STOCHASTIC ECONOMICS

Stochastic Processes, Control, and Programming

STOCHASTIC ECONOMICS
Stochastic Processes, Control, and Programming

GERHARD TINTNER
DISTINGUISHED PROFESSOR OF
ECONOMICS AND MATHEMATICS
UNIVERSITY OF SOUTHERN CALIFORNIA
LOS ANGELES, CALIFORNIA

JATI K. SENGUPTA
PROFESSOR OF ECONOMICS AND STATISTICS
IOWA STATE UNIVERSITY
AMES, IOWA

 1972

ACADEMIC PRESS   New York and London

Copyright © 1972, by Academic Press, Inc.
ALL RIGHTS RESERVED
NO PART OF THIS BOOK MAY BE REPRODUCED IN ANY FORM,
BY PHOTOSTAT, MICROFILM, RETRIEVAL SYSTEM, OR ANY
OTHER MEANS, WITHOUT WRITTEN PERMISSION FROM
THE PUBLISHERS.

ACADEMIC PRESS, INC.
111 Fifth Avenue, New York, New York 10003

United Kingdom Edition published by
ACADEMIC PRESS, INC. (LONDON) LTD.
24/28 Oval Road, London NW1 7DD

LIBRARY OF CONGRESS CATALOG CARD NUMBER: 72-75626

PRINTED IN THE UNITED STATES OF AMERICA

For Léontine, Phillip, Cindy, and Heidi
and
For Krishna and Rimi

TABLE OF CONTENTS

Preface   ix

1. The Stochastic View of Economics

1.1 What Is Stochastic?	1
1.2 Stochastic and Deterministic Economics	9
1.3 Stochastic Elements in Econometric Models	12
1.4 The Need for Stochastic Treatment of Economic Problems	21
1.5 Scope of Our Stochastic Analysis	24

2. Stochastic Models of Economic Development

2.1 Stochastic Systems	27
2.2 Discrete Stochastic Processes and Their Application to Trends of Economic Development	35
2.3 Multi-Dimensional Discrete Stochastic Process with Applications	59
2.4 Diffusion Processes Applied to Economic Development	65

3. Stochastic Control Theory with Economic Applications

3.1 Deterministic Control Theory and its Economic Applications	94
3.2 Stochastic Control: Methods and Applications	128
3.3 Economic Applications of Stochastic Control	150
3.4 Problems in Stochastic Control	180

4. Stochastic Programming Methods with Economic Applications

4.1 Programming Framework in Economic Analysis	187
4.2 The Approach of Stochastic Linear Programming: A Brief Summary	203

4.3 Economic Applications of Stochastic Programming	229
4.4 Stochastic Programming and Economic Decisions	259

References and Bibliography 269

Author Index	297
Subject Index	305

PREFACE

This book presents some aspects of economics from the stochastic or probabilistic point of view. After describing briefly in general terms the stochastic view of economics, it applies stochastic processes to the theory of economic development, stochastic control theory, and various aspects of stochastic programming.

The methodology used in this book is very much influenced by recent trends in probability theory, engineering mathematics, and operations research. But, in a sense, it also follows the much neglected American tradition of Evans (1930) and Roos (1934). It is our conviction that these methods are appropriate and should be utilized in the solution of problems of economic policy.

We originally planned a volume on the Econometrics of Development and Planning. But the literature on deterministic theories in this field has grown so enormously that we chose to confine ourselves to purely stochastic matters, which have been somewhat neglected. We may recommend some useful books in the deterministic theory of development: Allen (1968), Burmeister and Dobell (1970), and Morishima (1964, 1970), Sir John Hicks (1965). Naturally, our ideas about a stochastic theory of development draw very much upon the deterministic literature in this field.

In Chapter 1 we give a short survey of the stochastic view in economics. The second chapter treats stochastic models of economic development. Here we utilize tentatively discrete and continuous stochastic models.

Chapter 3 concerns itself with stochastic control, a decidedly modern approach to problems of economic policy. We owe a great deal in this field to the ideas of the Russian mathematician Pontryagin (1962), the American mathematician Bellman (1957), and the voluminous literature in control theory, published in many languages.

The last chapter is devoted to stochastic programming. Linear programming originates with the Russian mathematician Kantorovich (1939) and the American mathematician Dantzig (1948). But the data in a linear programming problem (the coefficients of the objective function, the coefficients of the inequalities involved) are assumed to be "sure" numbers. If more realistically we assume that these data are subject to a probability distribution, we confront the much more difficult problem of stochastic programming. We present in this chapter the theory and also some application of these ideas.

Throughout the book we have approached our subject in the econometric spirit. We present many tentative applications, and the models involved have been estimated by econometric methods. The empirical applications are intended to be illustrative of the following areas in particular: econometric estimation in nonstatic models with and without control variables, problems of resource allocation under risk, and the evaluation of alternative economic policies under dynamic models when parts of the latter are subject to a stochastic generating process.

The book is a cooperative product; the authors have influenced each other through discussions and co-authorship of joint articles which underlie some of its chapters. However the main lines of responsibility are as follows: Chapters 1 and 2 are by the first author and Chapters 3 and 4 by the second. A number of empirical illustrations which underlie Chapters 2–4 have been done in collaboration with some of our former students to whom we are extremely grateful.

The book is not a textbook in the conventional sense, as it includes numerous results and concepts arising from the authors' own research. The areas in which original results are reported include (a) the specification and estimation of stochastic process models applicable to dynamic models of growth, (b) the variety of models from the theory of stochastic control which are applicable to resource allocation problems in economics, (c) the theory of quantitative economic policy under various stochastic specifications, and (d) the use of several operational methods of probabilistic programming in static, dynamic, and other models.

We hope that the book will stimulate economists to consider economic problems without neglecting the stochastic point of view. We emphasize, however, that we have stressed in this book only those operational aspects of the stochastic framework which are applicable to economic and other empirical models; in this sense it is a book on stochastic methods rather than theory and it deals more with applications than with existence theorems and mathematical characterizations. The areas in which an economist would be primarily interested include (a) the implications of alternative decision rules under different stochastic schemes, (b) the limitations of deterministic models which are supposed to be on the average correct, and (c) the various operational methods of incorporating and analyzing the risk element in the framework of resource-allocation models of dynamic and static programming.

We believe that the stochastic methods treated in this book would also interest workers in fields other than economics, e.g., applied statistics and mathematics, operations research and management science, and systems engineering. The areas which would be of primary interest to the statisticians and applied mathematicians would include (a) empirical and computational problems of application as implied in our treatment of evolutionary time series (e.g., the small sample theory is yet undeveloped) and (b) the rise of nonparametric methods in stochastic programming, which are more or less free from assumptions of specific distributions and therefore more widely applicable.

Operations researchers and systems engineers will perhaps find here some ideas and approaches which might be helpful in their own fields. The point of view taken in this book, that econometric estimation should be viewed as a decision-making process under a stochastic framework, would not appear to them to be far fetched, although it might seem so to an economist. For these reasons the application of stochastic and mathematical methods to economic models and data would be most valuable in a multidisciplinary field. We do emphasize throughout that the unifying theme of our book is the need for integrating the stochastic methods—stochastic processes, control, and programming—with the specification of the deterministic parts, in particular when decision rules and economic policies are to be evaluated.

We believe this book offers considerable flexibility for use in any field which deals with methods for applying stochastic processes, control or programming. However in order to appreciate some of the econometric results on stochastic processes in Chapter 2, stochastic programming in Chapter 4 and stochastic control in Chapter 3 the students and research workers would find it very helpful to have prior background in some econometrics (e.g., at the level of Tintner (1952) or Johnston (1963)), some ideas on linear programming (e.g., at the level of Hadley (1962)), and some control theory applied to economic models (e.g., at the level of Intriligator (1971)).

Readers who are primarily interested in the applied economic aspects and implications of stochastic processes control and programming should preferably concentrate on the following chapters: 1.1–1.3, 2.1, 2.2, 3.3, and 4.3. Readers with interests in statistical and mathematical methods would probably find the following chapters more interesting: 2.3, 2.4, 3.2, and 3.4. Readers with a strong interest in mathematical economics with econometrics and operations research should proceed consecutively from Chapter 1 through Chapter 4.

We wish to express our gratitude to the National Science Foundation for suppoting most of the research upon which this book is based. Also we would like to express our sincere appreciation and thanks to Professor Karl Shell who went through the whole manuscript and offered several suggestions.

1 THE STOCHASTIC VIEW OF ECONOMICS

1.1 What Is Stochastic ?

Stochastic is a term now used universally as indicating something connected with probability. The problem of the meaning of probability is very disputed, but will not be discussed in this book (see Tintner, 1949, 1968). Whereas there is a very lively dispute about the meaning of probability, there is general agreement among mathematicians interested in probability and statisticians about the validity of the laws of probability. For an excellent introduction into probability theory see Feller (1966, 1968). Most texts on econometrics like Christ (1966), Goldberger (1964), Haavelmo (1944), Johnston (1963), Klein (1962), Malinvaud (1970), Theil (1961), Tintner and Millham (1970), and Wold and Jureén (1953) contain information on probability theory.

Probability theory as traditionally presented in statistics texts (see e. g., Mood and Graybill, 1963) is essentially static. We assume a given probability distribution and investigate its properties. Probability distributions may be discrete or continuous. Examples are the binomial distribution as a discrete distribution and the normal distribution as a continuous one.

In probability theory we are able to derive various laws of large numbers and different versions of the central limit theorem. The laws of large numbers investigate what happens (under certain conditions) to samples taken from given probability distributions as these samples become larger and larger. The central limit theorems investigate under what conditions certain functions of the samples of given populations tend, for example, to the normal distribution. For an excellent account of these matters and the results of modern research in this field the reader is referred to the two volumes of Feller (1966, 1968).

If we look at probability distributions from a more dynamic point of view, we obtain stochastic processes which may be defined as families of random variables depending upon a parameter (Fisz, 1963). A random variable is a (discrete or continuous) variable which can assume certain given values with definite probabilities. The parameter upon which the families of random variables depend is frequently time. Hence the reader may envisage a normal distribution, which can be characterized by just two magnitudes, its population mean (mathematical expectation) and population variance. In elementary probability theory we assume that the probability distribution is fixed, for example, we may sample from the same normal distribution with (unknown) population mean and variance. Think now of a family of normal distributions where the means and variances are (e. g., linear) functions of time. We then have a stochastic process, in this particular case a diffusion process. The statistical problems connected with such stochastic processes are much more difficult than traditional statistics which concerns itself with fixed given (but unknown) probability distributions. Many problems connected with the statistical treatment of stochastic processes (Bartlett, 1961; Fisz, 1963) are known under the name of time series analysis (see Tintner, 1952, 1968; Hannan, 1960; Davis, 1941; Granger and Hatanaka, 1964; Grenander and Rosenblatt, 1957; Quenouille, 1957; Whittle, 1963).

We give now short descriptions of some of the most important types of stochastic processes, given in the excellent little book of Takacs (1960).

Consider first Markov chains. Assume the existence of a system of mutually exclusive and exhaustive events: $E_1, E_2 \cdots E_j \cdots$. Define a random variable x_n $(n = 0, 1, 2 \cdots)$, $x_n = j$ if E_j is the outcome of the nth trial. Let P be probability

$$P(x_n = j \mid x_0 = i_0, x_1 = i_1, \cdots x_{n-1} = i_{n-1})$$
$$= P(x_n = j \mid x_{n-1} = i_{n-1}),$$

where $P(a \mid b)$ is the conditional probability of a, given that b has occurred. In other words, the probability distribution at step (time) n is completely determined by the outcome at step $n - 1$. We obtain a homogeneous Markov chain if

$$P(x_n = j \mid x_{n-1} = i) = P_{ij},$$

where P_{ij} is independent of n and is called the transition probability.

Up to now the state variables j and the time n were discrete. Consider now a sequence of real random variables $x_0, x_1, \cdots x_n \cdots$ and the relation

$$P(x_n \leq x \mid x_0 = y_0, x_1 = y_1 \cdots x_{n-1} = y_{n-1})$$

1.1 What Is Stochastic ?

$$= P(x_n \leq x \mid x_{n-1} = y_{n-1}).$$

This constitutes a Markov chain with continuous state space. This chain will be homogeneous if

$$P(x_n \leq x \mid x_{n-1} = y) = K(x, y),$$

where $K(x, y)$ is independent of time (n).

A sequence of real random variables x_n ($n = 0, \pm 1, \pm 2, \cdots$) is a stationary stochastic sequence if expectation and variance of x_n exist (are not infinite) and are independent of n (time) and the correlation coefficient between x_n and x_m depends only on the time difference $m - n$.

Define:

$$Ex_n = a,$$

the mathematical expectation;

$$E(x_n - a)^2 = \sigma^2,$$

the variance;

$$R(x_n, x_m) = E(x_n - a)(x_m - a)/\sigma^2 = R(m - n),$$

the correlation coefficient. We have $R(-n) = R(n)$.

Consider now Markov processes, characterized by a family of real random variables x_t

$$P(x_t \leq x \mid x_{t_1} = y_1, x_{t_2} = y_2 \cdots x_{t_n} = y_n) = P(x_t \leq x \mid x_{t_n} = y_n),$$

where $t_1 < t_2 < \cdots t_n < t$ ($n = 1, 2 \cdots$).

The transition probabilities are given by

$$P(x_t \leq x \mid x_s = y) = F(s, y; t, x), \qquad s < t.$$

Such a Markov process is called homogeneous in time if the transition probabilities depend upon x, y, and $t - s$. It is additive (has independent increments) if the transition probabilities depend upon $x - y$ and s and t.

A Markov process has a finite or denumerably infinite number of states if

$$P(x_t = j \mid x_s = i) = P_{ij}(s, t), \qquad s < t$$

and its transition probabilities are given by

$$F(s, y; t, x) = \sum_j{}' P_{ij}(s, t).$$

Let $c_j \Delta t$ be the (approximate) probability that a transition will occur if the system is in state E_j in the time interval $(t, t + \Delta t)$. Let $P_{jk}(t)$ be the conditional probability of the event of a transition from E_j to E_k in the time interval $(t, t + \Delta t)$. We derive the forward Kolmogorov equations

$$\partial P_{ik}(s, t)/\partial t = -c_k(t)P_{ik}(s, t) + \sum_{j \neq k} P_{ij}(s, t)c_j(t)P_{jk}(t).$$

The backward Kolmogorov equations are

$$\partial P_{ik}(s, t)/\partial s = c_i(s)P_{ik}(s, t) - c_i(s) \sum_{j \neq i} P_{ij}(s)P_{jk}(s, t).$$

Let

$$P(x_t = k) = P_k(t)$$

denote the probability distribution of the random variable x_t, and denote by $P_i(0)$ the initial distribution, then

$$P_k(t) = \sum P_i(0)P_{ik}(0, t),$$

where the $P_k(t)$ satisfy the following system of differential equations

$$dP_k(t)/dt = -c_k(t)P_k(t) + \sum_{j \neq k} c_j(t)P_{jk}(t)P_j(t), \qquad k = 0, 1, 2, \cdots.$$

If the Markov process is homogeneous $P_{jk}(t) = P_{jk}$ and $P_{ik}(s, t) = P_{ik}(t-s)$, $c_j(t) = c_j$.

Then the two systems of Kolmogorov equations become

$$dP_{ik}(t)/dt = -c_k P_{ik}(t) + \sum_{j \neq k} c_j P_{jk} P_{ij}(t),$$

$$dP_{ik}(t)/dt = -c_i P_{ik}(t) + c_i \sum_{j \neq i} P_{ij} P_{jk}(t).$$

Define now the matrices

$$P(t) = [P_{ij}(t)],$$

$$A = [a_{ij}], \qquad a_{ij} = c_i P_{ij}, \qquad (i \neq j), \qquad a_{ii} = -c_i.$$

We have the matrix equation

$$dP(t)/dt = P(t) \cdot A = A \cdot P(t); \qquad P(0) = I,$$

where I is the unit matrix. The solution is

1.1 What Is Stochastic?

$$P(t) = e^{At}.$$

The absolute probabilities $P(x_t = k) = P_k(t)$ satisfy

$$dP_k(t)/dt = -c_k P_k(t) + \sum_{j \neq k} c_j P_{jk} P_j(t).$$

If we have continuous transitions, let $x_t = x$ characterize the state of the system at time t. Define

$$F(s, y; t, x) = \int_{-\infty}^{x} f(s, y; t, x)\, dx,$$

where $f(s, y; t, x)$ is the transition probability density function.

$$f(s, y; t, x) \geq 0, \qquad \int_{-\infty}^{\infty} f(s, y; t, x)\, dx = 1.$$

The Chapman Kolmogorov equation is

$$f(s, y; t, x) = \int_{-\infty}^{\infty} f(s, y; u, z) f(u, z; t, x)\, dz.$$

Let $a(t, x)$ be the instantaneous mean and $b(t, x)$ the instantaneous variance, defined by appropriate limiting processes. The differential equations become now

$$\partial f/\partial s + a(s, y)(\partial f/\partial y) + (1/2)b(s, y)(\partial^2 f/\partial y^2) = 0,$$
$$(\partial f/\partial t) + [\partial a(t, x) f/\partial x] - (1/2)[\partial^2 b(t, x) f/\partial y^2] = 0.$$

If the process is homogeneous and additive,

$f(s, y; t, x) = g(t - s, x - y)$ and a, b are constants and we have just one differential equation

$$(\partial g/\partial t) + a(\partial g/\partial x) - (b/2)(\partial^2 g/\partial x^2) = 0.$$

Mixed cases, where transitions occur continuously or by jumps, are also possible.

For non-Markovian processes, consider recurrent processes. Let $T_1, T_2 \cdots T_n \cdots$ be the instants at which random events occur. Now assume that $T_n - T_{n-1}$ ($n = 1, 2 \cdots$; $T_0 = 0$) the time differences are identically distributed, independent, positive random variables whose common distribution function is $F(x)$. Let x_t be the number of events occurring in the time interval $(0, t)$.

T_1 might have a distribution different from $F(x)$ (general recurrent process). Let

$$M = \int_0^\infty x\, dF(x),$$

the finite mean value of $F(x)$,

$$F^*(x) = \left(\frac{1}{M}\right)\int_0^x [1 - F(y)]\, dy,$$

is the distribution function of T_1. Then x_t is a stationary recurrent process. Now define

$$W(t, n) = P(x_t \leq n).$$

Let $F_n(t)$ be the nth iterated convolution of the distribution $F(t)$ with itself; $F_0(t) = 1$ for $t \geq 0$, $F_0(t) = 0$ for $t < 0$.

$$W(t, n) = P(t \leq T_{n+1}) = 1 - P(T_{n+1} \leq t) = 1 - F_{n+1}(t).$$

The mean of x_t is given by

$$m(t) = Ex_t = \sum_{n=0}^\infty [1 - W(t, n)] = \sum_{n=1}^\infty F_n(t).$$

Let e_t be the distance between t and the instant of the next random event. Its distribution is

$$P(e_t \leq x) = \int_t^{t+x} [1 - F(t + x - u)]\, dm(u).$$

A stochastic process x_t is strictly stationary if the joint distribution of the random variables $x_{t_1+T}, x_{t_2+T}, \cdots x_{t_n+T}$ is identical with the joint distribution of the random variables $x_{t_1}, x_{t_2}, \cdots x_{t_n}$. A stochastic process is stationary in the wide sense if Ex_t and Ex_t^2 exist (are not infinite) and are independent of t; also $Ex_t x_{t+T}$ depends only on T.

A stochastic process is homogeneous if the joint distribution of all $x_{t_k} - x_{u_k}$ ($k = 1, 2, \cdots, n$) is invariant under a translation of time. The correlation coefficient

$$R(x_t, x_s) = R(t - s)$$

is the correlation function. $R(0) = 1$ if $R(t)$ is continuous.

Khinchine showed that a necessary and sufficient condition for a function $R(t)$ to be the correlation function of a continuous stationary stochastic process is the existence of

$$R(t) = \int_{-\infty}^\infty \cos tx\, dF(x),$$

1.1 What Is Stochastic?

where $F(x)$ is a probability distribution function (spectral distribution function).
A stochastic integral may be defined as a limit

$$I = \lim\left[I_n = \sum_{i=1}^{n} f(t_i)x(t_i)(t_i - t_{i-1})\right] = \int_a^b f(t)x(t)\,dt.$$

where $a = t_0 < t_1 < \cdots t_n = b$,

Consider a stationary stochastic process for which
$Ex_t = 0$; $Ex_t^2 = 1$ with correlation function $R(t)$.
Then

$$E\left[\int_a^b f(t)x(t)\,dt\right] = 0,$$

$$E\left[\int_a^b f(t)x(t)\,dt\right]^2 = \int_a^b \int_a^b R(t-s)f(t)f(s)\,dt\,ds,$$

if the integral exists.

Kolmogorov and Cramer proved for this process: it can be represented by a stochastic Stieltjes integral,

$$x(t) = \int_{-\infty}^{\infty} e^{iLt}\,dz(L),$$

where $z(L)$ is a complex valued stochastic process with $E\,z(L) = 0$ and:

$$E[Z(L_1 + \Delta L_1) - Z(L_1)][\bar{Z}(L_2 + \Delta L_2) - \bar{Z}(L_2)] = 0,$$

where the bar means complex conjugate. This is the spectral representation of the process. The spectral distribution function $F(x)$ satisfies

$$F(L + \Delta L) - F(L) = E[Z(L + \Delta L) - Z(L)]^2.$$

Assume now that the sample function of a stationary stochastic process $x(u)$ is known for $u < t$. We want to use this sample function in order to predict $x(t + T)$ for $T > 0$. Using as a measure of effectiveness mean square error we use linear prediction.

Assume $E\,x(t) = 0$ and $E\,x(t)^2 = 1$. We want to minimize

$$E[x(t + T) - L_T(t)]^2,$$

where $L_T(t)$ is the linear predictor. If

$$\lim_{n\to\infty} \sum_{k=1}^{n} a_k x(t - s_k) = L_T(t),$$

$$F_T(L) = \lim_{n\to\infty} \sum_{k=1}^{n} a_k \exp(-iLs_k).$$

The best linear predictor is given by

$$L_T(t) = \int_{-\infty}^{\infty} e^{iLt} F_T(L)\, dZ(L),$$

where $F_T(L)$ is the spectral characteristic function of the process $x(t)$. It may be determined from

$$\int_{-\infty}^{\infty} e^{isL}[e^{iTL} - F_T(L)]\, dF(L) = 0.$$

The minimum mean square error is

$$\sigma_T^2 = \int_{-\infty}^{\infty} \{1 - [F_T(L)]^2\}\, dF(L).$$

Let x_t be the number of random events occurring in the time interval $(0, t)$. The signal depends upon a random parameter. The magnitude of the signal is $f(u, x)$, where u is time measured from the instant of occurrence and the random parameter. Signals superimpose and e_t is the sum of amplitudes of signals at time t. Then e_t is a secondary stochastic process.

$$e_t = \sum_{0 < T_k \le t} f(t - T_k, c_k),$$

where T_k are the instances of occurrence of the random events in the interval $(0, t)$ and c_k the values of the random parameters. Assume now that the c_k have a distribution function $H(x)$ and are identically distributed independent random variables. The sequences c_k and T_k are mutually independent.

A stochastic integral for the process e_t is

$$e_t = \int_0^t f(t - u, c_u)\, dx_u.$$

Let

$$P(e_t \le x) = F(x).$$

If we have in the limit

$$\lim_{t\to\infty} P(e_t \le x) = F(x).$$

and

$$\lim_{t\to\infty} E(e_t) = M; \quad \lim_{t\to\infty} (e_t - M)^2 = D^2$$

are all finite, then we have a stationary process e_t^*

$$e_t^* = \sum_{T_k \leq t} f(t - T_k, c_k)$$

such that:

$$P(e_t^* \leq x) = F(x); \quad E(e_t^*) = M; \quad E(e_t^* - M)^2 = D^2$$

for all t. Also

$$R(e_t^*, e_{t+T}^*) = R(T)$$

depends only on T.

1.2 Stochastic and Deterministic Economics

The history of economics (Schumpeter, 1954) shows that this social science, which has attained a certain amount of maturity, has been profoundly influenced by contemporary developments in the natural sciences, especially physics (Tintner, 1966, 1968; Fels and Tintner, 1967).

Already the system of the physiocrats (Blaug, 1962) exhibits the potent influence of contemporary natural science. The classical system is very much influenced by Newtonian physics. Neoclassical economics cannot be understood without taking into account the potent influence of 19th century deterministic physics (Walras, 1954). As a matter of fact, the only economic model which is not a more or less felicitous imitation of classical deterministic physics is game theory (von Neumann and Morgenstern, 1944). Unfortunately, whereas the influence of game theory has been most profound in some important fields of economics, it has never fulfilled its promise of dealing successfully with the problems for which it was designed: Market organizations which lie (in a certain sense) between free competition and pure monopoly, like oligopoly, oligopsony, bilateral monopoly, and so on.

The foregoing remark should not be interpreted as a criticism of the remarkable achievements of game theory in fields other than for which it was originally designed. The introduction of measurable utility has revolutionized this field and yielded many important results for the problem of choice under uncertainty (Marschak, 1950). Methods connected with game theory have (somewhat paradoxically) been used in order to establish the existence of solutions of static competitive economic systems (Debreu, 1959). Also, game theory has been used in order to establish the existence of meaningful solutions for the famous dynamic system of von Neumann (1945) and its extensions (Kemeny *et al.*, 1956). Finally the most recent investigations of Aumann (1964) using game theoretical methods, have for the first time established valid conditions for the theoretical existence of competitive markets.

So this leaves us for practical purposes with neoclassical models (Marshall, 1948). Models of general equilibrium (Kuenne, 1963) are evidently still imitations of classical deterministic physical systems. Essentially they are restricted to statics, dynamic factors being introduced very much *ad hoc*. But even within static (timeless, Hicks, 1946) economics they are deficient by being restricted to free competition.

Models of economic development, which have been very popular recently (Hahn and Matthews, 1964) are at least dynamic, but still very much deterministic. The famous dynamic model of von Neumann (1945) is also based upon the idea of free competition. Other dynamic models (Morishima, 1964) are also deterministic.

It is not our intention to criticize economic deterministic models, just because they are (somewhat feeble) imitations of classical deterministic physics. It is perhaps unavoidable that all other sciences are powerfully influenced by the most successful sicence, and this was undoubtedly physics in the 19th century. Also, we might point with pride to the fact that in one memorable instance classical economics has influenced the development of biology. It is well known that Darwin derived his famous theory of evolution (struggle for life) uuder the strong impression of the Malthusian theory of population. Hence the influence has not always been one sided.

But if we contemplate contemporary rather than 19th century physics we must observe a great change. True enough, relativity theory is deterministic and constitutes perhaps a powerful ending of classical physics. But since the rise of quantum theory modern physics has otherwise become stochastic, that is, uses the results of probability theory. This makes the results of classical physics not useless. The law of large numbers assures us that, if the number of particles is large (and for macro-systems it is enormous), deviations from the mean values (mathematical expectations) will be small. This explains the continuing success of the methods of classical physics in engineering.

Similarly, we might argue that deterministic economics must be considered as dealing with the mean values (mathematical expectations) of the random variables, which really characterize the economic system. This explains perhaps the great concern of classical and neoclassical economic theory with free competition. Cournot showed already in 1838 that (under certain conditions) free competition will prevail if the number of competitors becomes infinite. Later investigations seem to indicate that the infinity involved is even of the power of the continuum (Aumann, 1964). In such cases we are certainly justified in concentrating on the mean values.

1.2 Stochastic and Deterministic Economics

But is this assumption realistic? We believe that there is a great deal to the criticism of Galbraith (1967). He emphasizes the empirical importance of noncompetitive market structures. The results of modern economics in dealing with market structures which are somehow between free competition and pure monopoly are not very impressive. The brilliant beginning of the analysis in the 1930's (Chamberlin, 1948; Robinson, 1938; Zeuthen, 1933) has not been followed up. Game theory, which is perhaps the first economic model not derived from physics, was particularly designed to deal with these problems. In spite of valiant efforts (Shubik, 1959), the results in dealing with oligopoly, oligopsony, bilateral monopoly, and other market organizations (which characterize great parts of the economy of the United States and Western Europe) rather than free competition have been very disappointing. Even in cases in which game theory has been somewhat successfully applied (Harsanyi, 1966), the results do not seem to go much beyond the achievements of the 1930's (Zeuthen, 1933).

We propose to follow the example of modern physics and to consider economic phenomena frankly from a stochastic point of view, that is, to treat economic variables as random variables. A random variable is a variable which can assume given values with definite probabilities. More specifically, we will try to use the theory of stochastic processes in this field. This may, among other things, meet the criticism of Hayek (1952) that in economic phenomena we cannot assume that we are sampling from a static probability distribution, which is fixed for all times. Instead, we will make the assumption that we are dealing with families of probability distributions which typically change their form in time. This, as indicated above, can be accomplished by using the theory of stochastic processes.

We might simply argue that stochastic processes might be used on an experimental basis in economics, that we might imitate the physicists or that a cultural lag of 50 years is long enough. However, we shall see that there are more compelling reasons for advocating the use of stochastic processes in the analysis of economic phenomena.

However, as we shall see, this is by no means an easy matter. Economic time series contain typically trends and irregular cycles. If we confine the use of stochastic processes to stationary processes, we start from extremely unrealistic assumptions. Moreover, we preclude ourselves from the analysis of economic phenomena which are connected with economic development. These phenomena were already in the forefront of the concern of the classical school of economics (e. g., Ricardo, Malthus, Marx; see Baumol, 1959) and recent economics has for obvious practical and theoretical reasons paid very much attention to these phenomena.

But the theory of evolutionary time series or stochastic processes is not yet very well developed. Especially, we frequeutly lack statistical methods which would enable us to deal with this type of phenomena (Tintner, 1968). We should not overlook the great mathematical and statistical difficulties which make such an approach challenging but also very laborious.

There are perhaps a number of other difficulties of a principal nature in our path. Economic variables cannot become negative, and modern methods of linear programming (Dantzig, 1963; Kantorovich, 1963; Kreko, 1965) take this into account. Hence, conventional methods which use the normal probability distribution or its generalization, e.g., the classical diffusion process must be used with extreme caution. We propose to use instead the lognormal distribution (Gibrat, 1931; Aitchison and Brown, 1957). For some empirical applications see Tintner and Patel (1966). On the other hand, there is the possibility that the Pareto distribution might be more appropriate (Pareto, 1927; Mandelbrot, 1960; Steindl, 1965). This would create a number of mathematical difficulties since these distributions have very thick tails and infinite variances (Feller, 1966; see also Gnedenko and Kolmogorov, 1968).

1.3 Stochastic Elements in Econometric Models

Let us define two sets of variables: $y_t = \{y_{1t}, y_{2t}, \cdots, y_{nt}\}$ is a column vector of n contemporaneous endogenous variables. Also $z_t = \{z_{1t}, z_{2t}, \cdots, z_{mt}\}$ is a column vector of m predetermined variables. These variables include truly exogenous variables and also lagged values of the endogenous variables, for example, y_{jt-m}. Now we contruct a linear system

$$A \cdot y_t + B \cdot z_t = u_t.$$

Here A is a matrix of constants of order $n \cdot n$; B is a matrix of constants of order $n \cdot m$. Further, $u_t = \{u_{1t}, u_{2t}, \cdots, u_{nt}\}$ is a column vector of random variables. See Klein (1953, 1962), Malinvaud (1966), Christ (1966), Menges (1961), Goldberger (1964), Johnston (1963), Tintner (1952, 1960), Leser (1966), and Fisk (1967).

The matrices A and B consist of the economically important parameters which constitute the system to be estimated. We are here mostly interested in the stochastic elements which enter into the system.

The observations consist of samples of the elements of the vectors y_t and z_t. Regarding their stochastic nature, it will frequently happen that $y_t = y_t^* + y_t^{**}$, $z_t = z_t^* + z_t^{**}$, where the elements of y_t^{**} and z_t^{**} are observational errors (Tintner, 1952). Morgenstern (1963) has emphasized the importance of errors of observations in many empirical economic time series which are utilized in practice in an effort to estimate the system described. But in most econometric investigations the assumption is made that errors of observations can be neglected.

Another problem which is closely related to the existence of errors of observations is multicollinearity or near multicollinearity in the data (Tintner, 1952; Winckler, 1966). Collinearity exists if we have linear dependence between two or more elements of the set z_t. We will obtain a situation which is almost collinear if these linear relations hold not strictly but only approximately. It may happen that in those cases the estimated relationships in the presence of errors of observations z_t^{**} will almost entirely depend upon these errors.

Because of the stochastic nature of the variables u_t the observations y_t are also random variables, because the functions of random variables are themselves random variables. In fact, if the matrix A is not singular, we might solve for y_t and obtain

$$y_t = C \cdot z_t + D \cdot u_t,$$

where $C = -A^{-1} \cdot B$ and $D = A^{-1}$. These reduced form equations, which are important for certain methods of estimation of the system, show clearly the dependence of the endogenous variables y_t upon the random variables u_t.

Making the very unrealistic assumptions that the random variables u_t are normally distributed and that u_t is independent of u_{t-m} ($m = 1, 2, \cdots$) various methods are available for the estimation of our system of equations if the equations are identified (Fisher, 1966). We might use the full information maximum likelihood method (Hood and Koopmans, 1953) which however necessitates numerical procedures, because the resulting equations are not linear. Alternatively, we might use the limited information method of Anderson and Rubin (1949). Let us assume that we want to estimate just one of the n equations of our system, which is written as

$$a y_t^{(1)} + b z_t^{(1)} = u_t^{(1)},$$

where $y_t^{(1)}$ is the vector of endogenous variables entering into the equation to be estimated; $z_t^{(1)}$ is the vector of the predetermined variables entering into our equation. The vectors a and b are the corresponding elements of the matrices A and B. The random variable is the disturbance in the equation to be estimated $u_t^{(1)}$.

We form the reduced form equations for the endogenous variables entering into our equation

$$y_t^{(1)} = E z_t^{(1)} + v_t^{(1)}$$

and denote by Q the variance–covariance matrix of the deviations $v_t^{(1)}$. On the other hand, from the complete reduced form system:

$$y_t = Cz_t + Du_t = Cz_t + v_t,$$

where $v_t = Du_t$, we derive the variance–covariance matrix of the v_t, which we call R.

The maximum likelihood estimates of the elements of the vector b are obtained by

$$(Q - LR) \cdot b = 0,$$

where L is the smallest root of the determinantal equation

$$|Q - LR| = 0.$$

Also

$$c = H \cdot b,$$

where H is the variance–covariance matrix of the elements of $z_t^{(1)}$. With a just identified equation we have just enough relations to compute the desired equation

$$ay_t^{(1)} + bz_t^{(1)} = u_t^{(1)},$$

by elimination of the elements of the vector z_t other than $z_t^{(1)}$ from the reduced form equations. This is called indirect least squares.

With the two stage least squares method of Theil (1961) and Basmann (1957) we must first normalize. We write the equation to be estimated as

$$y_t^{(2)} = a^{(2)} y_t^{(3)} + b^{(2)} z_t^{(1)} + u_t^{(1)},$$

where now $y_t^{(2)}$ is one member of the vector $y^{(1)}$; $y^{(3)}$ are the remaining members of the vector $y^{(1)}$.

We utilize the reduced form equations in order to estimate by the classical method of least squares the elements of $y_t^{(3)}$. Call these estimates $y_t^{(3)*}$. Then we estimate by the classical method of least squares the relation

$$y_t^{(2)} = a^{(2)} y_t^{(3)*} + b^{(2)} z_t^{(1)} + u_t^{(1)*}.$$

Finally, consider the case of recursive systems where the matrix A is lower triangular with -1 in the diagonals. Then we might write the system as follows

$$y_t = Fy_t + Bz_t + u_t$$

(Wold and Jureén, 1953). This assumes the existence of a causal structure. Here F is a lower triangular matrix with zero in the diagonal. Then, under certain circumstances, we might estimate the relations directly by the classical method of least squares.

We give now a short survey of some of the methods which have some practical use in dealing with stochastic processes in economics. This is frequently called time series analysis (Tintner, 1968).

The problem of dealing with time series has baffled economists from the very beginning. In other statistical applications the assumption of randomness can be easily justified. Hence, most elementary statistical textbooks (which seem to be written mostly for biological and industrial statisticians) neglect these problems. But it is evident that the assumption of statistical independence of consecutive observations cannot be satisfied in economics: If this year's price of a given good was independent of last year's price then we would certainly live in an entirely different economic universe than the one we know.

How are we to deal with this baffling problem? The methods sketched below (Tintner, 1966) are very frequently makeshift and *ad hoc*. What is missing is a valid theory of the statistical treatment of stochastic process (Bartlett, 1961; Anderson, 1964; Davis, 1941; Grenander and Rosenblatt, 1957; Fisz, 1963; Granger and Hatanaka, 1964; Hannan, 1970; Moran, 1959; Quenouille, 1957; Wold, 1954). This problem is complicated by a number of difficulties; Economic time series are typically not stationary (e. g., they contain a trend). Most of the few statistical methods proposed deal with stationary time series. But if we eliminate the trend, we eliminate the economic problems which were already recognized by the classical school to be decisive: economic development. It has also moved into prominence in economic theory recently. Another difficulty is the fact that economic time series are typically short. But small sample results in this field are very rare (R. L. Anderson, 1942; Rao and Tintner, 1962, 1963). The mathematical difficulties in this field are formidable.

The traditional method of dealing with economic time series is the analysis into trend, irregular cycle, fairly regular seasonal movement and remaining random component. Apart from the seasonal movement, where modern methods of spectral analysis are most promising (Hannan, 1960, 1963; Nerlove, 1964), this method is no longer very popular. But it is still sometimes applied because of the unavailability of more adequate methods.

If we deal with reasonably long time series, as we should if the ideas of economic development are taken seriously, polynomial (or exponential) trends are rather inadequate. These trends tend to infinity with increasing time. We might then compute logistic trends. This is a difficult problem, because the differential equation of the logistic is not linear (Hotelling, 1927). Anyway, time derivatives of economic time series are difficult to approximate, since typically economic data are only given at discrete intervals (months, years). It seems preferable to use a difference equation rather than a differential equation for the estimation of the logistic trend (Hotelling, 1927).

Let the equation of logistic trend be

$$X_t = k/(1 + b e^{-at}),$$

where k is the upper asymptote, a and b are positive constants. Transforming our variable, $z_t = 1/X_t$, we derive the first order linear difference equation (Tintner, 1960, p. 273)

$$z_{t+1} = (1 - e^{-a})/k + e^{-a}z_t.$$

This permits the estimation of $(1 - e^{-a})/k$ and e^{-a} and hence of a and k. If desired, the constant b might be estimated by the method of Rhodes (1940).

This method has been utilized by Tintner et al. (1961) for the estimation of a logistic trend for Indian agricultural income. In this paper there are also asymptotic formulae for the standard errors in the estimates.

Polynomial and logistic trends are especially important for purposes of prediction. Otherwise the method of moving averages might be preferable (Kendall and Stuart 1963, p. 366). The weights of moving averages might be determined by the idea that we fit a polynomial trend to $2m + 1$ consecutive observations, where the length of the moving average is $2m + 1$. The order of the polynomial might be determined by a method proposed by Tintner (1940, p. 100). But caution should be used in removing trends by moving averages. The moving average will introduce autocorrelations even into a pure random series (Tintner 1952, p. 203; Slutsky, 1927).

For tests of autocorrelation we have the von Neumann (1941) ratio available, which has been tabulated by Hart (1942). For a parametric test of autocorrelation, assuming normal distributions we introduce the following notations; The empirical (sample) autocorrelation coefficient r_L is the simple correlation between the series $X_1, X_2, \cdots, X_{N-L}$ and the series $X_{L+1}, X_{L+2}, \cdots, X_N$. Unfortunately, the distribution of the sample autocorrelation coefficient r_L is not known for small samples. To simplify the mathematical problem consider the circular autocorrelation coefficient r_L^*. This is the simple correlation coefficient between the series: $X_1, X_2, \cdots, X_N$ and $X_{L+1}, X_{L+2}, \cdots, X_n, X_1, X_2, \cdots, X_L$. Anderson (1942) derived the exact small sample distribution of r_L^* which might be used as an approximation for the test of the empirical sample autocorrelation coefficient r_L.

Suppose we fit a stochastic difference equation

$$X_t = a_0 + \sum_{i=1}^{p} a_i X_{t-i} + e_t,$$

where the a_i are constants and e_t are non-autocorrelated random disturbances with zero mean and constant variance. If the order of the stochastic difference equation (p) is unknown, a large sample test of goodness of fit by Quenouille (1947) is available in order to determine p.

Most economic time series are autocorrelated. Orcutt and James (1948) have proposed a method which shows that, in certain sense, the set of N autocorrelated observations is equivalent to a smaller number n' of independent observations. See also Bartlett (1935).

Very popular is the Durbin and Watson (1950, 1951) test for residuals from a least squares regression. Let z_t ($t = 1, 2, \cdots, N$) be the residual from a least squares regression with any number of independent variables. The Durbin–Watson ratio is

$$d = \sum_{t=2}^{N} (z_t - z_{t-1})^2 / \sum_{t=1}^{N} z_t^2.$$

No exact significance levels are available, but only lower and upper bounds, so the test is often not conclusive. But Theil and Nagar (1961) have provided an approximate test.

Sometimes we may want to transform our observations in order to use the more conventional methods applicable with sets of independent observations.

Consider the Variate Difference Method of Anderson (1929). See also Tintner (1940, 1952). Here the assumption is that the economic time series consists of a "smooth" function of time, the systematic part, and a purely random disturbance. The assumption of smoothness excludes sinusoidal oscillations with short periods (e.g., the seasonal variations).

If the smooth systematic part is a polynomial, it can be eliminated by taking enough differences. Any sufficiently smooth function of time, on the other hand, can be reduced by the same process.

The main object of the Variate Difference Method is to estimate the order of the difference which eliminates the systematic part and to estimate the variance of the random element, say σ^2. Suppose that we compute the (properly scaled) variances of the differences of our original observations, V_k. If k_0 is the order of the difference in which the systematic element is eliminated or sufficiently reduced we must have

$$V_{k_0} = V_{k_0+1} = \cdots$$

at least approximately. The variance of the random element σ^2 might then be estimated by V_{k_0}.

Some large sample tests were considered by. Anderson (1929) and Zaycoff (1937). A rather inefficient small sample test was suggested by Tintner (1940). But if we assume a circular population we might derive the exact small sample distribution of the circularly defined variance of the kth difference series (Tintner, 1955). An exact test is provided for the circular case in Rao and Tintner (1962). But the problem of finding k_0 is a multiple choice problem. Using some ideas of Anderson (1962), Rao and Tintner (1963) use the circular definition and provide a valid estimation procedure under this assumption.

H. Wold (Wold and Jureén, 1953) has provided an asymptotic method by which we may modify classical tests in regression analysis in the presence of autocorrelation of the residuals from the fitted regression. Consider the case of a simple least squares regression. Let us assume that we are dealing with first-order autoregressive schemes and designate the estimated first autocorrelation coefficient of the residuals from the regression by r_1 and the first autocorrelation coefficient of the independent variable in the regression equation by R_1. Then Wold shows that the adjustment factor for standard errors, tests of significance, and so on, is

$$\left(\frac{1 + r_1 R_1}{1 - r_1 R_1}\right)^{1/2}.$$

An important theorem of Frisch and Waugh (1933) states the following proposition. Suppose we are dealing with multiple regression and utilize deviations from linear trends. Instead of computing these deviations and then using the deviations in order to compute the multiple regression we might simply introduce time as an additional variable into our regression equation. This proposition has been generalized by Tintner (1952) to polynomial trends and more generally to any set of orthogonal functions.

Another way of dealing with autocorrelation is the use of autoregressive transformations, proposed by Cochrane and Orcutt (1949). Suppose we want to compute a linear multiple regression

$$X_{1t} = k_0 + k_2 X_{2t} + \cdots + k_p X_{pt} + u_t,$$

where u_t is a random element. But the sequence $u_1, u_2, \cdots, u_N$ is not a sequence of independent random variables, but the u_t follow, for example, a simple Markov process

$$u_t = A u_{t-1} + v_t,$$

where now the v_t are a set of independent random variables with mean zero and constant finite variance.

Let $|A| < 1$ and make the transformations

$$Y_{it} = X_{it} - A X_{it-1}.$$

1.3 Stochastic Elements in Econometric Models

Then the transformed relation

$$Y_{1t} = k_0(1 - A) + k_2 Y_{2t} + \cdots + k_p Y_{pt} + v_t$$

might now be fitted by classical methods, since the v_t are a set of independent random variables. Incidentally, if $A = 1$ we get the transformation $Y_{it} = X_{it} - X_{it-1}$ the first difference transformation which has been extensively used in econometric work (Stone *et al.*, 1954).

The difficulty is of course the estimation of A. Johnston (1963, p. 192) has proposed an iterative method for this purpose.

It has been mentioned above that one of the difficulties with economic time series is their evolutionary nature. Maximum likelihood estimates are given by Fisz (1963, p. 488) for the poisson process.

Let $N(t)$ be the number of events in the time period from zero to t. Let us assume that $N(t)$ has independent and homogeneous increments, that is, $N(t_2) - N(t_1)$ is independent of and has the same distribution as $N(t_2 + h) - N(t_1 + h)$ for all t_1, t_2 and all $h > 0$. The transition probability $P_{ji}(t_1, t_2)$ is the probability that $N(t_2) = i$ if $N(t_1) = j$ and is

$$P_{ji}(t_1, t_2) = [\exp -L(t_2 - t_1)]\frac{[L(t_2 - t_1)]^{j-1}}{(j-1)!}.$$

Here L, the parameter to be estimated, is the average number of events per unit of time. Assume we have observations $N(t_k)$ for the time points t_k ($k = 1, 2, \cdots, n$) and denote $N(t_k) = j_k$. The likelihood function is

$$P = \prod_{k=0}^{n-1} P_{j_k j_{k+1}}(t_k, t_{k+1}).$$

If this expression is maximized with respect to L, the maximum likelihood estimate of L is given by j_n/t_n.

One of the few results in the field of explosive (or evolutionary) processes is due to Anderson (1959). Consider the first-order stochastic difference equation

$$X_t = a X_{t-1} + u_t.$$

Assume that the initial value X_0 is constant and the u_t are non-autocorrelated normally distributed random variables with zero means and finite variance. The maximum likelihood estimator is

$$\hat{a} = \sum_{t=2}^{N} X_t X_{t-1} / \sum_{t=1}^{N} X_t^2.$$

If $|a| < 1$, the x_t are stationary and there is no trend. Then under wide conditions, the expression $(\hat{a} - a)\sqrt{N}$ is asymptotically normally distributed with mean zero.

If $a > 1$ we have an evolutionary process and an exponential trend in the x_t. The expression

$$(\hat{a} - a)|a|^N/(a^2 - 1)$$

has a Cauchy distribution. With a Cauchy distribution the mean and variance are infinite. But if $X_0 = 0$ and the u_t are normally distributed, then

$$(\hat{a} - a)\left(\sum_{t=1}^{N} X_{t-1}^2\right)^{1/2}$$

is a maximum likelihood estimator, is asymptotically normally distributed, and may be used for tests, confidence limits, and so on, in large samples.

Still less explored is the theory of multiple or vector stochastic processes, which is evidently of supreme importance in economic applications. We present as an interesting attempt to deal with these important matters the approach of Quenouille (1957, p. 70). See also Bartlett (1962).

Assume that we have a multiple Markov scheme, that is a system of first-order stochastic linear difference equations,

$$a_{11}X_{1t} + a_{12}X_{2t} + b_{11}X_{1t-1} + b_{12}X_{2t-1} = e_{1t},$$
$$a_{21}X_{1t} + a_{22}X_{5t} + b_{21}X_{1t-1} + b_{22}X_{2t-1} = e_{2t}.$$

We have a set of observations X_{1t}, X_{2t} $(t = 1, 2, \cdots, N)$. Let us assume that the e_{1t} and e_{2t} are random variables with zero means and constant and finite variances and that they are independent over time. Under certain conditions, the equations can be rewritten in the following form

$$X_{1t} = u_{11}X_{1t-1} + u_{12}X_{2t-1} + v_{11}e_{1t} + v_{12}e_{2t},$$
$$X_{2t} = u_{21}X_{1t-1} + u_{22}X_{2t-1} + v_{21}e_{1t} + v_{22}e_{2t}.$$

We define the simple covariances as

$$c_{ij} = (1/N)\sum_{t=1}^{N} X_{it}X_{jt} \qquad (i, j = 1, 2),$$

and the lagged covariances as

$$c'_{ij} = (1/(N-1))\sum_{t=2}^{N} X_{it}X_{jt-1} \qquad (i, j = 1, 2).$$

The maximum likelihood estimates of the u_{ij} are the solutions of the following system,

$$c_{11}\hat{u}_{11} + c_{12}\hat{u}_{21} = c'_{11},$$
$$c_{21}\hat{u}_{11} + c_{22}\hat{u}_{21} = c'_{21},$$
$$c_{11}\hat{u}_{12} + c_{12}\hat{u}_{22} = c'_{12},$$
$$c_{21}\hat{u}_{12} + c_{22}\hat{u}_{22} = c'_{22}.$$

We might also obtain asymptotic standard errors for the estimates $\hat{u}_{ij}$. But in general, the a_{ij} and b_{ij} cannot be estimated without further assumptions. (See F. M. Fisher, 1965; also Mann and Wald, 1943.)

1.4 The Need for Stochastic Treatment of Economic Problems

Why should stochastic ideas, for example, methods which are connected with probability, be introduced in a fundamental way into economic reasoning? A number of reasons for this proposal have been indicated above: The idea of keeping up with the Joneses, in our case the mathematical physicists. Physics, easily the most successful of all natural sciences (Georgescu-Roegen 1966) has in our time become almost completely stochastic, that is, based upon the idea of probability distributions. But similarly in biology probabilistic ideas have been most important, especially in genetics, perhaps the most successful branch of biology (Kempthorne). Incidentally, genetic improvements in agriculture (hybrid corn) have also had very profound economic implications for American apriculture (Griliches).

The proposal to introduce stochastic methods into economics can however be emphasized by contemplating the development of operations research. This, in our opinion, is nothing else but the econometrics of the enterprise, both private and public. As a matter of fact, some of the most important contributions to operations research have come out of the nationalized French industries (Massé, 1959; Lesourne, 1960).

The book by Bharucha-Reid (1960) is remarkable by giving a very broad survey of the application of the theory of Markov processes in many fields. Economics is here prominent by its absence. But many interesting applications of stochastic processes to biology, physics, astronomy, astro-physics, and chemistry are given. The survey of the applications of stochastic processes in operations research deals with the theory of queues. Here Bharucha-Reid deals with telephone traffic and servicing of machines. Similar methods are of course important for inventory theory (Arrow et al., 1958; Whitin, 1957; Dvoretskzky et al., 1952; Morse 1958; Arrow el at., 1952). Stochastic methods have also been utilized by Holt, *et al.*, (1960) for economic planning in the private enterprise. They deal with production, inventories, and the work force.

Theil (1964) shows that the treatment of problems in operations research (in a private enterprise) are formally equivalent to problems of economic policy. In the same chapter (p. 75) he deals with a simple anti-depression policy for the United States in the 1930's and a case of production and employment scheduling in a paint factory.

Probabilistic elements are also introduced into stochastic programming (Tintner, 1960; Sengupta *et al.*, 1963). An interesting recent book (Borch, 1968) shows the importance of stochastic considerations in the theory of enterprise in a most fundamental fashion.

Probabilistic considerations are hence not exactly foreign to modern economics. If we consider size distributions (Steindl, 1965) we realize that these problems are really very old. They can be traced to the famous pareto distribution of incomes (Pareto, 1927) and the voluminous literature concerned with these problems. Unfortunately, these ideas have never been very well integrated into the main body of economic theory.

As we shall see in the following chapters, the application of stochastic processes in economics faces still formidable difficulties. It is doubtful if the theory of evolutionary processes is well enough developed in order to apply these methods to problems of economic development. Statistical methods applicable to the type of processes we consider are virtually non-existent and even large sample methods (which do not seem to be particularly applicable in economics where we frequently have small samples) are relatively scarce and almost entirely confined to stationary processes (Grenander and Rosenblatt, 1957). In economic language, they might be applicable to seasonal fluctuations and perhaps the business cycle but leave us almost helpless to deal with problems of the trend (economic development). It will take a great deal of hard work and mathematical ingenuity to develop more adequate statistical methods in time series analysis.

There is moreover a fundamental challenge which has been put forward forcefully by Mandelbrot (1963). It is suggested on the basis of some empirical evidence connected with commodity prices that the distributions involved may belong to the family of stable indivisible destributions (Levy, 1925, Feller, 1966, p. 531). If this is the case we are typically dealing with distributions whose variance is infinite. This is indeed the case with the famous Pareto (1897) distribution, which has been widely used for approximating the income distribution, at least for relatively high incomes.

If Mandelbrot is right, this would indeed create a very difficult situation in economic statistics. Let us note that most biological and physical applications utilize the normal distribution, and most applications in operations research the Poisson distribution or some of its variants. All these distributions have of course

1.4 The Need for Stochastic Treatment of Economic Problems

finite variances. Hence, least squares methods and its variants have some desirable properties. The statistical problems connected with the Pareto and similar distributions are entirely different, and new methods of dealing with problems of estimation, confidence and fiducial limits, tests of hypotheses, and so on, seem now to be called for. Perhaps we might have to utilize nonparametric methods in this field.

Most econometricians probably think that the use of normal distributions might be justified (in a fashion) by the appeal to a kind of central limit theorem. This is indeed true for a great variety of statistical problems, and investigations have shown that traditional methods are relatively robust (Kendall and Stuart, 1961, p. 465). But this robustness and general adequacy of traditional statistical methods might ultimately prove an illusion. For example, it is easy to show that if we are sampling from a normal distribution, the median is relatively inefficient compared with the arithmetic mean. If we have a sample of 100 items it would take a sample of 157 items in order to obtain the same accuracy with the sample median as with the sample mean, if we measure accuracy by the variance.

This is true enough as long as the population from which we sample is normal. But suppose we sample from the Cauchy distribution. Then it is well known that the distribution of the sample mean is exactly the same as the distribution of each individual item in the sample. In this (we concede, extreme) case we gain nothing by computing the arithmetic mean of the sample. We might just as well have picked out one of the items in our sample at random.

If probabilistic considerations of the type discussed might make us skeptical about our statistical procedures, they are reinforced by the work of Basmann (1960). Econometric models are unfortunately not based upon large samples. It is extremely difficult to derive small sample distributions of econometric estimates. Basmann has succeeded in doing so and has discovered that, even if our errors or deviations have normal distributions, the distribution of econometric estimates is very complicated. But they are related to the Cauchy distribution and have infinite variances. Here Mandelbrot gets support from a statistician with extreme classical prejudices. The work of Basmann shows that even if we start with (in our opinion rather unrealistic) classical assumptions of normality, we might, with our empirical econometric estimates, end up with random variables with infinite variances. These variables belong again to the family studied by Levy and Mandelbrot.

Having presented the economic statistician and econometrician with some reflections about the way in which more useful ideas might be introduced into economic statistics, we turn now to the concern of the mathematical economist. In contradiction to many other sciences, economics, especially mathematical economics, has reached a certain maturity (Tintner, 1968). Our time has seen a certain decline of the neoclassical point of view (Fels and Tintner, 1967) but this has been more than compensated by the rise of linear methods, for example, linear programming, activity analysis, input–output analysis. With game theory economics has for the first time achieved a model which was not derived by analogy with the natural sciences. Many beautiful theorems have been derived, especially in connection with competitive economic systems. The challenging problem of economic development has been boldly attacked, and if the actual theoretical results are still meager they might be promising.

Just like 19th century deterministic physics has not been in practice replaced by 20th century stochastic physics, but is still used as a valid approximation in the majority of physical problems (especially by engineers), so we might also retain provisionally the somewhat less impressive results of deterministic economic theory, especially mathematical economics. They only have to be reinterpreted as concerning perhaps certain mean values of the economic variables concerned. Also, since it is much less certain that the law of large numbers applies in our field we might be more cautious than the classical physicist about the validity of our results, even as crude approximations.

1.5 Scope of Our Stochastic Analysis

Our view of stochastic analysis in economic models adopted in this book is based on three fundamental considerations. First, we emphasize the need for operational presentation of those mathematical results in stochastic processes, stochastic control, and stochastic programming, which have actual or potential usefulness in economic models. Second, we consider empirical applications primarily for illustrative purposes in order to show the complexities of calculation involved and the nature of pay-offs expected. In realistic applications the researcher should consider some of our stochastic methods as broad guidelines, although we feel that in near future the computational difficulties would be considerably reduced thanks to the recent trend of developments of computer algorithms in stochastic control and mathematical programming. Third, we emphasize in our applications economic models primarily in the macrodynamic fields of economic growth, development and investment planning; however, we have considered a few resource allocation models in the theory of the multiproduct firm to show that some of the stochastic methods like stochastic programming and control have significant scope of application here. In the fields of operations research stochastic methods, for example, stochastic programming and control, are most frequently applied to microeconomic fields like the firm, the industry and the projects; and considerable literature on applications (Hanssmann, 1962; Sengupta and Fox, 1969) exist in this area.

The plan of our book is as follows. Chapter one has briefly reviewed the use of stochastic concepts in natural and social sciences with some emphasis on the

1.5 Scope of Our Stochastic Analysis

need for stochastic treatment of economic phenomena in economic theory and its applications. Chapter Two analyzes the problems of specification, estimation, and empirical application of selected *stochastic processes* which are believed to be useful for economic systems. Chapter Three discusses the problems of *stochastic control* and its stability and optimality in relation to selected deterministic and stochastic models of economic policy, including some models of economic growth and stabilization. Chpater Four presents a comprehensive view of the recent operational methods of *stochastic programming* and discusses in great details their applications to static and dynamic economic problems. The latter applications include, on the one hand, the recent methods of geometric and reliability programming, extensions of chance-constrained, and stochastic programming, and, on the other, the economic problems of resource allocation within a firm, dynamic problems of investment planning under two or more sectors, lifetime portfolio allocation problems, and the limitations of the competitive equilibrium models under uncertainty.

The three facets (stochastic process, stochastic control, and stochastic programming) comprise in our view one integrated field of stochastic economics, of which the existing deterministic economics should be an important part. However, it is important to realize that the stochastic viewpoint has its unique contributions and applications independent of deterministic standpoints. Although we have presented the three facets of stochastic economics independently, we do emphasize the goal of stochastic economics to be one of combining and integrating the three facets for characterizing equilibrium and optimum conditions in economic theory and policy.

2 STOCHASTIC MODELS OF ECONOMIC DEVELOPMENT

In this chapter we discuss some stochastic models whose application to problems of economic development seems to us to hold some promise. Unfortunately the existing theory of stochastic processes is not always adequate for our purposes and many methods, especially those regarding evolutionary processes are still very inadequate from the point of view of statistical inference. Hence the results of this chapter cannot be considered anything but suggestive as far as the problems of economic development are concerned.

In recent applications of the statistical theory of time series to economic models, two methods have gained some prominence, they are, the spectral analysis (Jenkins and Watts, 1968) for identification of cyclical characteristics and the theory of stochastic control (Box and Jenkins, 1970) viewed as a modified method of forecasting. In our view the spectral theory, although useful in specific economic enquiries (e. g., analysis of stock market prices or short run investment behavior) is not very suitable for evolutionary time series containing nonstationary trend, particularly when multiple series and specific control rules are involved. The general theory of stochastic processes offers in our opinion a more satisfactory and generalized approach, particularly when models of economic growth are concerned. Like the recent theory of optimal economic growth under a neoclassical framework, which is more complete in specification for the one and two-sector cases, our applications of stochastic processes to economic growth models are motivated by the same spirit; we have indicated however the lines of extension to multivariate (i. e., multisector) cases, although many empirical problems of statistical estimation remain unsolved in this area. From an applied viewpoint, the two most important aspects of stochastic models are the specification of the stochastic process in relation to the economic trend and the statistical estimation of its parameters. These two aspects have been emphasized in this chapter with particular reference to models of economic growth. Although short run fluctuations are not specifically analyzed, it is our view that these could be incorporated in terms of the policy variables which are exogenous to the model.

A second aspect emphasized in our analysis is the specific use of decision rules (or control laws) in the stochastic process framework and its various implications. In particular, we have shown that in some cases this regulation problem reduces itself to a forecasting problem where the parameters of the stochastic process are to be properly estimated (Whittle, 1963; Aoki, 1967). It is in this sense that the estimation and decision problems are interrelated, as will be shown more fully in Chapter 3.

Our discussion in this chapter proceeds as follows: The general framework of stochastic systems and their relevance to various economic fields are mentioned in Section 2.1; this is followed by the specification and empirical identification of several discrete and multidimensional stochastic processes in Section 2.2, and Section 2.3 emphasizing in particular the methods and problems of statistical estimation of parameters and their application to simple decision rules. Section 2.4 discusses the theory and application of diffusion processes, which repesent in our view the most important tool yet to be fully analyzed and applied to economic models, although its relevance for diffusion of technology, spread of effects of media advertisement, and the problems of migration in spatial models are well known to the applied workers. As emphasized before our discussion has stressed only operational methods which we believe to be useful in various economic and other applications.

2.1 Stochastic Systems

The great tradition of theoretical economics runs from the physiocrats through the English classical school (Smith, Malthus, Ricardo, Jame, and John Stuart Mill; also Say and Marx) to the neoclassical school of Marshall and Pigou (Blaug, 1962). The marginalist contribution is important but does not perhaps today appear as vital as previously (Fels and Tintner, 1967). Certainly, the present concern with economic development is in the classical tradition (Baumol, 1959).

The ideas of the writers indicated above have been most vitally influenced by classical deterministic physics. This is particularly evident with the great models of general equilibrium developed by Quesnay and Walras and much elaborated in our own time (Wald, 1951; Debreu, 1959; Koopmans, 1957). It would be unjust to deny the great contribution these writers have made to our understanding of economic phenomena.

But in the meantime physics itself has changed. We may consider Einstein's relativity theory simultaneously as the culmination and the swan song of deterministic physics. Modern physics, especially micro-physics, has become stochastic, that is, probability considerations enter in a fundamental sense into theory construction. But the bulk of economic theorizing still persists in the construction of deterministic models, especially in the field of economic development with which we are here concerned (Morishima, 1964). There seems to be a cultural lag of more than 50 years between economics and physics.

Specialization, which is increasing in all the sciences, has brought about a certain isolation between economics and operations research, which is really the econometrics of the enterprise. In this specialized field, which is perhaps not too well known to most economic theoreticians (see however Baumol, 1965), probability considerations play most important part (Graves and Wolfe, 1963; Ackoff, 1963). It is true that certain methods related to linear propramming have been utilized by economic theoreticians (Dorfman *et al.*, 1958; Morishima, 1964). But probabilistic methods, utilized in operations research have had much less influence (see however Holt *et al.*, 1960; Whittle, 1963; Sengupta, 1967; Steindl, 1965; Borch, 1968; Murphy, 1965). These applications of stochastic processes to operations research concern queuing theory and its many applications to problems of the enterprise: Waiting line problems, inventories, machine repair, stochastic programming, chance constrained programming, and so on.

It cannot be denied that these problems are genuine economic problems of the enterprise. Probabilistic methods, especially methods connected with stochastic processes, have been applied with more or less success to these problems. One might well ask oneself if similar methods could not also be utilized in dealing with different economic problems, especially problems of economic development.

In econometrics stochastic variables make indeed an appearance. But there they are not included in a fundamental way, but only as errors or deviations, errors in the equations, errors in the variables, and so on. It is only perhaps in the more recent theory of distributed lags (Jorgenson, 1966) that stochastic processes appear in a slightly more fundamental way in systems of econometric relations.

How are we to interpret the theoretical structures of modern mathematical economics and the empirical studies of the econometricians? Perhaps the best interpretation is the following: They investigate relationships between the mathematical expectations (mean values) of the economic variables involved over time. However perhaps we should introduce stochastic ideas into economics in a slightly more fundamental way.

We have already mentioned inventory problems, where evidently stochastic considerations are of great importance. But there are also problems of production and the work force that involve fundamental stochastic considerations (Holt *et al.*, 1960). Stochastic and chance constrained programming (Charnes and Cooper, 1959) may be applied when programming problems are considered more realistically and seen to imply probability considerations (Tintner, 1960). Labor mobility has been studied from the stochastic point of view by Blumen *et al.* 1955). A similar study of econonically important demographic factors is due to Orcutt *et al.* (1961). It has been suggested that the theory of choice and production should be formulated fundamentally in stochastic terms (Luce and Raiffa, 1957; Tintner, 1941, 1942; Borch, 1968). Hence, examples are not lacking for the application of probability ideas in economics proper.

We should also mention in this connection a number of studies of price movements with the help of stochastic processes (Cootner, 1964; Granger and Morgenstern, 1963; Granger and Hatanaka, 1964). Similar ideas are applied to speculative markets by Murphy (1965). Dynamic programming (Bellman, 1957; Bellman and Dreyfus, 1962) involves sometimes stochastic considerations. See also Howard (1960). Simon (1957) has indicated the use of stochastic processes for certain economic distributions. These ideas are fully exploited in a recent book by Steindl (1965). See also Mandelbrot (1960, 1961).

The following theory due to Haavelmo (1954) explains economic development in stochastic terms. In what follows, $y(t)$ might for instance, be interpreted as national income at time t. The theory shows that even with most simple assumptions the introduction of stochastic (i. e., probability) considerations makes a lot of difference. It also demonstrates that it is at least possible that divergences in economic development might be explained in stochastic terms. Assume that

$$y(t) = ay(t-1) + a_0 \qquad (1)$$

represents a first-order constant coefficients difference equation for an economic time series. Then the solution is

$$y(T) = A a^T + B, \qquad T = 1, 2 \cdots, \qquad (2)$$

where $B = a_0/(1-a)$ and A is given by $Y(0) = A + B$. Now let $u(t)$ be a random variable. If it enters linearly we obtain $y(t) = ay(t-1) + a_0 + u(t)$ and the solution is now

$$y(T) = A a^T + B + \sum_{S=1}^{T-1} a^S u(T-s). \qquad (3)$$

Assume that the random variables $u(t)$ and $u(s)$ are stochastically independent, with zero mean and constant finite variance σ^2. Then if $a^2 < 1$, we have for the variance of $y(T)$

$$\sigma^2_{y(T)} = \sigma^2 \sum_{S=1}^{T-1} a^{2S} = \sigma^2 \left(\frac{1-a^{2T}}{1-a^2} \right). \qquad (4)$$

Now assume that observations are made not at interval i, but i/n. Then we have

$$y(t + i/n) = ay(t - 1 + i/n) + a_0, \qquad t = 1, 2, \cdots,$$
$$i = 0, 1, 2, \cdots n - 1,$$

The solution is now

$$y(T + i/n) = A_i a^T + B; \qquad T = 1, 2, \cdots; \qquad i = 0, 1, 2, \cdots, n - 1, \quad (5)$$

where

$$A_i = y(i/n). \qquad (6)$$

Shocks also occur at intervals i/n. Then we have

$$y(t + i/n) = a\, y(t - 1 + i/n) = a_0 + u(t + i/n) \qquad (7)$$

and the solution is now

$$y(T + i/n) = A_i a^T + B + \sum_{s=1}^{T-1} a^s\, u(T - s + i/n) \qquad (8)$$

and the variance of y is the same as before. It is constant in the interval $T - 1 \leq t < T$.

Assume lags in the shock. Then we have

$$y(t) = ay(t - 1) + a_0 + \sum_{i=1}^{n} u(T - 1 + i/n). \qquad (9)$$

The solution is

$$y(t) = A\, a^T + B + \sum_{s=0}^{T-1} [a_s \sum_{i=1}^{n} u(T - 1 - s + i/n)]. \qquad (10)$$

Let the shocks $u(t)$ be statistically independent with zero mean and constant finite variance σ^2. Then we obtain

$$\sigma^2_{y(T)} = n\sigma^2 \left[\sum_{s=1}^{T-1} a^{2s} \right], \qquad (11)$$

evidently, as $n \to \infty$ this becomes infinite. But if we replace $u(t)$ by $u^*(t) = u(t)/n$, then

$$\sigma^2_{y(T)} = \sigma^2 \left[\sum_{s=1}^{T-1} a^{2s}/n \right] \qquad (12)$$

and as $n \to \infty$, this variance becomes zero.

2.1 Stochastic Systems

Consider a differential equation $\dot{y}(t) = ay(t) + a_0;\ a > 0$, which has the solution

$$y(t) = (A + a_0/a)e^{at} - a_0/a \tag{13}$$

Now assume shocks which have lasting influence upon the paramater a_0, occurring at time $t = 1, 2, \cdots$. Also $u_0 = 0$. Then we have

$$y(T) = [(A + a_0/a)e^{aT} - a_0/a] + [(u_1/a)e^{a(T-1)} - u_1/a]$$
$$+ \cdots +](u_{T-1}/a)e^a - u_{T-1}/a] \tag{14}$$
$$= [(A + a_0/a)e^{aT} + a_0/a] + (1/a)\sum_{s=1}^{T-1} u_s(e^{a(T-s)} - 1). \tag{15}$$

Assume again that the u_t are independent with zero means and constant finite variance σ^2. Then we obtain for the variance of $y(T)$

$$\sigma^2_{y(T)} = \frac{\sigma^2}{a^2}\left[(T-1) - \left(\frac{2(e^{a(T-1)} - 1)}{1 - e^{-a}}\right) + \left(\frac{e^{2a(T-1)} - 1}{1 - e^{-2a}}\right)\right]. \tag{16}$$

This result corresponds to the following assumption
Let

$$\dot{y}(t) = ay(t) + b(t),$$

where

$$b(t) = a_0; \quad 0 \leq t \leq 1,$$
$$b(t) = a_0 + u_1 + u_2 + \cdots + u_j; \quad \begin{array}{l} j \leq t \leq j+1 \\ (j = 1, 2 \cdots). \end{array}$$

The solution is

$$y(t) = e^{at}\int_0^t b(s)e^{-as}\,ds + Ae^{at}. \tag{17}$$

This yields the above result. Now assume that shocks occur at a more frequent rate

$$b(t) = a_0, \quad 0 \leq t \leq i/n,$$
$$b(t) = a_0 + u(1/n) + u(2/n) + \cdots + u(j/n), \quad \begin{array}{l} j/n \leq t \leq (j+1)/n, \\ (j = 1, 2, \cdots). \end{array} \tag{18}$$

we have

$$y(T) = \left[(A + a_0/a)e^{aT} - a_0/a + \{(1/a) \cdot \sum_{i=1}^{Tn-1} u(i/n)(e^{a(T-i/n)} - 1)\}\right], \quad T = 1, 2, \cdots. \quad (19)$$

Then the variance of $y(T)$ goes to infinity as $n \to \infty$. Now assume that the effect of each shock lasts one unit of time. Assume $u_0 = 0$,

$$y(T) = (A + a_0/a)e^{aT} - a_0/a + (1/a)u_1(1 - e^{-a})e^{a(T-1)}$$
$$+ \cdots + (1/a)u_{T-1}(1 - e^{-a})e^a = (A + a_0/a)e^{aT} - a_0/a$$
$$+ (1/a)(1 - e^{-a})\sum_{s=1}^{T-1} u_s\, e^{a(T-s)}. \quad (20)$$

Assume again that the shocks are stochastically independent, with zero mean and finite variance σ^2. Then we obtain the variance of $y(T)$

$$\sigma^2_{y(T)} = \frac{\sigma^2(1 - e^{-a})(e^{2a(T-1)} - 1)}{a^2(1 + e^{-a})}. \quad (21)$$

Now define $b(t)$ as follows

$$\begin{aligned} b(t) &= a_0, & 0 \le t \le 1, \\ b(t) &= a_0 + u_j, & j \le t \le j+1, \end{aligned} \quad (22)$$

and we obtain the same solution. On the other hand, if

$$\begin{aligned} b(t) &= a_0, & 0 \le t \le 1/n, \\ b(t) &= a_0 + u(j/n); & j/n \le t \le (j+1)/n, \end{aligned} \quad (23)$$

we obtain

$$y(T) = \left\{(A + a_0/a)/e^{aT} - a_0/a + (1/a)(1 - e^{-a/n}) \cdot \left(\sum_{i=1}^{Tn-1} u(i/n) \cdot e^{a(T-i/n)}\right)\right\}.$$

The variance of $y(T)$ under the old assumptions is now

$$\sigma^2_{y(t)} = \frac{\sigma^2(1 - e^{-a/n})^2(\exp[2a(Tn-1)/n] - 1)}{1 + e^{-a/n}} \quad (24)$$

and this approaches zero as $n \to \infty$. Now consider the difference equation

2.1 Stochastic Systems

$$y(t + 1/n) = (1 + a/n)y(t) + u(t + 1/n). \tag{25}$$

Assume $y(0)$ as given. The solution is

$$y(T) = (1 + a/n)^{Tn}y(0) + (1/n)\sum_{i=0}^{Tn-1} u(T - i/n)(1 + a/n)^i \tag{26}$$

and the variance of $y(T)$ is

$$\sigma^2_{y(T)} = \frac{\sigma^2[1 - (1 + a/n)^{2Tn}]}{n^2[1 - (1 + a/n)^2]}. \tag{27}$$

The variance becomes zero as $n \to \infty$.

Consider a different set-up. Take a fixed unit of time, for example, a year. It is big shocks that matter. The probability of a shock occurring in a given interval is L/n, L constant. Let u_i be the shock in interval i

$$p(u_i < U^i) = \left(\frac{L}{n}\right) \int_{-\infty}^{U} p(u_i)\,du_i, \quad i = 1, 2, \cdots. \tag{28}$$

But the number of (nonzero) shocks occurring is a random variable

$$p_k = \binom{n}{k}\left(\frac{L}{n}\right)^k \left(1 - \frac{L}{n}\right)^{n-k}, \tag{29}$$

and as $n \to \infty$ we get

$$p_k = \frac{L^k e^{-L}}{k!}. \tag{30}$$

Let $\bar{\sigma}^2$ be the constant variance of shocks and assume shocks are independent. Then the variance of the sum of all shocks, u^* occurring during a year, from t to $t + 1$ is as $n \to \infty$

$$\sigma^2_{u^*} = L\bar{\sigma}^2. \tag{31}$$

Now consider two regions (or countries). Denote by y_1 and y_2 two variables which belong to the two regions. Assume $y_1(0) = y_2(0)$; u_1 are the shocks in region one and u_2, the shocks in region two. We also have

$$\begin{aligned}\dot{y}_1(t) &= ay_1(t) + b_1(t), \\ \dot{y}_2(t) &= ay_2(t) + b_2(t),\end{aligned} \tag{32}$$

where

$$b_1(t) = a_0, \quad 0 \leq t \leq 1,$$
$$b_1(t) = a_0 + u_1(1) + u_1(2) + \cdots + u_1(n), \quad n \leq t \leq n+1$$
$$b_2(t) = a_0, \quad 0 \leq t \leq 1, \quad (33)$$
$$b_2(t) = a_0 + u_2(1) + u_2(2) + \cdots + u_2(n), \quad n \leq t \leq n+1$$

Now assume that u_1 and u_2 are random variables with zero mean, no autocorrelation, no serial correlation and finite variance σ^2.

Let their covariance be σ_{12}. Then we compute for the variance of the difference

$$\text{var}[y_1(T) - y_2(T)] = \sigma^2_{y_1(T) - y_2(T)}$$
$$= \frac{2(\sigma^2 - \sigma_{12})}{a^2}\left[(T-1)\frac{-2(e^{a(T-1)}-1)}{1-e^{-a}} + \frac{e^{2a(T-1)}-1}{1-e^{-2a}}\right]. \quad (34)$$

If there is perfect positive correlation between the shocks, the variance of the difference is zero. The smaller *ceteris paribus* the covariance, the more probable is a large dissimilarity between the regions.

Now assume independence ($\sigma_{12} = 0$); if $a > 0$ and if $y_1(T) > y_2(T)$ at a given T, the difference can be expected to continue and increase with time. If $y_1(T)$ is much larger than $y_2(T)$, then it is very unlikely that the position will be reversed in the future.

This simple model due to Haavelmo tries to show that two structurally very similar models may diverge because of stochastic influences.

The deterministic models for economic development seem to be inadequate to deal with some of the fundamental problems. We follow the general ideas of Haavelmo (1954), who was the first to emphasize the fundamental importance of stochastic influences in this field. Indeed, we might consider the dederministic models discussed above as rough approximations, which deal with the mathematical expectations of the random variables involved.

As Bharucha–Reid (1960) has shown in his most valuable survey, the idea of stochastic processes has many applications. A stochastic process is a family of random variables, depending on a parameter (in our case time). The applications in operations research are most pertinent for economics. (Churchman *et al.*, 1957; Holt *et al.*, 1960; Kromphardt *et al.*, 1962; Lesourne, 1960; Massé, 1959; Saaty, 1961; Takacs, 1962; Whittle, 1963). After all, operations research is simply the econometrics of the enterprise.

If workers in operations research utilize stochastic processes in queuing theory, inventory theory, machine repair theory, and so on, it is difficult to see why similar methods are not used in economics proper. Most economic models are derived from classical physics. Wiener (1964, p. 88) puts it this way,

"I have found mathematical sociology and mathematical economics or econometrics suffering under a misapprehension of what is the proper use of mathematics in the social sciences. . ." In a recent book (Murphy, 1965) there is an attempt to use methods of statistical physics in economics, especially in connection with speculation and the capital market. See also Cootner (1964). Steindl (1965) applies ideas borrowed from the theory of stochastic processes to the growth of the firm. Similar methods are also to be found in Orcutt *at el.* (1961).

In the following pages we will utilize various stochastic processes essentially in an effort to deal with macroeconomic problems of economic development. In a sense, this might be considered as a continuation of the earlier model (Tintner, 1942, 1944, 1948). There, an effort was made to explain by the same model economic development (trend) and fluctuations (business cycle) with the help of an essentially deterministic model, where random elements play only a very subordinate part.

Mandelbrot (1960) has given some very good reasons for using the pareto distribution (Pareto, 1927) in econometric work. See also Champernowne (1953), Simon (1957, pp. 145—164), and Klein (1962, p. 140). Steindl (1965) also favors this distribution. Our work leads more or less to the lognormal distribution, which seems to explain better the main body of the distribution of income (O' Carroll, 1963). See also Quandt (1966).

2.2 Discrete Stochastic Processes and Their Application to Trends of Economic Development

We consider first a simple Poisson process (Tintner *et al.*, 1965). Let x be a certain event and contemplate its possible change in the small time interval between t and $t + \Delta t$. The probability of a transition from x to $x + 1$ is $L(\Delta t) + o(\Delta t)$. The probability of no change is $1 - L(\Delta t) + o(\Delta t)$; all other possible transitions are of order $o(\Delta t)$. L is a constant. This is a pure birth process. we obtain for the probability $p_x(t)$ that x has a given value at the time t, the following equation

$$p_x(t + \Delta t) = L(\Delta t)p_{x-1}(t) + [1 - L(\Delta t)]p_x(t). \tag{1}$$

If we go to the limit, that is for $\Delta t \to 0$, we obtain the differential-difference equation

$$dp_x(t)/dt = Lp_{x-1}(t) - Lp_x(t). \tag{2}$$

we introduce the generating function

$$F(s, t) = \sum_{x=0}^{\infty} p_x(t)s^x \tag{3}$$

and obtain from Eq. (2) the differential equation

$$dF(s, t)/dt = L(s - 1)F(s, t). \tag{4}$$

Integration gives

$$F(s, t) = k\, e^{L(s-1)t} \tag{5}$$

and from the condition that at all times t the sum of the probabilities is one

$$F(1, t) = \sum_{x=0}^{\infty} P_x(t) = 1, \tag{6}$$

we obtain $k = 1$, so that finally the generating function is

$$F(s, t) = e^{L(s-1)t}. \tag{7}$$

This is the generating function of a Poisson distribution and we obtain for the probabilities

$$p_x(t) = e^{-Lt}(Lt)^x/x\,!. \tag{8}$$

The mean and variance of this distribution are Lt.

If we consider applications to economic development we might construct a process, where transitions from x to $x + 1$, $x + 2 \cdots x + m$ are possible with not negligible probabilities. These are very complicated stochastic processes. Hence, we choose as a first approximation a process where the transition from y to $y + u$, $u > 0$ and no other transitions will be possible with not negligible probability. As a rough approximation we have a process where the relevant economic variable (say national income) can only assume the discrete values $0, u, 2u, 3u, \cdots$.

More formally, consider y some measure of economic development, for instance national income. The probability of a transition in the small time interval t to $t + \Delta t$ is envisaged. The probability that y will increase from y to $y + u$ is $L(\Delta t) + o(\Delta t)$; the probability of no change is $1 - L(\Delta t) + o(\Delta t)$; all other possible transition probabilities are $o(\Delta t)$.

Let $q_y(t)$ be the probability that y has a given value at point t. We obtain by similar methods as before the differential equation

$$dp_y(t)/dt = Lq_{y-u}(t) - Lq_y(t). \tag{9}$$

We introduce the new generating function

$$G(s, t) = \sum_{y=0}^{\infty} q_y(t) s^y, \qquad (y = 0, u, 2u, \cdots) \tag{10}$$

and obtain the differential-difference equation

$$dG(s, t)/dt = L(s^u - 1)G(s, t). \tag{11}$$

Integration gives

$$G(s, t) = \exp[L(s^u - 1)t]. \tag{12}$$

This is the generating function of a modified Poisson process, that is, a process where y can only assume the values $0, u, 2u, \cdots$. Hence, the probabilities are now

$$q_y(t) = e^{-Lt}(Lt)^{y/u}/(y/u)!, \qquad y = 0, u, 2u, \cdots. \tag{13}$$

The mean value of this modified Poisson distribution is a function of time

$$M_t = uLt \tag{14}$$

and its variance is also a function of time

$$V_t = u^2 Lt. \tag{15}$$

suppose that we observe the quantities $y_1, y_2, \cdots, y_N$ at discrete intervals $t = 1, 2, \cdots, N$. We estimate the mean value (M_t) by the method of least squares

$$\widehat{uL} = \sum_{t=1}^{N} ty_t \bigg/ \sum_{t=1}^{N} t^2. \tag{16}$$

Because of the obvious relationship

$$V_t = u M_t, \tag{17}$$

we define $\hat{V}_t$ and $\hat{M}_t$ as

$$\hat{V}_t = (y_t - uLt)^2, \qquad \hat{M}_t = uLt \tag{18}$$

and estimate u by

$$\hat{u} = \sum_{t=1}^{N} \hat{V}_t \bigg/ \sum_{t=1}^{N} \hat{M}_t. \tag{19}$$

We have utilized this theory in order to construct a simple stochastic model for the value of Pennsylvania manufacturing 1899–1959 in five year intervals (see Table 1).

2. Stochastic Models of Economic Development

TABLE 1
VALUE OF PENNSYLVANIA MANUFACTURING

Year	t	y_t	M_t
1899	1	691.58	629.91
1904	2	812.61	1259.82
1909	3	1044.18	1889.73
1914	4	1141.00	2519.64
1919	5	3029.98	3149.55
1924	6	3036.21	3779.46
1929	7	3430.61	4409.37
1934	8	1662.48	5039.28
1939	9	2476.86	5669.19
1944	10	5270.67	6299.10
1949	11	6941.47	6929.01
1954	12	9930.49	7558.92
1959	13	12643.65	8188.83

SOURCE: "Census of Manufacturers." Bureau of the Census, U. S. Department of Commerce, Washington, D. C.

Estimates with the help of Eq. (16) give

$$\widehat{uL} = 515897 \cdot 550/819.000 = 629.9115. \qquad (20)$$

Also we have from Eq. (19)

$$\hat{u} = 52359870.173/57321.810 = 913.4380. \qquad (21)$$

Hence, our estimates are

$$\hat{u} = 913.4380, \qquad \hat{L} = 0.6896. \qquad (22)$$

This has to be interpreted in the following way. Let y_t be the value of manufacturing of Pennsylvania in year t. Consider changes in a small time interval t to $t + \Delta t$. The probability of increase of y to $y + 913.4380$ is approximately $0.6896\ (\Delta t)$. The probability of no change is approximately $1 - 0.6896(\Delta t)$, all other possible transitions are negligible. Then the probability distribution of value of manufacturing in Pennsylvania is

$$q_y(t) = e^{-0.6896t}(0.6896t)^{y/913.4380}/(y/913.4380)!$$
$$y = 0, 913.4380, 1826.8760, \cdots$$

Finally, we predict for $t = 16$ (year 1974) a mean value of 10078.5760. The 95 % limits are in this case 4597.9840 and 15559.2040. The 90 % limits are given by 5511.3860 and 14645.7660.

2.2 Discrete Stochastic Processes

In the following we consider a very simplified system which yields a linear trend. We consider only steps upwards and no steps down in economic development (Mukherjee et al., 1964).

Consider a simple birth process, where only transitions from x to $x + u$, $u > 0$ are possible without negligible probabilities. More formally, we have a quantity x and consider its possible change in a small interval of time, from t to $t + \Delta t$. There are two possibilities. A transition from x to $x + u$ with probability $b(\Delta t) + o(\Delta t)$; or no change, with probability $1 - b(\Delta t) + o(\Delta t)$. All other transitions have probabilities of order $o(\Delta t)$.

The joint probability of k observations $y_1, y_2, \cdots, y_k$ is given by

$$P = \prod_{r=1}^{k} \frac{\exp\left[-b(y_r - y_{r-1})/u\right]}{\left[\frac{y_r - y_{r-1}}{u}\right]!}, \quad (23)$$

where $y_0 = a$.

The maximum likelihood estimates are given by

$$\frac{\partial \log P}{\partial b} = -k + \frac{y_k - a}{ub} = 0, \quad (24)$$

$$\frac{\partial \log P}{\partial a} = \frac{-1}{u} \cdot \log b + \frac{\partial}{\partial a} \log \left(\frac{y_1 - a}{u}\right)! = 0, \quad (25)$$

$$\frac{\partial \log g}{\partial u} = \frac{-(y_k - a)}{u^2} \log b + \sum_{r=1}^{k} \frac{\partial}{\partial u} \log \left(\frac{y_r - y_{r-1}}{u}\right)! = 0. \quad (26)$$

This method is not here very practical, and hence, we consider the following simpler, but perhaps not very efficient, methods. We have the mean

$$M_t = a + ubt. \quad (27)$$

Using the method of least squares and minimizing

$$\sum_{t=1}^{k} (y_t - M_t)^2, \quad (28)$$

we derive the following relations for the estimates of a and ub

$$k\hat{a} + \widehat{ub} \sum_{t=1}^{k} t = \sum_{t=1}^{k} y_t, \quad (29)$$

$$\hat{a} \sum_{t=1}^{k} t + \widehat{ub} \sum_{t=1}^{k} t^2 = \sum_{t=1}^{k} ty_t. \quad (30)$$

This yields the estimates $\hat{a}$ and $\widehat{ub}$. Further, we have the variance

2. Stochastic Models of Economic Development

$$V_t = u^2 bt = E[y_t - M_t]^2. \tag{31}$$

Hence, an estimate for $(u^2 b)$ is given by

$$u^2 b = \frac{1}{(k-2)} \sum_{t=1}^{k} \frac{1}{t}(y_t - a - ubt)^2, \tag{32}$$

where $k - 2$ is the factor for appropriate degrees of freedom. It makes allowance for the number of parameters estimated in the regression.

Applying the method of least squares indicated above to the data given in Table 2 on p. 43, the following trend emerges.

$$M_t = 78.0385 + 4.270t. \tag{33}$$
$$(0.454)$$

The standard error of the coefficient of t is indicated in brackets below. The correlation coefficient is 0.94. The estimate of u is $\hat{u} = 1.785$ and the estimate of $\hat{b} = 2.392$.

We try to predict, on the basis of our theory, Indian national income for 1964/65 ($t = 17$). We obtain from the above formula $M_{17} = 150.628$. National income is a Poisson variable, distributed with a parameter $bt = 40.664$. We use the normal approximation because of lack of tabulated values and determine the 95% confidence or fiducial limits for national income in India 1964/65 as 128.318 and 172.938.

If we make the trend also a function of government expenditure (G) we arrive at the following estimate of the trend

$$M_t = 68.6009 + 2.005 G_t + 1.952t. \tag{34}$$
$$(1.182) \quad (1.431)$$

The multiple correlation coefficient is here 0.95. The estimates for u and b are $\hat{u} = 2.820$ and $\hat{b} = 0.692$.

We try again to predict the trend of national income in India for 1964/1965 ($t = 17$). Using our formula, we investigate two hypotheses about government consumption in 1964/1965

(a) $G_t = 25$; (b) $G_t = 30$.

When $G_t = 25$, the estimated trend value is 151.910 and the 95% confidence limits are 132.952 and 170.868. If on the other hand $G_t = 30$, the estimated trend value is 161.935 and the 95% confidence limits are 142.977 and 180.893.

If we apply weighted regression to the problem of estimating the parameters, we minimize

2.2 Discrete Stochastic Processes

$$\sum_{t=1}^{k} \frac{(y_t - M_t)^2}{t}, \tag{35}$$

and the estimates are derived from the following system of equations

$$\hat{a} \sum_{t=1}^{k} \frac{1}{t} + \widehat{ub} \cdot k = \sum_{t=1}^{k} \frac{y_t}{t}, \tag{36}$$

$$k\hat{a} + \widehat{ub} \sum_{t=1}^{k} t = \sum_{t=1}^{k} y_t, \tag{37}$$

$$V = \widehat{u^2 b} = \frac{1}{(k-2)} \sum_{t=1}^{k} \frac{1}{t} (y_t - M_t)^2. \tag{38}$$

Applying this method to our data, we have for the simple linear trend

$$M_t = 82.0598 + 3.734t. \tag{39}$$

The correlation coefficient is 0.95 and the estimates of the parameters are $\hat{u} = 1.542$ and $\hat{b} = 2.421$.

Using this formula, we predict national income in India for 1964/65 ($t = 17$) as 145.378. Using again the normal approximation, we obtain for the 95% confidence limits, 125.989 and 164.767.

If we introduce government expenditure (G_t) into the trend we have

$$M_t = 73.3013 + 1.372 G_t + 2.455t. \tag{40}$$

The multiple correlation coefficient is 0.96. Our estimates are $\hat{u} = 2.128$ and $\hat{b} = 1.153$.

We may again investigate the aforementioned two hypotheses concerning government consumption. If $G_t = 25$, then the estimated value is 149.336 and the 95% confidence limits are 130.870 and 167.802. If on the other hand, government expenditure should rise to $G_t = 30$, we obtain a trend value of 156.196 and the following 95% confidence limits are 137.730 and 174.662.

Consider now a slight generalization of the Posisson process. We consider the situation at the time t. The probability of the transition from x to $x + u$ is $b(t)(\Delta t) + o(\Delta t)$. The probability of no change is now $1 - b(t)(\Delta t) + o(\Delta t)$. Note that now the birth rate $b(t)$ is a function of time. Also, as before, we assume that at time $t = 0$ we have $x = a$ with probability one.

The probability of x at time t is

$$p_x(t) = \left[\exp\left\{ -\int_0^t b(z)\,dz \right\} \right] \left[\int_0^t b(z)\,dz \right]^{(x-a)/u} \bigg/ [(x-a)/u]\,!,$$
$$x = a, a+u, a+2u, \cdots. \tag{41}$$

2. Stochastic Models of Economic Development

The mean of the stochastic process is

$$N_t = a + u \int_0^t b(z)\, dz \tag{42}$$

and the variance is

$$V_t = u^2 \int_0^t b(z)\, dz. \tag{43}$$

The third and fourth moments about the mean are given by

$$\mu_3 = u^3 \int_0^t b(z)\, dz, \tag{44}$$

$$\kappa_4 = \mu_4 - 3\mu_2^2 = u^4 \int_0^t b(z)\, dz. \tag{45}$$

Hence, the skewness is

$$\gamma_1 = \left(\int_0^t b(z)\, dz\right)^{-1}, \tag{46}$$

$$\gamma_2 = \left(\int_0^t b(z)\, dz\right)^{-1}. \tag{47}$$

It is evident, that the distribution tends towards normal if the integral becomes infinite.

As a special case, consider now the influence of government consumption G_t. We assume this influence linear and have

$$b(t) = b_0 + b_1 G(t). \tag{48}$$

We assume that $G(t)$ is a step function: $G(t) = G_r$ for $r - 1 \leq t < r$. We then have

$$\int_0^t b(z)\, dz = b_0 t + b_1 \sum_{r=1}^{t} G_r. \tag{49}$$

Using the simple least squares mothod, we compute the estimates for our stochastic process. The mean value is

$$M_t = 88.8026 - 3.141 t + 0.550 \sum_{r=1}^{t} G_r. \tag{50}$$

The multiple correlation coefficient is 0.98. other estimates are $\hat{a} = 88.8026$ and $\hat{u} = 2.289$. The function $b(t)$ which includes the linear influence of government consumption is

2.2 Discrete Stochastic Processes

$$b(t) = -1.3724 + 0.240G(t). \tag{51}$$

We have again estimated some results for 1964/1965 ($t = 17$). Assuming (a) $G_t = 25$ on the average in the three intervening years we have for our estimate of the trend of Indian national income 179.671. The 95% limits are 151.431 and 207.911.

If we assume that government consumption will be on the averge 30 in the three years, we have an estimate of the trend as 187.921. The 95% limits of this estimate are 158.426 and 217.416.

In this case we cannot apply easily the weighted regression approach as

$$V(y) = u^2 \left[b_0 t + b_1 \sum_{r=1}^{t} G_r \right] \tag{52}$$

depends upon two independent unknown constants $u^2 b_0$, and $u^2 b_1$. However, an analog can be had by using

$$V(y_t) \propto u^2 b_0 t \tag{53}$$

as a first approximation. To some extent this may be justified by the high correlation likely to be present between the expression $\sum_{r=1}^{t} G_r$ and t.

Applying this "approximate" weighted regression, we get the following results.

TABLE 2

Net National Output (Y_t) and Government Expenditures (G_t) For India At Current Prices: 1948-49 to 1961-62

Year	Y_t	G_t	t
1948-49	86.5	8.5	1
1949-50	90.1	8.1	2
1950-51	95.3	8.3	3
1951-52	99.7	8.8	4
1952-53	98.2	9.0	5
1953-54	104.8	9.8	6
1954-55	96.1	11.0	7
1955-56	99.8	12.9	8
1856-57	113.1	14.8	9
1957-58	113.9	17.5	10
1958-59	126.0	18.5	11
1959-60	129.5	17.4	12
1960-61	141.6	19.3	13
1961-62	146.3	23.4	14

Note: y and G are in billions of rupees.
Source: "Estimates of National Income," C.S.O., Government of India, April, 1963.

$$M_t = 85.7538 - 1.323t + 0.420 \sum_{r=1}^{t} G_r. \tag{54}$$

The multiple correlation is 0.98 and the other estimates are $\hat{a} = 85.7538$ and $\hat{u} = 1.244$. The function $b(t)$ now turns out to be

$$b(t) = -1.0641 + 0.338G(t). \tag{55}$$

Under our two hypotheses about the government expenditure, the trend value and the confidence limits of the Indian national income in 1964/65 are as follows. If $G = 25$ on the average for three intervening years, the trend of Indian national income is 173.429 and the 95% confidence limits are 152.948 and 193.910.

The assumption that $G_t = 30$ on the average during the three years gives 179.729 as the estimated trend value. The 95% confidence limits are 158.524 and 200.934.

We now apply a model based upon the idea of stages of economic development to British industry (Tintner and Thomas, 1963; Sengupta and Tintner, 1966).

The following theory is a combination of ideas of Rosenstein-Rodan (1943) and Rostow (1960) with Haavalmo (1954). Let us assume that, in the short run, the economic system can be represented by a system of linear stochastic difference equations (Tintner, 1961) It should be noted that the system is only an approximation to economic relations which are probably much more complicated (Klein and Goldberger, 1955; Duesenberry et al., 1965). Now, as Tinbergen (1939 has shown, it is possible to replace the system by a single stochastic linear difference equation of higher order. We have used the data of W. Hoffmann (1955) for British industrial production, 1700–1939, excluding building (base: 1913 = 100) in order to verify the theory. Also, for the sake of simplicity, we have replaced the stochastic difference equation by the simple one

$$x_t = a^{(k)} x_{t-1} + u_t. \tag{56}$$

This equation is of the first order and represents the short-term exponential trend. Here $a^{(k)}$ is a constant u_t, a random variable. Equation (56) holds only if

$$c_k \leq x_{t-1} \leq c_{k+1}, \tag{57}$$

If x_{t-1} falls outside the interval defined in Eq. (57), the constant $a^{(k)}$ has to be replaced by another constant $a^{(k-1)}$ or $a^{(k+1)}$. This method for estimating the switch-over points c_k is an adaptation of the method of Quandt (1958, 1960). All possible values of c_k are studied and the one chosen which gives maximum

2.2 Discrete Stochastic Processes

likelihood. The constants $a^{(k)}$ are computed by the method of maximum likelihood.

We give in the following Table 3 the empirical results for a number of regimes (Tintner and Thomas, 1963).

TABLE 3

Number of regimes	a_k	95% Confidence	Limits	Corresponding year	c_k
2	1.030234	1.020258	1.040210	1834	14.20
	1.010767	0.993004	1.028530		
3	1.023861	1.006460	1.041262	1791	4.13
	1.029882	1.018080	1.040684	1869	40.60
	1.009564	0.987216	1.031912		
4	1.010724	0.989705	1.031743	1777	2.47
	1.026710	1.008864	1.044556	1820	8.13
	1.030094	1.016952	1.043236	1869	40.60
	1.009564	0.987216	1.031912		

It should be remembered that the confidence limits given in this table are only approximate, since only the asymptotic theory is available.

Tests also show that the difference between the values of $a^{(k)}$ belonging to neighboring regimes are not significant, showing that the transition is a gradual one. But the tests again are only approximate. On the other hand, the variance of the u_t connected with different regimes are probably different, as statistical tests show. The general conclusion from this analysis seems to be that exponential trends resulting from a first-order stochastic difference equation seem to give a reasonably good short-term explanation of the trend of economic development. But more elaborate systems of equations must be used in order to give a complete description and explanation of economic development. Since these are nonlinear stochastic difference equations, the theoretical and practical difficulties of such an analysis are considerable.

In the theory of stochastic process, it is known that for linear models restricted to very short periods, the solution of a deterministic model can, under fairly general conditions, be shown to be identical with the mean solution of the corresponding stochastic model (Bharucha-Reid, 1960). However, such results do not hold for even very simple types of nonlinearities. This generalization has far reaching implications from an operational standpoint, because when a growth model is applied for predictive or policy purposes, one can formulate alternative decision rules for specifying the optimal course of policy under various conditions of uncertainty.

The stochastic considerations may be introduced at various levels into the structural equations of a growth model, for example, by considering the observed set of coefficients as the expected values of corresponding stochastic coefficients or by introducing equational errors as shocks to the otherwise deterministic model. But perhaps the most generalized specification would be to consider the basic variable (or variables) of a growth model to be subject to stochastic processes of different types. Let T denote a set of points on a time axis such that at each point t in T an observation is made of a random variable $X(t, w)$ occupying a point w in the entire state space Ω, then a stochastic process is completely specified by the family of random variables $\{X(t, w), t \in T, w \in \Omega\}$. Apart from the question of inferring the probability law of the stochastic process from a set of observations [or realizations of the process $X(t, w)$ for fixed w], we may consider the set of alternative solutions of a growth model when the assumptions behind the underlying stochastic process are varied in different ways. This would be useful, not only from the standpoint of predictive power, but also from that of policy applications and optimality intepretations. We shall consider the process of economic growth in terms of stochastic birth and death processes under simplifying assumptions, regarding the type of stochastic processes and the number of independent variables determining the process of overall growth. Apart from empirical applications, this will be followed by a discussion of a simple interdependent growth model, when probabilistic consideration are introduced at a simpler level through capacity and/or demand variations.

In order to show the relevance and implications of stochastic process in the aggregative theory of economic development, let us consider the simplest assumption that economic development is measured by a single variable $x(t)$ which may denote real national income overall or per capita (Sengupta and Tintner, 1964). This is here assumed to be a discontinuous variable. Let $p_x(t)$ denote the probability that $X(t)$ will have a given value x.

$$p_x(t) = \text{prob}\{X(t) = x\} \qquad \text{for } x = 0, 1, 2, \cdots. \tag{58}$$

We now make the following assumptions about the stochastic process $\{X(t), t \in T\}$, that is, about the possible changes in the value of $X(t)$ during a small time interval between t and $(t + \Delta t)$.

(i) Assumption about stationary independent increments which postulates the following conditions:

(a) If at time t the system $\{X(t), t \in T\}$ takes the value $x = 1, 2, 3, \cdots$ the probability of transition from x to $(x + 1)$ in the small interval $(t, t + \Delta t)$ is given by $\lambda_x \Delta t + o(\Delta t)$, where the symbol $o(\Delta t)$, denotes a value of smaller order of magnitude than Δt.

(b) If at time t, the system $\{X(t), t \in T\}$ takes the value x, $x = 1, 2, 3, \cdots$ the probability of transition from x to $(x - 1)$ in the small interval $(t, t + \Delta t)$ is $\mu_x \Delta t + o(\Delta t)$.

(c) The probability of transition between any two specific values (or states) of the system, x and $(x + s)$ is independent of the initial position.

(ii) The probability of no transition to a neighboring value (or state) is given by $1 - (\lambda_x + \mu_x)\Delta t + o(\Delta t)$.

(iii) The probability of a transition to a value other than a neighboring value is $o(\Delta t)$ which becomes negligible in the limit when t tends to zero.

(iv) The transition from $x = 0$ to $x = 1$ is not possible, that is, in the terminology of Markov chains the value (or state) $x = 0$ is an absorbing state from which no exit is possible.

In view of these assumptions we can derive a recurrence relation for the transition probabilities of this birth and death process as follows.

$$p_x(t + \Delta t) = \lambda_{x-1} p_{x-1}(t)\Delta t + [1 - (\lambda_x + \mu_x)\Delta t] p_x(t) \\ + \mu_{x+1} p_{x+1}(t) + o(\Delta t). \tag{59}$$

Taking the limit $\Delta t \to 0$ this relation leads to the following difference-differential equation for the probability $p_x(t)$.

$$\frac{dp_x(t)}{dt} = \lambda_{x-1} p_{x-1}(t) - (\lambda_x + \mu_x) p_x(t) + \mu_{x+1} p_{x+1}(t). \tag{60}$$

The initial conditions governing Eq. (60) are specified by

$$p_x(0) = \partial_{xx_0} = \begin{cases} 1 & \text{for } x = x_0, \\ 0 & \text{otherwise}, \end{cases} \tag{61}$$

when the system $\{X(t), t \in T\}$ takes the value $x = x_0$, $0 < x_0 < \infty$ at time zero. Let us further extend the notation for transition probabilities as

$$p_{i,k}(t) = \text{prob}\,[X(t + s) = k \mid X(s) = i], \tag{62}$$

so that $p_{ik}(t)$ denotes the conditional probability of $\{X(t + s)\}$ taking a specific value k, given that $\{X(s)\}$ has been observed at time point s to take the value i. Then it can be shown (Parzen, 1960) that the functions λ_x and μ_x given in Eq. (60) admit of the following interpretation.

$$\lim_{\Delta t \to 0} \frac{1}{\Delta t}[p_{x,x+1}(t + \Delta t)] = \lambda_x(t) \quad \text{for } x \geq 0,$$

$$\lim_{\Delta t \to 0} \frac{1}{\Delta t}[p_{x,x-1}(t + \Delta t)] = \mu_x(t) \quad \text{for } x \geq 1, \tag{63}$$

$$\lim_{\Delta t \to 0} \frac{1}{\Delta t}[1 - p_{x,x}(t + \Delta t)] = \lambda_x(t) + \mu_x(t) \quad \text{for } x \geq 0,$$

where we have for all $t \geq 0$ the plausible condition

$$\mu_0(t) = 0. \tag{63}$$

Now it must be apparent from the system of Eqs. (63) that $\lambda_x(t)$ and $\mu_x(t)$ specify the probabilities of transition from a specific value of national income x to $x + 1$ and $x - 1$, respectively. By making alternative assumptions about these arbitrary functions of time $\lambda_x(t)$ and $\mu_x(t)$, we may generate alternative probabilistic models of economic growth as follows.

(i) Linear birth process type growth model. Here we make the assumption that $\mu_x(t) = 0$ and $\lambda_x(t) = \lambda$ for all t and x. In this case the solution of Eq. (60) is known to be

$$p_x(t) = \binom{x-1}{x-j} \exp(-j\lambda t)(1 - \exp(-\lambda t))^{x-j} \quad \text{for } x > j \geq 1, \tag{64}$$

where j denotes the value of x at time zero. Since by appropriate choice of units one take $j = 1$, one may write this more simply as

$$p_x(t) = \exp(-\lambda t)(1 - \exp(-\lambda t))^{x-1} \quad \text{for } x = 1, 2, 3, \cdots, \infty. \tag{65}$$

where $\exp(m)$ is a notation for e^m. Denoting the expected value and variance of $X(t)$ by $M(t)$ and $V(t)$, respectively, we can compute their values from Eq. (64) as follows.

$$\begin{aligned} M(t) &= \sum_{x=0}^{\infty} x\, p_x(t) = j\, e^{\lambda t}, \\ V(t) &= \sum_{x=0}^{\infty} \{x - M(t)\}^2 p_x(t) = j\, e^{\lambda t}(2j e^{\lambda t} - 1). \end{aligned} \tag{66}$$

To consider the conomic meaning of the proportional growth rate (or birthrate) λ in terms of economic models, we interpret λ as the product of the two structural coefficients of the Harrod–Domar type growth model, that is, the marginal output-capital ratio (σ) and the saving coefficient (s). By writing $\lambda = (s\sigma)$ we can specify the deterministic growth model of the Harrod–Domar type as

$$(dx/dt) = \lambda x. \tag{67}$$

Solution: $x(t) = je^{\lambda t}$, where $j = x(t)$ at $t = 0$.

By comparing Eqs. (66) and (67) two interesting results become readily apparent. In the first place, the mean value function of the stochastic model, Eq. (64), is exactly identical with the solution of the deterministic growth model

Eq. (67) of the Harrod–Domar type. Secondly, it is apparent that the linearity assmption for $\lambda_x = \lambda$ makes the two models applicable to very restrictive situations (i. e., short periods). For applications to long-run framework, however, we may still retain the linearity assumption by postulating that the entire long- run time scale T is divided into several subperiods or regimes, for each of which a linearized sequence of birthrates (or growth rates) may be a good approximation. Thus, the new assumption would be

$$\lambda_x = \lambda^{(k)} x(t) \quad \text{for} \quad c_k \le x(t) \le c_{k+1}, \tag{68}$$
$$(k = 1, 2, \cdots, K),$$

where c_k and c_{k+1} are arbitrary constants such that the growth rate parameter $\lambda^{(k)}$ is a good approximation for the kth regime when the total number of regimes is K (Tintner and Thomas, 1963). As soon as $x(t)$ moves outside the region defined in Eq. (68) we must make a new approximation, that is, we must replace the constants $\lambda^{(k)}$ by new constants, for example, $\lambda^{(k+1)}$ or $\lambda^{(k-1)}$ and so on. However, it is not difficult to show that the idea of regime changes defined by Eq. (68) is implicit in the growth model formalized by Domar (1946), since he makes a distinction between the actual (or observed) output-capital ratio and the potential maximum of output capital ratios, the gap between the two being caused by misdirection of investment or in-optimal decision making under various conditions of uncertainty and lack of perfect knowledge.

(ii) Linear birth and death process type growth model. Here we make the assumption that $\mu_x(t) = \mu x(t)$ and $\lambda_x(t) = \lambda x(t)$, that is the proportional rates of birth (λ) and death (μ) are constant parameters characterizing the stochastic distribution of national income over time. From an economic viewpoint the birthrate λ reflects the average increase in real national product (or income) resulting from an additional dose of real capital formation. By the same interpretation the death rate μ represents an average decrease in real national product (or income) resulting from any decrease in investment caused by the increased size of the total capital stock. Any misdirection of investment, nonoptimal depreciation policy or sub-optimal capacity variations which result in a reduced proportional rate of growth of real national income may thus be subsumed under the rate μ. The formulas (64) through (67) now hold as before, except that the net birthrate ($\bar{\lambda}$) has to be defined as $\bar{\lambda} = \lambda - \mu$. Hence, we have to replace by $\bar{\lambda}$ in the above formulas and write the variance function as

$$V(t) = [(\lambda + \mu)(\lambda - \mu)^{-1} \exp(t(\lambda - \mu)) - \{\exp(t(\lambda - \mu)) - 1\}]. \tag{69}$$

Similarly, one may define a set of changes of regime, both by conditions like Eq. (68) on the birthrate and similar conditions on the death rate parameters

μ. The deterministic version of this model would now be equivalent to the Duesenberry (1949) type of modification into the conventional Domar-type growth model. For instance, denoting the kth regime values of the two parameters as $\lambda^{(k)}$ and $\mu^{(k)}$, $k = 1, 2, 3, \cdots, K$, it is easy to see that the mean value function $M^{(k)}(t)$ for the kth regime may show a negatively proportional rate of growth or even no growth at all, if $\mu^{(k)} > \lambda^{(k)}$, or $\mu^{(k)} = \lambda^{(k)}$. Now, if the concept of regime is identified with a succession of short periods into a long-run framework, the cyclical fluctuations due to the negative and positive discrepancy between $(\lambda^{(k)} - \mu^{(k)})$ for different $k = 1, 2, \cdots, K$ may persist under different conditions characterizing the long-run trend of national income. For instance, if λ_0 and μ_0 be defined to be the average values of $\lambda^{(k)}$ and $\mu^{(k)}$ for all $k \in K$ (when the set K is finite), then the long-run trend of the mean national income $M(t)$ would follow a strictly increasing (decreasing) exponential time-path accordingly, as $\lambda_0 \gg \mu_0$, $(\mu_0 \gg \lambda_0)$, where the symbol $\gg$ is an abbreviation for "sufficiently greater than." If we make the further assumption that the subperiods classified by the sequence of regimes are equi-spaced and the set K is enumerably infinite, then even a mild restriction $\lambda_0 > \mu_0$ (or $\mu_0 > \lambda_0$) would ensure a positive (or negative) proportional rate of growth of the mean national income.

At this stage a relevant question of economic policy may be raised. Suppose on the basis of the above probabilistic growth model we found that the mean growth path is an increasing exponential function of time, then what sort of policy variables i. e., instrument variables (Tinbergen, 1956) should be chosen so as to ensure that the actual growth path converges to the mean growth path. Such convergence may imply a type of stabilizing policy pursued by the national policy-maker (or the planning board). This question may be answered on two different levels. In the first place, we may consider the problem of estimating the parameter $\lambda = s\sigma$ as a part of decision theory, which requires as a datum of the problem the specification of the loss for a given difference between the estimate and the true value of the unknown parameter. The optimality criterion in such a case is naturally the minimization of expected loss. Alternative linear decision rules can be formulated by using the minimum expected squared error and other criteria disussed elsewhere, in order to specify the optimal choice of the estimates of s and σ. Secondly, the paramter λ occurring in the mean value function $M(t)$ in Eq. (66) may be interpreted in such a way that the mean size of national income $je^{\lambda t}$ may vary from one subperiod (or regime) to another due to the variations of the output-capital coefficient σ. The mean value function $M(t)$ can now be written, if λ is regarded as the observed value of a random variable L with distribution function $F(\lambda)$.

$$E[X(t) \mid L = \lambda] = M(t) = je^{\lambda t}. \tag{70}$$

This, of course, would lead to compound distributions (Takacs, 1965) because in effect we are now assuming a mixture of two or more homogeneous statistical populations over subperiods, each with the conditional probability

$$\text{prob}[X(t) = x \mid \lambda = L] = p_x(t).$$

Although this step is a slight generalization of our original assumption, this permits an easier policy application when the variations in λ can be associated with variations in the output coefficient σ over a given period. For instance, let us consider now the aggregative coefficient σ to be weighted average $(q_1\sigma_1 + q_2\sigma_2)$ of the sectoral output-capital coefficients where the weights q_1, q_2 may denote for example, the proportion of total national investment allocated to two mutually exclusive sectors such that $q_1 + q_2 = 1$. Now, we may specify a quadratic loss function (L_0) similar to that used by Theil (1958) and others.

$$L_0 = [X(t) - E\{X(t) \mid \lambda = L\}]^2 \tag{71}$$

and select the optimal set of instrument variables $q_1 = 1 - q_2$ by minimizing the expected value of L_0 in Eq. (71).

(iii) Homogeneous linear growth model. In a strictly homogeneous model we make the assumption that $\lambda_x(t)$ and $\mu_x(t)$ do not depend on t for all x. A quasi-homogeneous model is defined by the conditions that at least one of the two parameters $\lambda_x(t)$, $\mu_x(t)$ does not depend on t for all values of x. For instance, with reference to economic models it may be plausible to assume

$$\lambda_x = ac \quad \text{and} \quad \mu_x = cx, \tag{72}$$

where a, c are positive constants independent of x, such that $\sum_{x=0}^{\infty} p_x(t) = 1$, that is, the constants a, c are so normalized that $p_x(t)$ [or the probability that $x(t)$ will have a given value x at time t] represents an honest probability distribution.

The assumptions, in Eq. (72) imply that the impact of productivity changes and technology, for example, on the time rate of change of national product is a constant independent of the size of the present national product, while the growth retarding effects [e. g., misdirection or lack of realization of investment (Frisch, 1961) or inoptimal capacity variations, etc.] are proportional to the size of the present national product (through the retarding effects of the present volume of capital stock on current realized investment and hence, on national product). The implications of these assumptions may be further clarified by means of $x(t)$ which is now non-stochastic by definition.

$$dx/dt = ac - cx. \tag{73}$$

Solution: $x(t) = a - (a - j)e^{-ct}$, where $j = x(t)$ at $t = 0$.

This solution indicates that national income (x) grows in an exponential fashion from the lower asymstotic initial level $x(0) = j$ to the upper asymptotic level given by $a = x(t)$ at $t = \infty$. The proportional rate of growth of national income $dx/(x\,dt)$ is, however, no longer a constant as in the model of Eq. (67), but varies from $c(a - j)/j$ at $t = 0$ to zero at $t = \infty$.

It may be easily shown that the stochastic analog of the deterministic model of Eq. (73) is given by the Eq. (60) and (72) combined. In this case the solution of the diffierence-differential equation (60) subject to Eq. (72) is given by the Poisson distribution,

$$p_x(t) = e^{-h(t)}[h(t)]^x/x\,!. \tag{74}$$

The mean value function $M(t)$ for this distribution is given by $h(t) = M(t)$ as follows

$$M(t) = E(X(t), t \geq 0) = \sum_{x=0}^{\infty} x\,p_x(t) = [a - be^{-ct}] = V(t), \tag{75}$$

where b is a positive constant of integration such that $b = a - j$, where $j = h(t)$ at $t = 0$ and $V(t)$ is the variance function for the stochastic process $(X(t), t \geq 0)$.

It is easily seen that the mean value function $M(t)$ in Eq. (75) for the stochastic model is exactly identical with the solution of the deterministic model of Eq. (74) But the stochastic model is much more general in at least three different respects. In the first place, if the stochastic process holds good, it enables us to estimate the confidence limit for the realization of a given value of national income, which may be set as a target by a national policy-maker. Secondly, an estimating procedure like the maximum likelihood method, when applied to the stochastic model of Eqs. (60) and (73) would give consistent estimates for the parameters a, b and c, which are likely to obtain more information than any direct estimates from the deterministic model of Eq. (74) without the use of the corresponding stochastic model. Thirdly, it is easy to show that the stochastic model possesses a "long-run statistical distribution," since from Eq. (74)

$$\lim_{t \to \infty} (p_x(t)) = e^{-a}(a)^x/x\,!. \tag{76}$$

Hence, no matter what the initial unconditional probability distribution $p_x(0)$, the unconditional probability $p_x(t)$ always tends to a limiting probability of Eq. (76) which is independent of t. This last condition implies that probability distribution function is "infinitely divisible" (Feller, 1966). Hence, the stochastic model may be said to have a statistical equibrium which corresponds to the economic equilibrium of the deterministic model. The idea of change of regime and consequent switch-over of the parameters a, b, and c may again be applied here.

2.2 Discrete Stochastic Processes

An immediate theoretical generalization of this linear stochastic process model would be to define λ_x and μ_x as

$$\lambda_x = \lambda_0 + \lambda_1 x, \qquad \mu_x = \mu_0 + \mu_1 x, \tag{77}$$

where λ_0, λ_1, μ_0, μ_1 are positive constants satisfying the regularity conditions (Feller, 1968) that $p_x(t)$ must fulfill the conditions of an honest probability distribution function. Combining Eqs. (60) and (70) we get a difference-differential equation, whose solution is given as follows:

$$p_x(t) = \left[\left(\frac{\lambda_0 + \lambda_1 h(t)}{\lambda_0}\right)^{-\lambda_0/\lambda_1} \left\{\frac{\lambda_0(\lambda_0 + \lambda_1)\cdots(\lambda_0 + (x-1)\lambda_1)}{x!}\right.\right.$$
$$\left.\left.\cdot\left(\frac{h(t)}{\lambda_0 + \lambda_1 h(t)}\right)^x\right\}\right], \tag{78}$$

where $h(t) = M(t) = E\{X(t)\} = a - b e^{-ct}$.

From this Eq. (78) it is easy to show that by making different assumptions about the limiting value of the ratio $\lambda_3 = \lambda_0/\lambda_1$ we may generate different probability schemes. For instance, when $\lambda_3 \to \infty$ we get the Poisson type distribution given in Eq. (73) whereas if $\lambda_3 \to -1$ we get a geometric distribution with the probability distribution function as:

$$p_x(t) = (1 + h(t))^{-1}[h(t)/(1 + h(t))]^x, \qquad h(t) > 0, \tag{79}$$

where $h(t) = a - b e^{-ct}$ as before.

Similarly, other discrete distributions like the negative binomial, Pascal, and so on, could be derived as special cases.

(iv) *Nonlinear stochastic growth model.* A generalized form of a stochastic growth model, which has a very close relation with the deterministic form of a generalized logistic-type model of economic growth formalized by Haavelmo (see also Tintner, 1961; Tintner *et al.*, 1961) is obtained from the system of Eq. (60), when we introduce nonlinearities in the birth (λ) and death rate (μ) as follows:

$$\begin{aligned}\lambda_x &= \lambda x \quad \text{and} \quad \lambda = \alpha(k_2 - x),\\ \mu_x &= \mu x \quad \text{and} \quad \mu = \beta(x - k_1), \qquad k_1 < k_2,\end{aligned} \tag{80}$$

where α, β, k_1, and k_2 are absolute constants such that $x(t)$ at $t = 0$ lies in the closed interval $[k_1, k_2]$. A combination of Eqs. (80) an (60) leads to a system of difference-differential equations, which has not yet been explicitly solved so that the explicit form of the probability function $p_x(t)$ is not known. However, it has been shown by Feller (1968), Kendall (1949), and Bartlett (1961) that the mean value function $M(t)$ in this case satisfies the following differential equation:

2. Stochastic Models of Economic Development

$$dM(t)/dt = (\alpha k_2 + \beta k_1)M(t) - (\alpha + \beta)m_2(t), \tag{81}$$

where $m_2(t)$ is an unknown function representing the second moment about the origin for the process $\{X(t)\}$. Denoting the variance of the process $\{X(t)\}$ by $V(t)$ we can write Eq. (81) as

$$\frac{dM(t)}{dt} = \left[(\alpha + \beta)\left\{\frac{(\alpha k_2 + \beta k_1)}{\alpha + \beta}M(t) - M^2(t)\right\} - (\alpha + \beta)V(t)\right]. \tag{82}$$

Now the deterministic analog of model in Eq. (81) can be written in terms of the nonstochastic variable $y(t)$ representing national income total or per capita as follows;

$$\frac{dx(t)}{dt} = (\alpha + \beta)\left[\frac{(\alpha k_2 + \beta k_1)}{\alpha + \beta}x(t) - x^2(t)\right]. \tag{83}$$

By comparing Eqs. (82) and (83) it is readily apparent that for a nonzero value of the variance function $V(t)$, the mean solution of the logistic stochastic model given in Eq. (82) would be less than the deterministic solution given in Eq. (83). This rather surprising result (originally discovered by Feller, 1968) has two important consequences for the operational application of such models for policy purposes. In the first place, when we linearize a nonlinear model of Eq. (80) at k stages (or facets) by assuming

$$\lambda_x = \lambda^{(k)}x, \qquad \mu_x = \mu^{(k)}x, \qquad k = 1, 2, \cdots, K \tag{84}$$

as in the Eq. (68), where $\lambda^{(k)}$, $\mu^{(k)}$ are constants for the kth stage, we may introduce a specification bias (or error) of a large order and hence the linearly optimal decision rules for each linearized stage may not necessarily be the optimal for the complete nonlinear model.

The second interesting result of this nonlinear stochastic process model is that it leads to an *inverse* problem of the following kind. Given a deterministic model of Eq. (83), what are the conditions under which a *similar* stochastic model may be formulated such that the mean value function for the latter is exactly identical with the solution of the deterministic model.

For instance, it has been shown (Takashima, 1956) that if the birth (λ) and death (μ) rates are defined to be linearly dependent on the inverse of the present size of national income x, that is,

$$\lambda_x = \lambda x \quad \text{and} \quad \lambda = \alpha\left[\frac{(k_2 - 1)}{x}\right], \tag{85}$$

$$\mu_x = \mu x \quad \text{and} \quad \mu = \left[\beta\left\{1 - \left(\frac{k_1}{x}\right)\right\}\right], \tag{86}$$

2.2 Discrete Stochastic Processes

$$k_1 < k_2, \quad 0 < k_1 \leq x \leq k_2, \tag{87}$$

then this nonlinear model has a mean value function $M(t)$ which is identical with the solution of the following deteministic model defined in terms of the nonstochastic variable $x(t)$:

$$dx(t)/dt = (\alpha k_2 + \beta k_1) - (\alpha + \beta)x(t). \tag{88}$$

Since in economic growth models we have different kinds of nonlinearities, it would be a significantly new line of work to investigate the generality and applicability of the inversion hypothesis satisfied for example, by Eq. (88), which is sometimes referred to as the Prendiville Process.

Now we consider an application of a linear homogeneous stochastic growth model for which the probability distribution function has been given in Eq. (72). We have used the long-run annual index of total output for the United Kingdom (1700–1940) constructed by Hoffmann, where the latter index is defined as the sum total of the indices of output of consumer goods and producer goods industries with the base year 1913. Denoting the index of total output x_t and taking yearly values for the whole period (1700–1940), a first-order autoregressive equation fitted on the basis of least squares and its approximate solution turn out to be as follows:

$$x_t = 2.096715 + 0.997297 x_{t-1}$$
$$(0.4157) \quad (0.0112)$$

Solution:

$$x_t = 775.7125 - 772.8825 e^{-0.0027t} \quad \text{(approximately)},$$

where the standard errors are specified in respective parentheses. The overall mean of x_t for the whole period is about 99.3731. Using this as a normalizing factor, such that $h(t)$ in Eq. (79) is redefined as $h(t) = x_t/99.3731$ we obtain:

$$h(t) = \text{Mean value } M(t) = 7.8061 - 7.7776 e^{-0.0027t}.$$

Hence, the difference-differential equation characterizing the process of British economic development for this period becomes

$$dp_x(t)/dt = 0.0211 p_{x-1}(t) - (0.0211 + 0.0027x)$$
$$\cdot p_x(t) + 0.0027(x+1) p_{x-1}(t) \tag{89}$$

This equation has the following interpretation. The probability of x changing from a given value x to $x - 1$ in the interval t to $t + \Delta t$ is approximately $0.0211(\Delta t)$. The probability of x not changing in the same time interval is approximately $[0.9786 - 0.0027x]\,\Delta t$ and the probability of x changing from x to $x + 1$ in the same interval is about $0.0027x(\Delta t)$.

By following the same method we may consider some subperiods of the overall period and the normalized index of total output now turns out to be as follows:

$$1846 - 1880: \quad M(t) = h(t) = 11.5898 - 11.0154e^{-0.0024t},$$
$$1881 - 1908: \quad M(t) = 3.0976 - 2.3211e^{-0.0082t},$$
$$1909 - 1940: \quad M(t) = 1.0233 - 0.0711e^{-0.2653t}.$$

The corresponding differential-difference equations for the probability $p_x(t)$ can be easily derived. Statistical tests are available for the hypothesis that the underlying stochastic process is of the Poisson-type against the simple alternative, that it is not.

From an analytic standpoint, however, it may be more interesting to characterize the process of economic development in terms of more than one variable, for instance, capital stock, population, and total national output. However, the time series on these three variables are highly serially and temporarily interdependent. Hence, the assumption of mutual independence underlying the solution of a multiple difference-differential equation cannot be empirically justified. We may, however, apply the method of principal components. As an illustration we consider the time-series on population (x_{1t}), real value of fixed capital at 1913 prices (x_{2t}), and the index of total output at 1913 prices for the United Kingdom over the period 1870–1940. Since the figures of capital stock constructed by Cairncross (1953) are not available prior to 1870, we restrict ourselves to this period. This series x_{1t}, x_{2t}, x_{3t} along with the standardized series are presented in Table 4.

On the basis of the biennial values for three variables for the period 1870–1940 we compute the correlation matrix

$$\begin{array}{c} \\ x_{1t} \\ x_{2t} \\ x_{3t} \end{array} \begin{bmatrix} x_{1t} & x_{2t} & x_{3t} \\ 1.0000 & 0.9778 & 0.7314 \\ 0.9778 & 1.0000 & 0.9324 \\ 0.7312 & 0.9324 & 1.0000 \end{bmatrix}.$$

The system of linear equations which gives the coefficients of the first and the largest principal component is given by

$$1.0000k_{11} + 0.9778k_{21} + 0.7312k_{31} = \lambda k_{11},$$
$$0.9778k_{11} + 1.0000k_{21} + 0.9324k_{31} = \lambda k_{21}, \qquad (90)$$
$$0.7312k_{11} + 0.9324k_{21} + 1.0000k_{31} = \lambda k_{31},$$

2.2 Discrete Stochastic Processes

TABLE 4

POPULATION, CAPITAL STOCK, AND THE INDEX OF
TOTAL OUTPUT, UNITED KINGDOM
(Base Year 1913)

Year	Population		Capital Stock		Index of Output	
	x_{1t}	z_{1t}	x_{2t}	z_{2t}	x_{3t}	z_{3t}
1870	22.45	−8.165	69.36	−3.785	88.3	−2.522
1872	23.04	−7.716	69.98	−3.655	96.6	−2.203
1874	23.69	−7.222	70.66	−3.513	101.9	−1.999
1876	24.34	−6.728	71.35	−3.368	101.3	−2.022
1878	24.99	−6.234	72.08	−3.215	98.7	−2.122
1880	25.64	−5.740	72.78	−3.068	110.2	−1.680
1882	26.27	−5.261	73.61	−2.894	114.4	−1.519
1884	26.57	−5.033	74.53	−2.701	121.8	−1.234
1886	27.17	−4.577	75.40	−2.519	114.4	−1.519
1888	27.77	−4.120	76.24	−2.343	120.7	−1.276
1890	28.37	−3.664	77.24	−2.133	134.1	−0.761
1892	29.35	−2.919	78.25	−1.922	130.7	−0.892
1894	30.05	−2.387	79.19	−1.725	127.4	−1.019
1896	30.75	−1.855	80.20	−1.513	138.6	−0.588
1898	31.45	−1.323	81.32	1.278	148.6	−0.203
1900	32.15	−0.791	82.51	−1.029	155.1	0.046
1902	32.88	−0.236	83.72	−0.775	153.5	−0.015
1904	33.58	−0.296	84.97	−0.513	154.2	0.011
1906	34.28	0.829	86.27	−0.241	166.5	0.484
1908	34.98	1.361	87.68	0.054	173.4	0.750
1910	35.69	1.893	89.00	0.331	169.2	0.588
1912	36.30	2.364	90.44	0.633	185.6	1.219
1914	36.75	2.706	92.00	1.169	196.7	1.646
1916	37.20	3.048	93.47	1.268	178.9	0.961
1918	37.65	3.391	94.92	1.572	164.0	0.388
1920	38.10	3.733	96.34	1.869	176.8	0.880
1922	38.57	4.090	97.38	2.087	136.5	−0.669
1924	38.99	4.409	98.90	2.406	175.6	0.834
1926	39.41	4.729	100.16	2.670	180.5	1.023
1928	39.83	5.048	101.85	3.024	202.5	1.869
1930	40.25	5.367	103.61	3.993	208.2	2.088
1932	40.63	5.648	105.00	3.684	185.9	1.230
1934	40.24	5.892	106.68	4.036	202.1	1.853
1936	41.26	6.135	108.61	4.441	201.2	1.819
1938	41.58	6.378	110.51	4.839	209.8	2.149
1940	41.90	6.622	111.01	4.944	216.3	2.399

NOTE: Population (in mill.), capital stock (in £100 m. at 1913 prices).

for which the three characteristic roots are $\lambda_1 = 2.5657$, $\lambda_2 = 0.3525$, and $\lambda_3 = 0.0818$. Since the total variance of the three standardized variables z_{1t}, z_{2t},

z_{3t} is evidently 3, the first principal component explains about 85.5 per cent, the second about 11.8 per cent, and the third about 2.7 per cent of the total variance. Taking the largest principal component (λ_1) we compute its unit normal eigenvector as

$$\begin{bmatrix} k_{11} \\ k_{21} \\ k_{31} \end{bmatrix} = \begin{bmatrix} 0.614647 \\ 0.649990 \\ 0.446902 \end{bmatrix}.$$

Imposing the condition $\sum_{i=1}^{3} k_{i1}^2 = \lambda_1$ we get the standardized set of coefficients as $k_{11}^* = 0.984531$, $k_{21}^* = 1.041143$ and $k_{31}^* = 0.715840$. Similarly, the standardized eigenvectors corresponding to second (λ_2) and third (λ_3) principal components are obtained. Then we express the standardized variates $z_{it} = (x_{it} - \bar{x}_i)/\sigma_i$, ($i = 1, 2, 3$) in terms of the principal components u_{1t}, u_{2t}, u_{3t} as

$$z_{1t} = 0.984531 u_{1t} - 0.249248 u_{2t} - 0.228515 u_{3t},$$
$$z_{2t} = 1.041143 u_{1t} + 0.063501 u_{2t} + 0.134201 u_{3t}, \quad (91)$$
$$z_{3t} = 0.715840 u_{1t} + 0.217304 u_{2t} + 0.107493 u_{3t}.$$

We obtain the principal components by inversion of the non-singular coefficient matrix as

$$\begin{bmatrix} u_{1t} \\ u_{2t} \\ u_{3t} \end{bmatrix} = \begin{bmatrix} 0.376325 & 0.385237 & 0.319074 \\ 0.267029 & -4.539113 & 6.234596 \\ -3.045979 & 6.610685 & -5.425522 \end{bmatrix} \begin{bmatrix} z_{1t} \\ z_{2t} \\ z_{3t} \end{bmatrix}.$$

Now we fit a first-order autoregressive equation to each of the principal components u_{it} ($i = 1, 2, 3$), the solutions of which turn out to be as follows.

$$u_{1t} = 8.950987 - 14.2860 e^{-0.032971 t},$$
$$u_{2t} = -0.455001 + 1.249905 e^{-0.24901 t},$$
$$u_{3t} = -2.168802 + 15.701004 e^{-0.181828 t}.$$

The differential-difference Eq. (89) corresponding to the first principal component (u_1) now reduces to

$$dp_{u_1}(t)/dt = 0.295123 p_{u_1-1}(t) - (0.295123 + 0.032971 u_1) p_{u_1}(t)$$
$$+ 0.032971 (u_1 + 1) p_{u_1+1}(t).$$

The other principal components, which are mutually orthogonal by construction, may be similarly used to compute the multiple differential-difference equation.

An alternative approach (Tintner, 1961) in this case would be to define an interdependent deterministic model between the three variables and construct a corresponding stochastic model. Then on the basis of the mean value, variance function, and other characteristics of the stochastic process for these three variables, we can derive various types of inference having useful policy implications.

2.3 Multi-Dimensional Discrete Stochastic Process with Applications

We propose a multi-dimensional stochastic process taking into account the interdependence among the variables (Tintner and Narayanan, 1966). Evidently, such models are more appropriate when we characterize economic development in terms of more than one variable, for example, national income, capital stock, employment, and so on. In what follows, we shall confine our attention to the case of three veriables only, though the method is perfectly general and can be extended to any number of variables.

The equation $X_t = (X_{1t}, X_{2t}, X_{3t})$ stands for a three-dimensional discrete stochastic process, where each component takes values in the space of integers $S: (0, 1, 2, \cdots \infty)$. Let us denote the three- dimensional Cartesian product space $S \cdot S \cdot S$ by S^3. We write

$$P_x(x) = \text{prob}(X_t = x) \quad \text{if} \quad x = (x_1, x_2, x_3) \in S^3 \tag{1}$$

and

$$P_x(t) = 0 \quad \text{if} \quad x \notin S^3. \tag{2}$$

The transition probabilities during a small time interval $(t, t + \Delta t)$ are as follows.

$$\text{prob}(X_{t+\Delta t} = x + \delta \mid X_t = x) = \lambda_\delta(\Delta t) + o(\Delta t), \tag{3}$$

where λ_δ is a constant for each $\delta = (\delta_1, \delta_2, \delta_3) \neq 0$; $\delta_i = 0$ or 1 for $i = 1, 2, 3$ and

$$\text{prob}(X_{t+\Delta t} = x \mid X_t = x) = 1 - \lambda_\delta(\Delta t) + o(\Delta t). \tag{4}$$

The other transition probabilities are of $o(\Delta t)$ and we have further the condition

$$\sum_{\delta \neq 0} \lambda_\delta = \lambda_0 \tag{5}$$

in order to make the sum of all the transition probabilities equal to unity. These assumptions lead us to the system of difference-differential equations:

$$\frac{\partial p_x(t)}{\partial t} \sum_{\delta \neq 0} [P_{x-\delta}(t) - P_x(t)] \tag{6}$$

for $x \in S^3$. In order to solve this system of equations we introduce the probability generating function.

$$F(s_1, s_2, s_3, t) = \sum_{x_i} s_1^{x_1} \cdot s_2^{x_2} \cdot s_3^{x_3} P_x(t). \tag{7}$$

In terms of this function, the system of mixed difference-differential equations in Eq. (6) can be expressed as

$$\frac{\partial F}{\partial t}(s_1, s_2, s_3, t) = F(s_1, s_2, s_3, t) \sum_{\delta_i} [(s_1^{\delta_1} \cdot s_2^{\delta_2} \cdot s_3^{\delta_3} - 1)\lambda_\delta], \tag{8}$$

$$F(s_1, s_2, s_3, t) = C(s_1, s_2, s_3) \exp[\sum_{\delta_i}(s_1^{\delta_1} \cdot s_2^{\delta_2} \cdot s_3^{\delta_3} - 1)\lambda_\delta t], \tag{9}$$

where $C(s_1, s_2, s_3)$ is an arbitrary function of (s_1, s_2, s_3) to be determined from the initial conditions. Let us suppose that initially

$$\text{prob}(X_0 = 0) = 1 \tag{10}$$

so that

$$F(s_1, s_2, s_3, 0) = 1. \tag{11}$$

Under this boundary condition the solution of Eq. (9) becomes

$$F(s_1, s_2, s_3, t) = \exp[\sum_{\delta_i}(s_1^{\delta_1} \cdot s_2^{\delta_2} \cdot s_3^{\delta_3} - 1)\lambda_\delta t], \tag{12}$$

which shows that the marginal distribution of each component X_{it} has a Poisson distribution with parameter (mean $= \sum \delta_i \lambda_\delta t$). It may be noted in passing that for the time-dependent, nonhomogeneous case where $\lambda_\delta t$ is a function of time for each δ, the solution can be obtained by changing $\lambda_\delta(t)$ to $\int_0^t \lambda_\delta(t)\,dt$ in Eq. (12).

In order to apply the above process to economic data we transform the state space of each component X_{it} by writing $Y_{it} = a_{it} + u_{it}X_{it}$, where a_{it} and u_{it} are functions of time. The generating function of $Y_t = (Y_{1t}, Y_{2t}, Y_{3t})$ is given by

$$G(s_1, s_2, s_3, t) = E(s_1^{Y_{1t}} \cdot s_2^{Y_{2t}} \cdot s_3^{Y_{3t}})$$
$$= s_1^{a_{1t}} \cdot s_2^{a_{2t}} s_3^{a_{3t}} \exp[\sum_{\delta_i} s_1^{\delta_1 u_{1t}} \cdot s_2^{\delta_2 u_{2t}} \cdot s_3^{\delta_3 u_{3t}} - 1)\lambda_\delta t]. \tag{13}$$

The means, variances covariances, and other higher moments of Y_i can be obtained through repeated partial differentiation of the generating function. We obtain the following results without difficulty.

2.3 Multi-Dimensional Discrete Stochastic Processes

$$M_{it} = E(Y_{it}) = a_{it} + u_{it} \sum_{\delta_i}(\delta_i \lambda_\delta t), \quad i = 1, 2, 3; \tag{14}$$

$$V(Y_{it}) = u_{it}^2 t \sum_{\delta_i}(\delta_i \lambda_\delta t), \quad i = 1, 2, 3; \tag{15}$$

$$\text{cov}(Y_{it}, Y_{jt}) = u_{it} \cdot u_{jt} \sum_{i,j}(\delta_i \delta_j \lambda_\delta t), \quad i, j = 1, 2, 3 \text{ and } i \neq j \tag{16}$$

$$E \prod_{i=1}^{3}(Y_{it} - EY_{it}) = u_{1t} \cdot u_{2t} \cdot u_{3t} \lambda_{111} t. \tag{17}$$

It can also be seen as t tends to infinity that Y_t approaches a three-dimensional normal distribution with asymptotic means, variances, and covariances as in Eqs. (14)–(16).

We now present an empirical illustration of fitting such an interdependent multiple stochastic process for postwar U. S. data. In this model we consider only two variables, Y_{1t} = gross national product, and Y_{2t} = private consumer expenditures, both at constant (1954) prices, Further, for simplicity, we assume that all the quantities—a_{it}, u_{it}, $\lambda_\delta(t)$—are constants, independent of time. Equations (14)–(16) become

$$M_{1t} = E(Y_{1t}) = a_1 + u_1(\lambda_{10} + \lambda_{11})t, \tag{18}$$

$$M_{2t} = E(Y_{2t}) = a_2 + u_2(\lambda_{01} + \lambda_{11})t, \tag{19}$$

$$V(Y_{1t}) = u_1^2(\lambda_{10} + \lambda_{11})t, \tag{20}$$

$$V(Y_{2t}) = u_2^2(\lambda_{01} + \lambda_{11})t, \tag{21}$$

and

$$\text{cov}(Y_{1t}, Y_{2t}) = u_1 u_2 \lambda_{11} t. \tag{22}$$

The estimates for $a_1, a_2, u_1, u_2, (\lambda_{10} + \lambda_{11})$ and so on are obtained by minimizing

$$\sum_{t=1}^{k} \frac{1}{t}[Y_{1t} - a_1 - u_1(\lambda_{10} + \lambda_{11})t]^2 \tag{23}$$

and

$$\sum_{t=1}^{k} \frac{1}{t}[Y_{2t} - a_2 - u_2(\lambda_{01} + \lambda_{11})t]^2. \tag{24}$$

λ_{11} is estimated from

$$\hat{u}_1 \hat{u}_2 \hat{\lambda}_{11} = \frac{1}{(k-2)} \sum_{t=1}^{k} \frac{1}{t}(Y_{1t} - \hat{M}_{1t})(Y_{2t} - \hat{M}_{2t}), \tag{25}$$

where k is the sample size. The empirical reaults are as follows.

$$\hat{M}_{1t} = \hat{E}(Y_{1t}) = 263.7535 + 11.9058t, \tag{26}$$

$$\hat{a}_1 = 263.7535, \quad \hat{u}_1 = 1.3483, \tag{27}$$

$$\hat{M}_{2t} = \hat{E}(Y_{2t}) = 180.3709 + 7.4728t, \tag{28}$$

$$\hat{a}_2 = 180.3709. \quad \hat{u}_2 = 0.7276, \tag{29}$$
$$\hat{\lambda}_{10} = 4.1135, \quad \hat{\lambda}_{01} = 5.5532. \quad \hat{\lambda}_{11} = 4.7170. \tag{30}$$

The hat signs in the above expressions denote that the quantities are the corresponding estimates. The λ's have the following interpretation. Given that at time t, $Y_{1t} = y_1$, and $Y_{2t} = y_2$, the transition probabilities (conditional) during an infinitesimal time interval $(t, t + \Delta t)$ are, correct to first order of (Δt), given by

$$\text{prob}(Y_{1,t+\Delta t} = y_1 + 1.3483, Y_{2,t+\Delta t} = y_2) = 4.1135(\Delta t), \tag{31}$$
$$\text{prob}(Y_{1,t+\Delta t} = y_1, Y_{2,t+\Delta t} = y_2 + 0.7276) = 5.5532(\Delta t), \tag{32}$$
$$\text{prob}(Y_{1,t+\Delta t} = y_1 + 1.3483, Y_{2,t+\Delta t} = y_2 + 0.7276) = 4.7170(\Delta t), \tag{33}$$

and

$$\text{prob}(Y_{1,t+\Delta t} = y_1 Y_{2,t+\Delta t} = y_2) = 1 - 14.3837(\Delta t). \tag{34}$$

We may utilize the above estimates in order to predict the trend values of Y_{1t}, and Y_{2t} for any particular year. We may also employ the property of asymptotic normality to derive joint-confidence ellipse for the distribution of (Y_{1t}, Y_{2t}). For example, putting $t = 20$ we obtain from Eqs. (26) and (28) the trend values of Y_{2t} and Y_{2t} for the year 1965 as 501.8695 and 329.8269 respectively. The variance-covariance of (Y_{1t}, Y_{2t}) for the same year is estimated as using Eqs. (20)–(22) as follows.

$$\begin{bmatrix} \text{var}(Y_{1t}) = 321.0611 & \text{cov}(Y_{1t}, Y_{2t}) = 92.5494 \\ & \text{var}(Y_{2t}) = 108.7409 \end{bmatrix} \tag{35}$$

Therefore the quadratic form

$$Q(Y_{1t}, Y_{2t})$$
$$= \begin{bmatrix} Y_{1t} - 501.8695 \\ Y_{2t} - 329.8269 \end{bmatrix}' \begin{bmatrix} 321.0611 & 92.5494 \\ & 108.7409 \end{bmatrix} \begin{bmatrix} Y_{1t} - 501.8595 \\ Y_{2t} - 329.8269 \end{bmatrix}$$
$$= [0.004127(Y_{1t} - 501.8695)^2 - 0.007026(Y_{1t} - 501.8695)(Y_{2t} - 329.8269)$$
$$+ 0.019640(Y_{2t} - 329.8269)^2] \tag{36}$$

has, approximately, a chi-square distribution with two degrees of freedom. Hence an approximate 95% confidence ellipse for the values of Y_{1t} and Y_{2t} for the year 1965 is given by the inequality

$$Q(Y_{1t}, Y_{2t}) \leq 5.99 \quad \text{(the upper 5\% point of the chi-square distribution with two degrees of freedom).} \tag{37}$$

2.3 Multi-Dimensional Discrete Stochastic Processes

We present some results of fitting a three-dimensional process to U. S. data of economic development (Table 5). We use the same symbols and notations as earlier and we consider the following three series for our model.

Y_{1t} = private consumer expenditures in U. S at constant (1954) prices;
Y_{2t} = gross private domestic investment expenditures in U. S. at constant (1954) prices; and
Y_{3t} = government expenditures in U. S. at constant (1954) prices.

TABLE 5

GROSS NATIONAL PRODUCT (Y_{1t}) AND PRIVATE CONSUMER EXPENDITURES (Y_{2t}) FOR UNITED STATES AT CONSTANT (1954) PRICES FOR THE PERIOD 1946-63
(Billions of Dollars)

Year	Y_{1t}	Y_{2t}	t
1946	192.3	282.5	1
1947	195.6	282.3	2
1948	199.3	293.1	3
1949	204.3	292.7	4
1950	216.8	318.1	5
1951	218.5	341.8	6
1952	224.2	353.5	7
1953	235.1	369.0	8
1954	238.0	363.1	9
1955	256.0	392.7	10
1956	263.7	402.2	11
1957	270.3	407.0	12
1958	273.2	401.3	13
1959	288.9	428.6	14
1960	298.1	439.9	15
1961	303.6	447.7	16
1962	317.6	474.8	17
1963	328.9	492.9	18

SOURCE: "Historical Statistics of the United States: Colonial Times to 1957; A Statistical Supplement." Bureau of the Census, U. S. Department of Commerce, Washington, D.C. Statistics for the recent period are taken from "Statistical Abstract of the United States 1964." Bureau of the Census, U. S. Department of Commerce Washington D.C.

We follow the same estimation method as earlier and our empirical results are as follows.

$$Y_{1t} = 180.3709 + 7.4728t,$$
$$\hat{a}_1 = 180.3709, \quad \hat{u}_1 = 0.7276.$$
(39)

$$Y_{2t} = 41.3268 + 1.3833t,$$
$$\hat{a}_2 = 41.3268, \quad \hat{u}_2 = 3.2972.$$
(39)

2. Stochastic Models of Economic Development

$$Y_{3t} = 41.9968 + 3.0538t,$$
$$\hat{a}_3 = 41.9968, \quad \hat{u}_3 = 3.3356. \tag{40}$$

$$\lambda_{100} + \lambda_{110} + \lambda_{101} + \lambda_{111} = 10.2702. \tag{41}$$

$$\lambda_{110} + \lambda_{111} = 0.1730. \tag{42}$$

$$\lambda_{101} + \lambda_{111} = -0.8290. \tag{43}$$

$$\lambda_{010} + \lambda_{011} + \lambda_{110} + \lambda_{111} = 0.4195. \tag{44}$$

$$\lambda_{011} + \lambda_{111} = -0.0197. \tag{45}$$

$$\lambda_{001} + \lambda_{011} + \lambda_{101} + \lambda_{111} = 0.9979. \tag{46}$$

$$\hat{\lambda}_{111} = 0.1592. \tag{47}$$

The estimate for λ_{101} and λ_{011} from Eqs. (43), (45) and (47) turn out to be negative. Since they are proportional to certain probabilities, their estimate cannot be negative and we set them equal to zero. We get the follwing estimates for

TABLE 6

STATISTICS OF PRIVATE CONSUMPTION EXPENDITURES (Y_{1t}), GROSS DOMESTIC INVESTMENT EXPENDITURES (Y_{2t}), AND GOVERNMENT EXPENDITURES (Y_{3t}) FOR UNITED STATES AT CONSTANT (1954) PRICES, 1946–63 (Billions of Dollars)

Year	Y_{1t}	Y_{2t}	Y_{3t}	t
1946	192.3	42.4	43.9	1
1947	195.6	41.5	37.2	2
1948	199.4	49.8	42.1	3
1949	204.3	38.5	47.2	4
1950	216.8	55.9	45.1	5
1951	218.5	57.7	63.3	6
1952	224.2	50.4	77.7	7
1953	235.1	50.6	84.3	8
1954	238.0	48.9	75.3	9
1955	256.0	62.5	73.2	10
1956	263.1	63.1	72.9	11
1957	270.3	57.8	75.0	12
1958	273.2	49.0	79.3	13
1959	288.9	61.7	80.1	14
1960	298.1	60.2	79.9	15
1961	303.6	57.5	84.3	16
1962	317.6	65.2	90.2	17
1963	328.9	67.7	93.7	18

SOURCE: "Historical Statistics of the United States: Colonial Times to 1957, A Statistical Supplement." Bureau of the Census, U. S. Department of Commerce, Washington, D. C. Statistics for the recent period are taken from "Statistical Abstract of the United States, 1964." Bureau of the Census, U. S. Department of Commerce, Washington, D. C.

the parameters of the transition probabilities.

$$\hat{\lambda}_{100} = 10.0972, \quad \hat{\lambda}_{110} = 0.0138, \quad \hat{\lambda}_{101} = 0,$$
$$\hat{\lambda}_{010} = 0.2465, \quad \hat{\lambda}_{011} = 0,$$
$$\hat{\lambda}_{001} = 0.8387,$$
$$\hat{\lambda}_{111} = 0.1592.$$

The above estimates have the following interpretation. For example, given that private consumer expenditures are equal to y_1 and gross private domestic investment expenditures are equal to y_2, the probability that these two expenditures rise by 0.7276 billion and 3.2672 billion respectively in a small interval of Δt (government expenditures remaining the same) is given by 0.0138 (Δt), and so on.

2.4 Diffusion Processes Applied to Economic Development

The following (Tintner and Patel, 1965) tries to apply a log-normal diffusion process to the development of the Indian economy. The theory of diffusion processes has been very important in many applications (Bharucha–Reid, 1960; Fisz, 1963; Bartlett, 1961; Bailey, 1964; Takacs, 1960). The use of the lognormal rather than the normal or the Pareto distribution (Mandelbrot, 1963; Pareto, 1927; Davis, 1941) is suggested by a number of investigations which seem to point to its usefulness as a convenient approximation in a number of fields of economics (Aitchison and Brown, 1957; Granger and Morgenstern, 1963; Sprekle, 1961; Gibrat, 1931; Orcutt et al., 1961; Prais and Houthakker, 1955).

Let $X(t)$, national income at time t, be a random variable of a continuous process $\{X(t), t \geq 0\}$ of the Markovian type defined on the interval $(0, \infty)$ of real line, that is $X(t)$ is a Markovian random variable depending on a continuous time parameter t, which assumes values in the state space $H: 0 < x < \infty$.

We assume the transition probability density

$$f(\tau, x; t, y) = \Pr[X(t) = y; X(\tau) = x], \quad 0 < y, x < \infty \quad (1)$$

exists for every τ and t, where $0 \leq \tau < t$, and satisfies the backward and forward Kolmogorov equations. We also assume the random variable $X(t)$ continuous with probability one.

Consider the coefficients $b(t, x)$ and $a(t, x)$, the infinitesimal mean and variance of the change in $X(t)$ during a small interval Δt of time, that characterize a particular process of the diffusion type as

2. Stochastic Models of Economic Development

$$b(t, x) = b_t \cdot x \quad \text{and} \quad a(t, x) = a_t \cdot x^2 > 0, \tag{2}$$

where b_t and a_t are functions of time.

We take $b_t = b_0$ and $a_t = a_0$, where a_0, b_0 are constants, $a_0 > 0$. We thus assume that the expected change and its variance, in national income are proportional to the instantaneous size of it. With $a(t, x) > 0$, it is certain that some change in national income will take place in any interval Δt and it will be small if Δt is small. This seems to well describe the plausible characteristics of an economy.

With the coefficients as defined above, the backward and forward Kolmogorov equations are, respectively,

$$-\frac{\partial f}{\partial \tau} = \frac{1}{2} a_0 x^2 \frac{\partial^2 f}{\partial x^2} - b_0 x \frac{\partial f}{\partial x}$$

and

$$\frac{\partial f}{\partial t} = \frac{1}{2} a_0 y^2 \frac{\partial^2 f}{\partial y^2} + (2a_0 - b_0) y \frac{\partial f}{\partial y} + (a_0 - b_0) f. \tag{3}$$

The probability density function satisfying these diffusion equations Eq. (3) is then given by the log-normal density function,

$$f(\tau, x; t, y) = \left[\frac{1}{y[2\pi\gamma(t-\tau)]^{1/2}} \right.$$
$$\left. \cdot \exp\left[\left(-\frac{1}{2\gamma(t-\tau)} \right) \{ \log y - \log x - \beta(t-\tau) \}^2 \right] \right], \tag{4}$$

where $\gamma = a_0$ and $\beta = (b_0 - a_0/2)$.

The characteristics of this distribution are easily derived from its moments (Aitchison and Brown, 1957),

$$E[X(t)]^k = (X(\tau))^k \cdot \exp\left(k\beta_0 + k^2 \cdot \frac{\gamma}{2} \right)(t - \tau), \tag{5}$$
$$k = 1, 2, 3, \cdots.$$

Therefore,

$$\text{Mean,} \quad E[X(t)] = X(\tau) \cdot \exp\left(\beta_0 + \frac{\gamma}{2} \right)(t - \tau) \tag{6}$$
$$= X(\tau) \cdot \exp b_0(t - \tau),$$
$$\text{Variance,} \quad V[X(t)] = X(\tau)^2 \cdot \exp 2b_0(t - \tau) \cdot (\exp a_0(t - \tau) - 1). \tag{7}$$

2.4 Diffusion Processes

If we take $\tau = 0$ and assume $\Pr[X(0) = x_0] = 1$, then

$$\text{Mean,} \quad E[X(t)] = x_0 \cdot \exp b_0 t, \tag{8}$$

which is an exponential trend; and

$$\text{Variance,} \quad V[X(t)] = x_0^2 \cdot \exp 2b_0 t(\exp a_0 t - 1), \tag{9}$$

which is also an exponential function of time.

The coefficients of skewness and kurtosis are

$$\gamma_1 = \frac{e^{3\gamma t} - 3e^{\gamma t} + 2)}{(e^{\gamma t} - 1)^{3/2}},$$

$$\gamma_2 = \frac{e^{6\gamma t} - 4e^{3\gamma t} - 6e^{\gamma t} - 3}{(e^{\gamma t} - 1)^2}. \tag{10}$$

2.4.1 MAXIMUM LIKELIHOOD ESTIMATES

The estimates of the parameters β and γ, and hence of a_0 and b_0 by the method of maximum likelihood are obtained as follows. Let $x_0, x_1, x_2, \cdots, x_n$ be observed states of $x(t)$ at time $t_0, t_1, t_2, \cdots, t_n$ respectively, and let $P[x(0) = x_0] = 1$. The joint probability of the observations is

$$L = \left[\left(\frac{1}{(2\pi\gamma)^{1/2}}\right)^n \prod_{j=1}^{n} \frac{1}{x_j[t_j - t_{j-1}]^{1/2}} \right.$$
$$\left. \cdot \exp\left\{-\frac{1}{2\gamma(t_j - t_{j-1})}[\log x_j - \log x_{j-1} - \beta(t_j - t_{j-1})]^2\right\}\right]. \tag{11}$$

The maximum likelihood estimates are given by solving the equations,

$$\frac{\partial \log L}{\partial \beta} = \left(\frac{1}{\gamma}\right) \sum_{j=1}^{n} [\log x_j - \log x_{j-1} - \beta(t_j - t_{j-1})] = 0$$

and

$$\frac{\partial \log L}{\partial \gamma} = \frac{1}{n} \sum_{j=1}^{n} \left[\frac{1}{(t_j - t_{j-1})}\{\log x_j - \log x_{j-1} - \beta(t_j - t_{j-1})\}^2\right] = 0. \tag{12}$$

The estimates on solving Eq.(12) are

$$\hat{\beta} = \sum_{j=1}^{n} (\log x_j - \log x_{j-1}) \Big/ \sum_{j=1}^{n} (t_j - t_{j-1})$$

and

$$\hat{\gamma} = \frac{1}{n} \sum_{j=1}^{n} \left[\frac{\{\log x_j - \log x_{j-1} - \hat{\beta}(t_j - t_{j-1})\}^2}{(t_j - t_{j-1})}\right]. \tag{13}$$

2. Stochastic Models of Economic Development

Hence the maximum likelihood estimates a_0 and b_0 are

$$\hat{a}_0 = \hat{\gamma} \quad \text{and} \quad \hat{b}_0 = \hat{\beta} + \hat{\gamma}/2. \tag{14}$$

The dispersion matrix of the estimates $(\hat{\beta}, \hat{\gamma})$ is given by the inverse of the information matrix I,

$$I = \frac{n}{\gamma} \cdot \begin{bmatrix} (t - \tau) & 0 \\ & 1/2\gamma \end{bmatrix}. \tag{15}$$

Therefore,

$$V(\hat{\beta}) = \frac{\gamma}{n} \cdot \frac{1}{(t - \gamma)},$$
$$V(\hat{\gamma}) = 2\gamma^2/n, \tag{16}$$
$$\text{cov}(\hat{\beta}, \hat{\gamma}) = 0.$$

If we assume that the n observations are taken at equal intervals of time of unity, that is, $(t_j - t_{j-1}) = 1$, $j = 1, 2, \cdots, n$ and $t_0 = 0$, then the estimates and their sampling variances are

$$\hat{\beta} = \sum_{j=1}^{n} \frac{\{(\log x_j - \log x_{j-1})\}}{n},$$
$$\hat{\gamma} = \sum_{j=1}^{n} \frac{\{(\log x_j - \log x_{j-1})\}}{n} - \hat{\beta}^2 \tag{17}$$
$$V(\hat{\beta}) = \frac{\gamma}{n} \cdot \frac{1}{t} \to 0 \quad \text{as} \quad t \to \infty,$$
$$V(\hat{\gamma}) = 2\gamma^2/n,$$

and

$$V(a_0) = 2\gamma^2/n$$
$$V(b_0) = \frac{\gamma^2}{2n} + \frac{\gamma}{n} \cdot \frac{1}{t} \to \frac{\gamma^2}{2n} \quad \text{as} \quad t \to \infty. \tag{18}$$

We now fit the above model to the data of real national income of India for the period 1848–49 to 1960–61. Here $X(t)$ represents the real national income of India in the year t, in billions of rupees, and Pr $[X(0) = 86.5 = 1]$.

The maximum likelihood estimates and their variances are found as

2.4 Diffusion Processes

$$\hat{\beta} = 0.322 \quad \text{and} \quad \hat{\gamma} = 0.5947 \times 10^{-3},$$
$$V(\hat{\beta}) = 0.4956\left(10^{-4} \cdot \frac{1}{t}\right), \quad V(\hat{\gamma}) = 0.5895(10^{-7}), \tag{19}$$

and hence

$$b_0 = 0.0325 \quad \text{and} \quad a_0 = 0.5947 \times 10^{-3},$$
$$V(b_0) = 0.1474(10^{-7}) - 0.4956\left(10^{-4} \cdot \frac{1}{t}\right), \tag{20}$$

and

$$V(a_0) = 0.5895(10^{-7}).$$

The trend is given by

$$E[X(t)] = 86.5 e^{0.0325 t}, \quad t = 0, 1, 2, \cdots. \tag{21}$$

We try to predict the values of national income for 1965–66 (end year of Third five-year Plan), and for selected years of the Fourth five-year Plan period (1966–67 to 1970–71). The results of prediction based on our theory are given in Table 7. We observe a 3.35% growth rate during Fourth Plan on the basis of our model.

TABLE 7

ESTIMATES OF THE NATIONAL INCOME OF INDIA FOR SELECTED YEARS

Year	t	$E\,X(t)$
1965–66	17	150.3
1967–68	19	160.4
1969–70	21	171.2
1970–71	22	176.8

The influence of an exogeneous factor like government expenditure on economic growth can be considered by taking the parameters of the process as functions of such variables. Let us assume a linear effect of government expenditure $G(t)$ on β_t, that is,

$$\beta_t = \beta_0 + \beta_1 \cdot G(t). \tag{22}$$

We have then

$$\int_\tau^t \beta_t\, dt = \beta_0(t - \tau) + \beta_1 \int_\tau^t G(t)\, dt = \beta_0(t - \tau) - \beta_1 \cdot (G)_\tau^t,$$

where

$$\int_\tau^t G(t)dt = (G)_\tau^t,$$

say, with infinitesimal mean and variance of change in $X(t)$ defined as

$$b(t, x) = b_t \cdot x \quad \text{and} \quad a(t,x) = a_t \cdot x^2 > 0, \tag{23}$$

where $a_t = a_0 > 0$ and $b_t = b_0 + b_1 \cdot G(t)$, where $a_0, b_0, b_1 \neq 0$ are constants, we have the probability density function for $X(t)$ satisfying the Kolmogorov equations as

$$f(\tau, t; t, y) = \left[\frac{1}{y(2\gamma(t-\tau)\pi)^{1/2}}\right.$$
$$\left. \cdot \exp\left\{-\frac{1}{2\gamma(t-\tau)}\{\log y - \log x - \beta_0(t-\tau) - \beta_1(G)_\tau^t\}^2\right\}\right], \tag{24}$$

where $\gamma = a_0$, $\beta_0 = (b_0 - a_0/2)$ and $\beta_1 = b_1$.

The maximum likelihood estimates for γ, β_0 and β_1 are obtained as

$$\hat{\beta}_1 = \sum_{j=1}^n \left[\frac{(G)_{t_{j-1}}^{t_j} \log(x_j/x_{j-1})}{(t_j - t_{j-1})}\right]$$
$$- \frac{\frac{\sum_{j=1}^n [\log(x_j/x_{j-1})] \cdot \sum_{j=1}^n [(G)_{t_{j-1}}^{t_j}]}{\sum_{j=1}^n (t_j - t_{j-1})}}{\sum_{j=1}^n [\{(G)_{t_{j-1}}^{t_j}\}^2] - \frac{\{\sum_{j=1}^n [(G)_{t_{j-1}}^{t_j}]\}^2}{\sum_{j=1}^n (t_j - t_{j-1})}} \tag{25}$$

$$\hat{\beta}_0 = \frac{\sum_{j=1}^n [\log(x_j/x_{j-1})] - \hat{\beta}_1 \sum_{j=1}^n [(G)_{t_{j-1}}^{t_j}]}{\sum_{j=1}^n (t_j - t_{j-1})},$$

and

$$\hat{\gamma} = \frac{1}{n} \sum_{j=1}^n \left[\frac{\{\log(x_j/x_{j-1}) - \hat{\beta}_0(t_j - t_{j-1}) - \hat{\beta}_1(G)_{t_{j-1}}^{t_j}\}^2}{(t_j - t_{j-1})}\right].$$

If observations are made at equal intervals of time of unity, that is, $(t_j - t_{j-1}) = 1, j = 1, 2, \cdots, n$ and if we assume $\hat{G}(t)$ a step function

$$G(t) = G_j, \quad \text{for} \quad t_{j-1} \leq t \leq t_j, \quad j = 0, 1, 2, \cdots n, \tag{26}$$

so that

$$\int_{t_{j-1}}^{t_j} G(r) \, dr = G_j \cdot (t_j - t_{j-1}),$$

2.4 Diffusion Processes

then our estimates are

$$\hat{\beta}_0 = \sum_{j=1}^{n} \frac{(\log x_j - \log x_{j-1})}{n} - \beta_1 \cdot \sum_{j=1}^{n} (G_j),$$

$$\hat{\beta}_1 = \sum_{j=1}^{n} \frac{[(G_j)\log(x_j/x_{j-1})] - \dfrac{\sum\limits_{j=1}^{n} \log(x_j/x_{j-1}) \cdot \sum\limits_{j=1}^{n} [G_j]}{n}}{\sum\limits_{j=1}^{n} (G_j^2) - \dfrac{\{\Sigma(G_j)\}^2}{n}}, \quad (27)$$

and

$$\hat{\gamma} = \frac{1}{n} \cdot \sum_{j=1}^{n} [\log(x_j/x_{j-1}) - \beta_0 - \beta_1 G_j]^2.$$

The dispersion matrix of the estimates $(\beta_0, \beta_1, \gamma)$ is given by the inverse of the information matrix I,

$$I = (n/\gamma) \begin{bmatrix} t & \sum\limits_{j=1}^{t} G_j & 0 \\ & (\Sigma_j G_j)^2/t & 0 \end{bmatrix}. \quad (28)$$

Therefore

$$V(\hat{\beta}_0) = \frac{\gamma}{n} \cdot \frac{1}{t} \to 0 \quad \text{as } t \to \infty,$$

$$V(\hat{\beta}_1) = \frac{\gamma}{n} \cdot \frac{1}{(\Sigma_1^t G_j)^2/t} \to 0 \quad \text{as } t \to \infty, \quad (29)$$

when G_j is constant or increasing;

$$V(\hat{\gamma}) = 2\gamma^2/n.$$

Hence,

$$V(a_0) = \frac{2\gamma^2}{n},$$

$$V(b_0) = \frac{\gamma}{n} \cdot \frac{1}{t} + \frac{\gamma^2}{2n} \to \frac{\gamma^2}{2n} \quad \text{as } t \to \infty, \quad (30)$$

$$V(b_1) = \frac{1}{n} \cdot \frac{1}{(\Sigma_1^t G_j)^2/t} \to 0 \quad \text{as } t \to \infty,$$

when G_j is constant or increasing.

2. Stochastic Models of Economic Development

The probability density function of $X(t)$ with $t_0 = 0$ and $\text{pr}[X(0) = x_0] = 1$ is

$$f(x_0; t, x) = \frac{1}{x\sqrt{2\pi\gamma t}} \exp\left[-\frac{1}{2\gamma t}\left\{\log x - \log x_0 - \beta_0 t - \beta_1 \sum_{j=1}^{t} G_j\right\}^2\right]. \quad (31)$$

The moments of the distribution are given by

$$E[X(t)]^k = X_0^k \cdot \exp\left\{k\left[\left(\beta_0 - k\frac{\gamma}{2}\right)t + \beta_1 \sum_{j=1}^{t} G_j\right]\right\},$$

$$k = 1, 2, 3, \cdots. \quad (32)$$

Hence, Mean

$$E[X(t)] = X_0 \cdot \exp\left(b_0 t + b_1 \cdot \sum_{1}^{t} G_j\right) \quad (33)$$

and, Variance

$$V[X(t)] = X_0^2 \cdot \exp\left[2\left(b_0 t + b_1 \cdot \sum_{1}^{t} G_j\right)\right] \cdot [\exp a_0 t - 1]. \quad (34)$$

Also, the coefficients of skewness and kurtosis are

$$\gamma_1 = \frac{(e^{3\gamma t} - 3e^{\gamma t} + 2)}{(e^{\gamma t} - 1)^{3/2}},$$

$$\gamma_2 = \frac{(e^{6\gamma t} - 4e^{3\gamma t} + 6e^{\gamma t} - 3)}{(e^{\gamma t} - 1)^2}. \quad (35)$$

We now fit the generalized model with government expenditure G_t to the same Indian data.

The maximum likelihood estimates and their variances are obtained as

$$\hat{\beta}_0 = 0.01118, \quad \hat{\beta}_1 = 0.001623, \quad \hat{\gamma} = 0.5494 \times 10^{-3}$$

$$V(\hat{\beta}_0) = 0.4578 \times 10^{-4} \cdot 1/t,$$

$$V(\hat{\beta}_1) = 0.4578 \times 10^{-4} \cdot \frac{1}{(\sum_{1}^{t} G)^2/t}, \quad (36)$$

$$V(\hat{\gamma}) = 0.503 \times 10^{-7},$$

and hence

$$\hat{b}_0 = 0.01146, \quad \hat{b}_1 = 0.001623, \quad \hat{a}_0 = 0.5494 \times 10^{-3}$$

$$V(\hat{b}_0) = 0.1257 \times 10^{-7} + 0.4578 \times 10^{-4} \cdot 1/t \quad (37)$$

$$V(\hat{b}_1) = \frac{0.4578 \times 10^{-4}}{(\sum_{1}^{t} G_j)^2/t}, \quad V(\hat{a}_0) = 0.503 \times 10^{-7}.$$

The trend is given by

$$E[X(t)] = 86.5 \exp(0.01146t + 0.001623 \sum_1^t G_j), \quad t = 0, 1, 2, \cdots. \quad (38)$$

The fitted trend values for the national income $X(t)$ are given in Table 8, where the fit is found better than the one without the influence of government expenditure (Table 9, below).

TABLE 8

Real National Income $X(t)$ (at Constant Prices for 1948-49) and Government Expenditures $G(t)$ (at Current Prices) for India for 1948-49 to 1960-61 (Billions of Rupees)

Year	t	G(t)	X(t)	E[X(t)]
1948-49	0	8.5	86.5	86.5
1949-50	1	8.1	88.2	88.6
1950-51	2	8.3	88.5	90.9
1951-52	3	8.8	91.0	93.3
1952-53	4	9.0	94.6	95.7
1953-54	5	9.8	100.3	98.4
1954-55	6	11.0	102.8	101.3
1955-56	7	12.9	104.8	104.6
1956-57	8	14.8	110.0	108.4
1957-58	9	17.5	108.9	112.8
1958-59	10	18.5	116.9	117.6
1959-60	11	17.4	118.8	122.4
1960-61	12	19.3	127.3	127.7
1961-62	13	23.4	—	—

Source: "Estimates of National Income," C.S.O., Government of India, April, 1963.

TABLE 9

Estimates of the National Income of India for Selected Years

Year	t	(a)	(b)	$E[X(t)]$ (c)	(d)
1965-66	17	167.7	168.5	168.6	168.5
1967-68	18	190.2	192.5	193.1	193.4
1969-70	21	217.8	222.8	224.8	226.5
1970-71	22	234.0	240.9	244.0	246.8

We now try to predict the values of national income for the same years chosen earlier under following different hypotheses of G_t:

(a) yearly 1.5 billion increase in G during Third and Fourth five year Plan period,

(b) yearly 2.0 billion increase in G during Third and Fourth Plan period,

(c) yearly 2.5 billion increase in G during Third and Fourth Plan period,
(d) yearly 2.0 billion increase in G during Third Plan and 3.0 billions increase in G during Fourth Plan.

The results are obtained as shown in the following table.

We observe the following growth rates under the different hypotheses of government expenditure and levels of income at the end of Fourth five-year Plan (1970–71)

EXPECTED GROWTH RATES DURING FOURTH PLAN

Period	Hypothesis about $G(t)$			
	(a)	(b)	(c)	(d)
First two years	6.7	7.1	7.2	7.3
Next two years	7.2	7.8	8.2	8.5
Last year	7.4	8.1	8.5	9.0

Thus, depending on what growth rate is desired, the level of government expenditure may be maintained.

Consider a stochastic variable, say real per capita agricultual production in India (Tintner and Patel, 1966). Then the transition probability that this variable will be y at time t if it were x at time s is given by

$$f(s, x; t, y) = \frac{\exp\{-(\log y - \log x - b(t-s))/2c(t-s)\}}{y[2\pi c(t-s)]^{1/2}}. \tag{39}$$

This is a lognormal diffusion process which satisfies the forward and backward Kolmogorov equations. In other words, it is the logarithm of real agricultural production per capita in India which is normally distributed and follows a conventional diffusion process.

Assume now that with probability one the variable x has the value x_0, at the point $t = 0$, then we have for the mean

$$Ey(t) = x_0 \exp(b + \tfrac{1}{2}c)t. \tag{40}$$

The variance is

$$\sigma^2_{y(t)} = [(Ey(t))^2(\exp(ct) - 1)]. \tag{41}$$

We give in the following tables the data from which real per capita agricultural production and real per capita government expenditure on agriculture has been computed.

Assume now that $x = x_0$ with probability one at the point $t = 0$, also that the observations are evenly spaced from $t = 0$ to $t = n$ ($n + 1$ observations).

2.4 Diffusion Processes

The maximum likelihood estimates of the parameters of the process are then given by the fomulas

$$\hat{b} = \left(\sum_{t=1}^{n} \log x_t - \log x_{t-1}\right) \Big/ n, \tag{42}$$

$$\hat{c} = \left[\sum_{t=1}^{n} (\log x_t - \log x_{t-1})^2\right] \Big/ n - \hat{b}^2. \tag{43}$$

The asymptotic variances of the estimates are given by

$$\sigma_b^2 = c/nt, \tag{44}$$

$$\sigma_c^2 = 2c^2/n. \tag{45}$$

Since both variances go to zero as n tends to infinity, the estimates are seen to be consistent.

Fitting a lognormal diffusion process to the Indian agricultural data by the methods indicated we obtain the following fitted values and predictions, where the latter are denoted by single asterisk in Table 10.

TABLE 10

REAL PER CAPITA AGRICULTURAL PRODUCTION AND REAL PER CAPITA GOVERNMENT EXPENDITURE ON AGRICULTURE IN INDIA[a]

t	X_t	X_t^*	$\log X_t$	$\log X_t^*$	G_t
0	1.0347	1.0350	0.0149	0.0149	0.0211
1	1.2430	0.0380	0.0945	0.0121	0.0278
2	1.3063	1.0220	0.1158	0.0093	0.0338
3	1.4821	1.0150	0.1620	0.0065	0.0516
4	1.3870	1.0090	0.1421	0.0037	0.0619
5	1.2488	1.0020	0.0966	0.0009	0.0360
6	1.1347	0.9956	0.0551	1.9981	0.0379
7	1.1842	0.9893	0.0734	1.9953	0.0405
8	1.1076	0.9929	0.0445	1.9925	0.0417
9	1.0881	0.9766	0.0366	1.9897	0.0437
10	1.0647	0.9703	0.0273	1.9869	0.0452
11	1.0000	0.9640	0.0000	1.9841	0.0529
12	0.9401	0.9579	1.9731	1.9813	0.0548
13		0.9517		1.9785	
14		0.9456		1.9757	
15		0.9396		1.9729	
16		0.9335		1.9701	
17		0.9274		1.9673	
18		0.9215		1.9645	

[a] Based on third Five-year Plan statistics and the index of wholesale prices (base year 1952–53) published by the Reserve Bank of India and single asterisks denote predicted values.

The predicted value of x^*_{18}, that is, real per capita agricultural income for 1969–70 is 0.9215. Using the estimated variance, we might compute 95% confidence or fiducial limits, which are 0.708 and 1.113.

Consider now the influence of an external factor, for example, real per capita public expenditure in the year t, G_t. Define

$$H_t = G_0 + G_1 + \cdots + G_t, \tag{46}$$

the total expenditure in the period zero to t.

The effect of this expenditure will only be felt after p years. Then the transition probability will be

$$f(s, x; t, y)$$
$$= \left\{\exp[-\{\log y - \log x - b_0(t-s)b_1(H_{t-p} - H_{s-p})\}^2/2c(t-s)\right.$$
$$\left. \cdot \left(\frac{1}{y[2\pi c(t-s)]^{1/2}}\right)\right\}. \tag{47}$$

This is the conditional probability, that the variable in question (e. g. , real per capita agricultural production) will have the value y at the time t if it had the value x at the time s.

Assuming again that x has the value x_p with probability one at the time $t = p$, we derive the following for the mean value

$$E[y(t)] = x_p \exp[b_0 + \tfrac{1}{2}c)(t-p) = b_1 H_{t-p}] \tag{48}$$

and for the variance:

$$\sigma^2_{y(t)} = [Ey(t)]^2[\exp(ct) - 1]. \tag{49}$$

The maximum likelihood estimates of the parameters are now as follows

$$\hat{b} = \frac{\{\sum_{t=1}^{n-p} G_t(\log x_{t+p} - \log x_{t+p-1}) - [\sum_{t=1}^{n-p}(\log x_{t+p} - \log x_{t+p-1}) - b_1 \sum_{t=1}^{n-p} G_t]\}(n-p)^{-1}}{\sum_{t=1}^{n-p} G_t^2 - \frac{[\sum_{t=1}^{n-p} G_t]^2}{n-p}} \tag{50}$$

$$\hat{b}_0 = \frac{\sum_{t=1}^{n-p}(\log x_{t+p} - \log_{t+p-1}) - b_1 \sum_{t=1}^{n-p} G_t}{n-p}, \tag{51}$$

$$\hat{c} = \sum_{t=1}^{n-p}(\log x_{t-p} - \log x_{t+p-1} - b_0 - b_1 G_t)^2/(n-p). \tag{52}$$

Now the large sample variances of the estimates are given by

$$\sigma^2_{\hat{b}_0} = c/(n-p)t, \tag{53}$$

$$\sigma^2_{\hat{b}_1} = ct/(n-p) \sum_{t=1}^{n-p} G_t, \tag{54}$$

$$\sigma^2_c = 2c^2/n. \tag{55}$$

It is again seen that these variances tend to zero as t and n tend to infinity and G_t is constant or increasing. Hence under these circumstances the estimates are consistent. The fitted values of a lognormal diffusion process under the assumptions indicated are given in Table 11, where $p = 3$.

TABLE 11

t	log X'	log X'
4	0.1727	1.4880
5	0.1465	1.4020
6	0.1233	1.3280
7	0.1091	1.2850
8	0.1002	1.2600
9	0.0782	1.1980
10	0.0571	1.1400
11	0.0373	1.0900
12	0.0181	1.0420
13	1.9999	0.9997
14	1.9825	0.9605
15	1.9690	0.9311
16	1.9565	0.9046

We have also made a forecast for the year 1969–70 under the following alternative assumptions: (1) same real per capita government expenditure in the intervening years as in 1963–64, (2) an increase by one-half, (3) a doubling, (4) two and a half times, (5) three times the expenditure of 1963–64. The forecasts and 95% limits are given in Table 12.

TABLE 12

Assumption	Ext X'_{18}	95% limits
(1)	0.8541	0.5952–1.1130
(2)	0.9107	0.6347–1.1862
(3)	0.9712	0.6768–1.2656
(4)	1.0350	0.7212–1.3488
(5)	1.1040	0.7694–1.4386

Let $X(t)$, the yield per hectare of an agricultural crop at time t, be a random variable of a continuous stochastic process of Markovian type, depending on a continuous time parameter t.

The transition probability that this variable will be y at time t if it were x at time τ is assumed to exist for every τ and t where $0 < \tau < t$ and satisfies the backward and forward Kolmogorov equation (Bharucha-Reid 1960, pp. 130ff); $b(t, x)$ and $a(t, x)$ are considered as infinitesimal mean and variance of change in $X(t)$ during the small interval Δt of time, that characterized the process.

$$E[\Delta x(t)] = b(t, x) = b_t x \quad \text{and} \quad V[\Delta x(t)] = a(t, x) = a_t s^2 > 0,$$

where a_t and b_t are functions of time. We assume $a_t = a > 0$ and $b_t = b$; a and b are constants. These assumptions yield the lognormal diffusion process (Tintner and Patel, 1965, 1966).

Taking the initial time as zero, that is, $\tau = 0$ and assuming $P[X(0) = x_0] = 1$ (P denotes probability) the mean and variance of $X(t)$ can be expressed as

$$E[X(t)] = x_0 \exp(bt),$$
$$V[X(t)] = x_0^2 \exp(2bt)[\exp(at) - 1].$$

They are exponential functions of time.

When the observations are taken at equal time intervals the maximum likelihood estimates of the parameters and their variances are given by Fisz (1963, p. 489).

We applied this model to the yields from 1951–52 to 1963–64 of a few important agricultural crops of India namely, wheat, rice, barley, and sugar cane. The expressions for the trends in the yields and variances based on the above estimates are shown in Table 13.

TABLE 13

The Trend and Variance of The Yields of Wheat, Rice, Barley and Sugarcane in India (1951-52 to 1963-64)

Crop	Trend and Variance
	Trend: $E[X(t)]$
Wheat	$(6.63)\exp[(.02170)t]$
Rice	$(10.60)\exp[(.036742)t]$
Barley	$(7.63)\exp[(.008436t]$
Sugarcane	$(31.79)\exp[(.03417)t]$
	Variance: $V[X(t)]$
Wheat	$(6.63)^2 \exp[(.04340)t]\exp[(.01359)t-1]$
Rice	$(10.60)^2 \exp[(.07284)t]\exp[.01033)t-1]$
Barley	$(7.63)^2 \exp[(.016871)t]\exp[(.004067)t-1]$
Sugarcane	$(31.79)^2 \exp[(.06934)t]\exp[(.008661)t-1]$

Note: Units for wheat, rice and barley are 100 kgs and for sugarcane metric tons.

2.4 Diffusion Processes

Having observed a reasonably good fit of the trends we attempt to predict the yields for the next few years (1964–1976) by extrapolating the estimated trends in yields and applying an extended model which considers the influence of an exogenous variable. This can be done by considering the parameters of the process as the function of the exogenous variable.

Consider one of the important measures, irrigation. Let $I(t)$ denote proportion of area irrigated under the crop to total area under the crop at time (t) and $\Delta I(t) = I(t) - I(t-1)$.

The infinitesimal mean and variance of change in the yield $X(t)$ is

$$b(t, x) = b_t x \quad \text{and} \quad a(t, x) = a_t x^2,$$

where

$$a_t = a_0 > 0 \quad \text{and} \quad b_t = b_0 + b_1 \Delta I(t).$$

The mean and variance of $X(t)$ are given by

$$E[X(t)] = x_0 \exp(b_0 t + b_1 \sum_{j=1}^{t} \Delta I_j),$$

$$E[X(t)] = x_0^2 \exp[2(b_0 t + b_1 \sum_{j=1}^{t} \Delta I_j)][\exp(a_0 t) - 1].$$

TABLE 14

The Trend and The Variance of the Yield Per Hectare $X(t)$ of Wheat, Rice, Barley Sugarcane in India (1951-1963)

Crop	Trend and Variance
	Trend: $E[X(t)]$
Wheat	$6.63 \exp\left[.03414t + 1.4130 \sum_{j=1}^{t} \Delta I(t)\right]$
Rice	$10.60 \exp\left[.2429t + .8156 \sum_{j=1}^{t} \Delta I(t)\right]$
Barley	$7.63 \exp\left[.02450t + .1941 \sum_{j=1}^{t} \Delta I(t)\right]$
Sugarcane	$31.79 \exp\left[.04013t + .5798 \sum_{j=1}^{t} \Delta I(t)\right]$
	$V[X(t)]$
Wheat	$(6.63)^2 \exp\left[.6828t + 2.8260 \sum_{j=1}^{t} I(t)\right][\exp(.01163t) - 1]$
Rice	$(10.60)^2 \exp\left[.04584t + 1.6342 \sum_{j=1}^{t} I(t)\right][\exp(.01049t) - 1]$
Barley	$(7.63)^2 \exp\left[.0490t + .3882 \sum_{j=1}^{t} I(t)\right][\exp(.006375t) - 1]$
Sugarcane	$(31.79)^2 \exp\left[.08026t + 1.1596 \sum_{j=1}^{t} I(t)\right][\exp(.005705t) - 1]$

80   2. Stochastic Models of Economic Development

The expression for the mean and variances of $X(t)$, considering $I(t)$ as an exogenous variable, are shown in Table 14. We computed the coefficient of variation, that is, $(V[X(t)]^{1/2})/E[X(t)]$ in order to compare the fitted trends employing different models. With the extended model we got in most of the cases a smaller coefficient of variation which shows that this model gives better estimates than those given by the simple model. Having observed a better fit, we attempt to predict the values of the yields for the next few years (1963–64 to 1975–76) under the following hypothesis about $I(t)$.

Consider the following different hypotheses regarding $I(t)$ for rice.

(1) A yearly increase of 0.007 during the last three years of the Third Plan and of 0.019 and 0.014 during the Fourth and Fifth Plan periods respectively, which will raise the proportion of irrigated area under rice to total area under rice, $I(t)$ from 38.05 per cent in 1962–63 to 39.63, 49.33, and 56.33 per cent by the end of the Third, Fourth, and Fifth Plans respectively.

(2) A yearly increase of 0.006 during the last three years of the Third Plan and of 0.015 and 0.013 during the Fourth and Fifth Plan periods respectively —this will raise the level of $I(t)$ to 39.83, 47.44, and 53.83 percent by the end of the Third, Fourth, and Fifth Plans respectively.

(3) A yearly increase of 0.007 during the last three years of the Third Plan, and of 0.015 and 0.016 during the Fourth and Fifth Plan periods respectively, which will raise the level of $I(t)$ to 40.13, 47.63, and 55.63 per cent, by the end of the Third, Fourth, and Fifth Plans respectively.

(4) A yearly increase of 0.01 in $I(t)$ during the whole period under con-

TABLE 15

Estimates for Yield Per Hectare of Rice, $X(t)$ for 1963–65 to 1975–76

Year	t	$E[X(t)]$ Hypothesis Number About $I(t)$			
		(1)	(2)	(3)	(4)
1963–64	12	14.95	14.95	14.96	15.00
1964–65	13	15.39	15.39	15.42	15.49
1965–66	14	15.85	15.85	15.88	16.00
1966–67	15	16.49	16.44	16.48	16.53
1967–68	16	17.16	17.05	17.09	17.07
1968–69	17	17.85	17.68	17.72	17.64
1969–70	18	18.58	18.34	18.36	18.22
1970–71	19	19.33	19.02	19.07	18.82
1971–72	20	20.04	19.70	19.79	19.44
1972–73	21	20.76	20.40	20.55	20.08
1973–74	22	21.52	21.12	21.33	20.74
1974–75	23	22.30	21.87	22.14	21.43
1975–76	24	23.11	22.65	22.98	22.14

sideration, that is, during 1963–76 which will raise the level of $I(t)$ to 41.93, 46.03, 51.03 per cent by the end of the Third, Fourth, and Fifth Plans respectively.

The estimates for the yield per hectare of rice based on the above alternative hypothesis about $I(t)$ during the period of 1963–1976 are shown in Table 15.

Consider the following Hypothesis about $I(t)$ for wheat.

(1) A yearly increase of 0.005 during the last three years of the Third Plan and 0.02 and 0.015 during the Fourth and the Fifth Plan periods respectively, which will raise the proportion of $I(t)$ from 34.13 per cent in 1962–63 to 35.63, 45.63 and 54.63 per cent by the end of the Third, Fourth, and Fifth Plans respectively.

(2) A yearly increase of 0.005 during the last three years of the Third Plan and of 0.017 during the Fourth and Fifth Plan periods, respectively, which will raise the level of $I(t)$ to 35.63, 44.13, and 51.63 per cent by the end of the Third, Fourth, and Fifth Plans respectively.

(3) A yearly increase of 0.006 during the last three years of the Third Plan and of 0.016 and 0.017 during the Fourth and Fifth Plan periods respectively, which will raise the level of $I(t)$ to 35.93, 43.93, and 52.43 per cent by the end of the Third, Fourth, and Fifth Plans respectively.

(4) A yearly increase of 0.01 during the whole period under consideration, that is, 1963–76, which will increase the level of $I(t)$ to 37.13, 42.13, and 47.13 pen cent by the end of the Third, Fourth, and Fifth plans respectively. The

TABLE 16

ESTIMATES FOR YIELD PER HECTARE OF WHEAT $X(t)$ FOR 1963–64 TO 1975–76 (in Hundreds of Kgs)

Year	t	$E[X(t)]$ Hypothesis Number About $I(t)$			
		(1)	(2)	(3)	(4)
1963–64	12	9.89	9.89	9.91	9.96
1964–65	13	10.31	10.31	10.34	10.46
1965–66	14	10.74	10.74	10.79	10.97
1966–67	15	11.43	11.39	11.42	11.52
1967–68	16	12.17	12.07	12.09	12.09
1968–69	17	12.96	12.79	12.79	12.68
1969–70	18	13.79	13.56	13.54	13.31
1970–71	19	14.68	14.37	14.33	13.97
1971–72	20	15.51	15.19	15.19	15.39
1972–73	21	16.40	16.05	16.10	15.39
1973–74	22	17.33	16.97	18.08	16.15
1974–75	23	18.32	17.83	18.08	16.95
1975–76	24	19.36	18.95	19.17	17.78

estimates for the yield per hectare of wheat based on the above hypothesis are shown in Table 16.

Similarly, different hypotheses regarding $I(t)$ for barley and sugarcane can be incorporated in the above model.

Now we extend the model to consider influence of two exogenous variables. This is done by considering the mean value of the process as a function of two exogenous variables, say $\Delta I(t)$ and $\Delta F(t)$. $\Delta I(t)$ denotes, as in the previous section, the change in proportion of irrigated area under the crop to total area under the crop $I(t)$ at time (t) over time $(t-1)$. $F(t)$ denotes the change in amount of nitrogenous fertilizer used in kilograms per hectare $F(t)$ at time (t) over time $(t-1)$. With infinitesimal mean and variance of change in $X(t)$ defined as

$$b(t, x) = b_t x \quad \text{and} \quad a(t, x) = a_t x^2 > 0,$$

as before, we have

$$a_t = a_0 > 0 \quad \text{and} \quad b_t = b_0 + b_1 \Delta I(t) + b_2 \Delta F(t),$$

where a_0, b_0, b_1, and b_2 are constants.

The mean and variance of $X(t)$ are

$$E[X(t)] = x_0 \exp\left(b_0 t + b_1 \sum_{j=1}^{t} \Delta I_j + b_2 \sum_{j=1}^{t} \Delta F_j\right),$$

$$V[x(t)] = x_0^2 \exp\left[2\left(b_0 t + b_1 \sum_{j=1}^{t} \Delta I_j + b_2 \sum_{j=1}^{t} \Delta F_j\right)\right]$$

$$\cdot [\exp(a_0 t) - 1].$$

Let us apply the extended model to yield of wheat, $X(t)$ in consideration of the proportion of irrigated area under wheat to total area under wheat $I(t)$ and nitrogenous fertilizer used per hectare for wheat $F(t)$. Under the similar assumption of $P[X(0) = x_0] = 1$ and $I(t)$ and $F(t)$ as step functions, the expressions for mean and variance of X(t) are the following

$$E[X(t)] = 6.63 \exp\left[10^{-2}(0.2154)t + 0.8108 \sum_{j=1}^{t} \Delta I_j + 0.0798 \sum_{j=1}^{t} \Delta \bar{F}_j\right],$$

$$V[X(t)] = (6.63)^2 \exp\left[\frac{1}{2} 10^{-2}(0.2154)t + 0.8108 \sum_{j=1}^{t} \Delta I_j + 0.07980 \sum_{j=1}^{t} \Delta F_j\right]$$

$$\cdot [\exp(0.01039) - 1].$$

This model gave smaller coefficients of variation which shows that these estimates are better than those of the previous model. The positive values 0.8108 and 0.07980 show that an increase in the proportion of irrigated area under wheat to total area under wheat, and increase in the amount of nitrogenour fertilizer used per hectare have a positive effect on the yields of wheat.

We extrapolate the trend in the yield of wheat for the next few years with some specific hypotheses regarding the change in the $I(t)$ and $F(t)$ for wheat. Consider the following sets of hypotheses about $I(t)$ and $F(t)$ for wheat.

(1) A yearly increase of 0.005 in $I(t)$ during the last three years of the Third Plan and of 0.9 in $F(t)$ during the last two years of the Third Plan. A yearly increase of 0.02 in $I(t)$ and of 1.05 in $F(t)$ during the Fourth Plan period, and for the Fifth Plan period a yearly increase of 0.015 in $I(t)$ and of 1.00 in $F(t)$.

(2) A yearly increase of 0.005 in $I(t)$ during the last three years of the Third Plan and of 0.9 in $F(t)$ during the last two years of the Third Plan; a yearly increase of 0.017 in $I(t)$ and 1.02 in $F(t)$ during the Fourth Plan, and a yearly increase of 0.015 in $I(t)$ and 1.00 in $F(t)$ during the Fifth Plan period.

(3) A yearly increase of 0.006 in $I(t)$ during the last three years of the Third Plan, and 0.9 in $F(t)$ during the last two years of the Third Plan; a yearly increase of 0.016 in $I(t)$ and of 1.04 in $F(t)$ during the Fourth Plan Period, and 0.017 in $I(t)$ and of 1.09 in $F(t)$ during the Fifth Plan period.

(4) A yearly increase of 0.010 in $I(t)$ during 1963–64 to 1975–76, and of 1.00 in $F(t)$ during 1964–65 to 1975–76.

The estimates for the yield per hectare of wheat based on the above alternative hypotheses about $F(t)$ during the period 1963–76 are shown in Table 17.

TABLE 17

ESTIMATES FOR YIELD PER HECTARE OF WHEAT $X(t)$ FOR 1963–64 TO 1975–76 (in Hundreds of Kgs)

Year	t	$E[X(t)]$ Hypothesis Number About $I(t)$ and $F(t)$			
		(1)	(2)	(3)	(4)
1963–64	12	10.23	10.23	10.23	10.27
1964–65	13	11.06	11.06	11.11	11.24
1965–66	14	11.96	11.96	12.06	12.30
1966–67	15	13.24	13.18	13.31	13.46
1967–68	16	16.25	14.53	14.68	14.72
1968–69	17	16.25	16.01	16.19	16.11
1969–70	18	17.99	17.65	17.86	17.63
1970–71	19	19.93	19.46	19.71	19.30
1971–72	20	21.90	21.38	21.84	21.12
1972–73	21	24.06	23.49	24.21	23.11
1973–74	22	20.05	28.36	29.75	27.67
1975–76	24	31.92	31.16	32.98	30.28

Thus the empirical results show that by increasing the proportion of irrigated area under wheat to the total area under wheat, and by increasing the use of nitrogenous fertilizer for wheat, a significant increase in the yield of wheat is accomplished.

It is then our purpose now to extend the univariate lognormal diffusion process to a multivariate process which takes into account the interdependence among the major economic magnitudes, such as national income, investment, labor force, and so on. In what follows we shall consider first a general case of k variables and then illustrate it by an empirical example involving four variables.

Let $\mathbf{X}(t)$ be a vector of k continuous variables of a stochastic process $\mathbf{X}(t)$, $t \geq 0$ of the Markovian type such that each component $X_i(t)$, $i = 1, 2, \cdots, k$, assumes values in the interval $(0, \infty)$ of the real line and represents an economic variable. Assume that the transition probability density function f for $\mathbf{X}(t)$ is

$$f = f[\gamma, \mathbf{x}(\gamma); t, \mathbf{y}(t)] = [\text{Pr. } \mathbf{X}(t) = \mathbf{y} | \mathbf{X}(\gamma) = \mathbf{x}]; \quad 0 \leq \gamma < t < \infty \quad (56)$$

satisfying the backward and forward Kolmogorov diffusion equations. Consider the coefficients $b_i[t, \mathbf{x}(t)]$ and $a_{ij}[t, \mathbf{x}(t)]$, the so-called infinitesimal means and variance-covariances of the changes in $\mathbf{X}(t)$ during a small interval Δt of time which characterize the particular type of diffusion process as follows

$$b_i[t, \mathbf{x}(t)] = b_i(x_i), \quad i = 1, 2, \cdots, k, \quad (57)$$

$$a_{ij}[t, \mathbf{x}(t)] = a_{ij}(x_i x_j), \quad i, j = 1, 2, \cdots, k, \quad (58)$$

where b_i, a_{ij} are constants such that the matrix A of the coefficients a_{ij} is positive semidefinite. We thus assume that the expected changes in $\mathbf{X}(t)$ and their variance-covariances are proportional to the present values of the variables $\mathbf{X}(t)$. Assuming A positive semidefinite, it would imply that the economic variables $\mathbf{X}(t)$ do assume changes in the small interval of time Δt though the magnitudes of the changes would be small for small Δt. Looking to the constantly changing nature of the economy through the interactions of a complex of economic activities and the diffusion amongst them, the above specification for the stochastic process $\{\mathbf{X}(t), t \geq 0\}$ seems to describe well the real situation of the evolution on the economy.

With the coefficients so defined in Eqs. (57) and (58), the backward and forward Kolmogorov equations become

$$-\frac{\partial f}{\partial \gamma} = \frac{1}{2} \sum_i \sum_{j=1}^k a_{ij}(x_i x_j) \frac{\partial^2 f}{\partial x_i \partial x_j} + \sum_{i=1}^k b_i(x_i) \frac{\partial f}{\partial x_i}, \quad (59)$$

$$\frac{\partial f}{\partial t} = \frac{1}{2} \sum_i \sum_{j=1}^k \frac{\partial^2 [a_{ij}(y_i y_j) f]}{\partial y_i \partial y_j} - \sum_{i=1}^k \frac{\partial [b_i(y_i) f]}{\partial y_i}, \quad (60)$$

respectively. The probability density function f satisfying these diffusion equations, Eqs. (59) and (60), is given by the multivariate lognormal density function

2.4 Diffusion Processes

$$f(\tau, \mathbf{x}; t, \mathbf{y}) = 1 \Big/ \sum_{i=1}^{k} y_i [(2\pi)^{1/2}]^k (t-\tau)^{k/2} |\Sigma|^{1/2} \qquad (61)$$
$$\cdot \exp\{-[2(t-\tau)]^{-1} Q\},$$

where

$$Q = [\log \mathbf{y} - \log \mathbf{x} - \boldsymbol{\beta}(t-\tau)]' \Sigma^{-1} [\log \mathbf{y} - \log \mathbf{x} - \boldsymbol{\beta}(t-\tau)],$$

where

$$\Sigma = (\sigma_{ij}) = (a_{ij}) = A, \qquad i, j = 1, 2, \cdots, k$$

and

$$\boldsymbol{\beta} = (\mathbf{b} - \tfrac{1}{2}\mathbf{a}), \qquad \mathbf{a} = \{a_{ii}\}, \qquad i = 1, 2, \cdots, k.$$

The characteristics of this distribution given by Eq. (61) are easily derived from those of the well-known multivariate normal distribution. Denoting $\log \mathbf{X}(t)$ by $\mathbf{Z}(t)$, the density function for $\mathbf{Z}(t)$ is the multivariate normal with mean vector $\mathbf{Z}(\tau) + \boldsymbol{\beta}(t-\tau)$ and variance covariance matrix $V = \Sigma(t-\tau)$, whose moment genereration is then given by

$$E\{\exp[\boldsymbol{\theta}'\mathbf{Z}(t)]\} = \exp\{\boldsymbol{\theta}'[\mathbf{Z}(\tau) + \boldsymbol{\beta}(t-\tau)] + \tfrac{1}{2}\boldsymbol{\theta}' V \boldsymbol{\theta}\}. \qquad (62)$$

Therefore, the moments of $\mathbf{X}(t)$ are easily obtained from Eq. (62).

If we consider the initial time τ as zero and assume that $P[\mathbf{X}(0) = \mathbf{x}(0)] = 1$, then means and variance-covariances are given by

$$E[X_i(t)] = X_i(0) \exp(b_i t), \qquad i = 1, 2, \cdots, k, \qquad (63)$$
$$V[X_i(t)] = X_i^2(0) \exp(2b_i t)[\exp(a_{ii} t) - 1], \qquad i = 1, 2, \cdots, k \qquad (64)$$
$$C[X_i(t), X_j(t)] = X_i(0) \exp(b_i t) X_j(0) \exp(b_j t)$$
$$\cdot [\exp(a_{ij} t) - 1], \qquad i = 1, 2, \cdots, k, \qquad (65)$$
$$R[X_i(t), X_j(t)] = [\exp(a_{ij} t) - 1)/(\exp(a_{ii} t) - 1)^{1/2}]$$
$$\cdot (\exp(a_{jj} t) - 1)^{1/2}, \qquad i, j = 1, 2, \cdots, k. \qquad (66)$$

We observe that the dynamic path of the economy as characterized by the random variables $X_i(t)$, $i = 1, 2, \cdots, k$ is given by the exponential trends Eq. (63) for each of the variables. Also, the variances Eq. (64) show exponential functions of time. Given the initial conditions, namely values of $X_i(t)$ at $t = 0$, and the values of the parameters b_i's and a_{ij}'s, one can evaluate the future course of development of the economic variables from the above expression. The values of the parameters can be taken as the estimates obtained from empirical data. We derive the estimates using the maximum likelihood method.

The estimates of the parameters $\boldsymbol{\beta}$ and $\boldsymbol{\Sigma}$ are obtained as follows. Let $\mathbf{x}_0$, $\mathbf{x}_1, \cdots, \mathbf{x}_n$ be the observed states of the random vector $\mathbf{X}(t)$ at times $t_0, t_1, \cdots, t_n$ respectively and let $P[\mathbf{X}(0) = \mathbf{x}(0)] = 1$. The joint probability of the observations, known as the likelihood function, is

$$L = (2\pi)^{-nk/2} |\Sigma|^{-n/2} \left\{ \prod_{\alpha=1}^{n} \prod_{i=1}^{k} (x_{1\alpha})^{-1} (t_\alpha - t_{\alpha-1})^{-k/2} \right.$$
$$\left. \cdot \exp([-2(t_\alpha - t_{\alpha-1})^{-1}]Q) \right\}, \tag{67}$$

where

$$Q = [\log \mathbf{x}_\alpha - \log \mathbf{x}_{\alpha-1} - \boldsymbol{\beta}(t_\alpha - t_{\alpha-1})]' \Sigma^{-1}$$
$$\cdot [\log \mathbf{x}_\alpha - \log \mathbf{x}_{\alpha-1} - \boldsymbol{\beta}(t_\alpha - t_{\alpha-1})].$$

Therefore,

$$\log L = (-nk/2) \log (2\pi) + (n/2) \log |\Sigma|^{-1}$$
$$- \frac{k}{2} \sum_{\alpha=1}^{n} \log (t_\alpha - t_{\alpha-1}) - \sum_{\alpha=1}^{n} \sum_{i=1}^{k} \log x_{i\alpha}$$
$$- \frac{1}{2} \sum_{\alpha=1}^{n} (t_\alpha - t_{\alpha-1})^{-1} [\log \mathbf{x}_\alpha - \log \mathbf{x}_{\alpha-1} - \boldsymbol{\beta}(t_\alpha - t_{\alpha-1})]'$$
$$\cdot \Sigma^{-1} [\log \mathbf{x}_\alpha - \log \mathbf{x}_{\alpha-1} - \boldsymbol{\beta}(t_\alpha - t_{\alpha-1})]. \tag{68}$$

Differentiating Eq. (68) with respect to the parameters $\boldsymbol{\beta}, \Sigma$ and equating the results to zero, we obtain the following maximum likelihood estimates for b_i and a_{ij}

$$\hat{b}_i = \hat{\beta}_i + (1/2)\hat{\sigma}_{ii}, \quad i = 1, 2, \cdots, k \tag{69}$$

$$\hat{a}_{ij} = \hat{\sigma}_{ij}, \quad i, j = 1, 2, \cdots, k. \tag{70}$$

If we assume that the observations are taken at equal intervals of time of unity, that is, $(t_\alpha - t_{\alpha-1}) = 1$ for $\alpha = 1, 2, \cdots, n$ and $t_0 = 0$, then the estimates are obtained as

$$\hat{\beta}_i = \sum_{\alpha=1}^{n} (\log x_{i\alpha} - \log x_{i\alpha-1})/n, \quad i = 1, 2, \cdots, k, \tag{71}$$

$$\hat{\sigma}_{ij} = \sum_{\alpha=1}^{n} (\log x_{i\alpha} - \log x_{i\alpha-1} - \hat{\beta}_i)$$
$$\cdot (\log x_{j\alpha} - \log x_{j\alpha-1} - \hat{\beta}_j)/n, \quad i, j = 1, 2, \cdots, k. \tag{72}$$

2.4 Diffusion Processes

The sampling variance-covariances of the estimates $\hat{\beta}$ and $\hat{\Sigma}$ are easily obtained as

$$V(\hat{\beta}_i) = \sigma_{ii}/nt \qquad i = 1, 2, \cdots, k, \tag{73}$$

$$C(\hat{\beta}_i, \hat{\beta}_j) = \sigma_{ij}/nt \qquad i \neq j = 1, 2, \cdots, k, \tag{74}$$

$$V(\hat{\sigma}_{ij}) = (\sigma_{ij}^2 + \sigma_{ii}\sigma_{jj})/n \qquad i, j = 1, 2, \cdots, k, \tag{75}$$

$$C(\hat{\sigma}_{ij}, \hat{\sigma}_{i'j'}) = (\sigma_{ii'} + \sigma_{jj'} + \sigma_{i'j} + \sigma_{ji'})/n, \qquad i, j, i', j' = 1, 2, \cdots, k. \tag{76}$$

We note that the variance-covariance of the estimates tend to zero for large n and t, and hence the estimates $\hat{\beta}$ and $\hat{\Sigma}$ are consistent estimates.

The variances of the estimates $\hat{b}$ and $\hat{A}$ can be easily obtained from those of the estimates β and Σ as

$$V(\hat{b}_i) = a_{ii}/nt + a_{ii}^2/2n, \qquad i = 1, 2, \cdots, k \tag{77}$$

$$V(\hat{a}_{ij}) = (a_{ij}^2 + a_{ii}a_{jj})/n, \qquad i = 1, 2, \cdots, k, \tag{78}$$

and these also tend to zero for large n and t, and hence $\hat{b}$ and $\hat{A}$ are also consistent estimates.

Further, using the property of asymptotic normality, we can derive joint confidence ellipse for the distribution of $X(t)$ as follows,

The quadratic from

$$Q[X(t)] = \{X(t) - E[X(t)]\}'C^{-1}\{X(t) - E[X(t)]\}, \tag{79}$$

where C is the covariance matrix of $X(t)$, follows approximately a chi-square distribution with k degrees of freedom. Hence an approximate confidence ellipse of confidence coefficient $(1 - \alpha)$ for the values of $X(t)$ is given by the inequality

$$P\{Q[X(t)] \leq \chi_\alpha^2\} \geq 1 - \alpha, \tag{80}$$

where χ_α^2 is the upper $\alpha\%$ point of the chi-square distribution with k degrees of freedom.

We now present an empirical illustration of fitting the above model of an interdependent multivariate stochastic process of Markovian type to the data of U. S. for the period 1946–63. Here we consider four variables representing national income by $X_1(t)$, private consumption expenditure by $X_2(t)$, gross domestic investment by $X_3(t)$, and government expenditure by $X_4(t)$ at time t measured in constant prices of 1954.

The maximum likelihood estimates of the parameters and their variances are found as

2. Stochastic Models of Economic Development

$$\hat{\beta} = \begin{bmatrix} \hat{\beta}_1 \\ \hat{\beta}_2 \\ \hat{\beta}_3 \\ \hat{\beta}_4 \end{bmatrix} = \begin{bmatrix} 0.032743 \\ 0.031570 \\ 0.027526 \\ 0.044599 \end{bmatrix}, \tag{81}$$

$$\hat{\Sigma} = ((\hat{\sigma}_{ij})) = \begin{bmatrix} 0.001103 & 0.000344 & 0.003851 & 0.000876 \\ & 0.000447 & 0.001963 & -0.000454 \\ & & 0.024637 & -0.003793 \\ & & & 0.013111 \end{bmatrix}, \tag{82}$$

and hence

$$\hat{b} = \begin{bmatrix} b_1 \\ b_2 \\ b_3 \\ b_4 \end{bmatrix} = \begin{bmatrix} 0.033294 \\ 0.031794 \\ 0.039844 \\ 0.051155 \end{bmatrix}, \tag{83}$$

$$\hat{A} = ((\hat{a}_{ij})) = 10^{-3} \begin{bmatrix} 1.103 & 0.344 & 3.851 & 0.876 \\ & 0.477 & 1.963 & -0.454 \\ & & 24.638 & -3.793 \\ & & & 13.111 \end{bmatrix}, \tag{84}$$

$$V(\hat{\beta}) = \frac{10^{-4}}{t} \begin{bmatrix} 0.678 \\ 0.263 \\ 14.500 \\ 7.710 \end{bmatrix}, \quad t = 1, 2, \cdots, \tag{85}$$

$$V(\hat{\sigma}_{ij}) = 10^{-6} \begin{bmatrix} 0.1431 & 0.0365 & 1.5980 & 0.8957 \\ & 0.0235 & 0.8745 & 0.3568 \\ & & 71.4200 & 19.8500 \\ & & & 20.2200 \end{bmatrix}. \tag{86}$$

Hence the variances of $\hat{b}$ and $\hat{A}$ are obtained as

$$V(b) = \frac{10^{-4}}{t} \begin{bmatrix} 0.678 \\ 0.263 \\ 14.500 \\ 7.701 \end{bmatrix} + 10^{-6} \begin{bmatrix} 0.03578 \\ 0.00588 \\ 17.85240 \\ 5.05583 \end{bmatrix}, \quad t = 1, 2, \cdots, \tag{87}$$

TABLE 18

Gross National Product $X_1(t)$, Private Consumption $X_2(t)$, Domestic Investment $X_3(t)$, and Government Expenditure $X_4(t)$ for U. S. at Constant Prices of 1954 for the Period 1946–63 (in Billions of Dollars)

Year	t	$X_1(t)$	$E[X_1(t)]$	$X_2(t)$	$E[X_2(t)]$	$X_3(t)$	$E[X_3(t)]$	$X_4(t)$	$E[X_4(t)]$
1946	0	282.5	282.50	192.3	193.30	42.4	42.40	43.9	43.90
1947	1	282.3	292.06	195.6	198.51	41.5	44.12	37.3	46.20
1948	2	293.1	301.95	199.3	204.92	49.8	45.92	42.1	48.63
1949	3	292.7	312.17	204.3	211.54	38.5	47.78	47.2	51.18
1950	4	318.1	322.74	216.8	218.38	55.9	49.72	45.1	53.87
1951	5	341.8	333.67	218.5	225.43	57.7	51.75	63.3	56.69
1952	6	355.5	344.96	224.2	232.72	50.4	53.85	77.7	59.67
1953	7	369.0	356.64	235.1	240.23	50.6	56.04	84.3	62.80
1954	8	363.1	368.72	238.0	247.99	48.9	58.32	75.3	66.10
1955	9	392.7	381.20	256.0	256.01	62.5	60.69	73.2	69.57
1956	10	404.2	394.11	263.7	264.28	63.1	63.15	72.9	73.22
1957	11	407.0	407.45	270.3	272.81	57.8	65.72	75.0	77.06
1958	12	401.3	421.24	273.2	281.63	49.0	68.59	79.3	81.11
1959	13	428.6	435.50	288.9	290.72	61.7	71.17	80.1	85.36
1960	14	430.0	450.25	298.1	300.12	60.2	74.07	79.9	89.84
1961	15	447.6	465.49	303.6	309.81	57.5	77.08	84.3	94.56
1962	16	474.8	481.25	317.6	319.82	65.2	80.21	90.2	99.52
1963	17	492.9	492.54	328.9	330.15	67.7	83.47	93.7	104.75

Source: "Historical Statistics of the United States: Colonial Times to 1957, A Statistical Supplement." Bureau of the Census, United States Department of Commerce, Washington, D. C. Statistical data for the recent period are taken from "Statistics Abstract of the United States, 1964. "Bureau of the Census, U. S. Department of Commerce, Washington, D. C.

$$V(a_{ij}) = 10^{-6} \begin{bmatrix} 0.1431 & 0.0365 & 1.5980 & 0.8957 \\ & 0.0235 & 0.8745 & 0.3568 \\ & & 71.4200 & 19.8500 \\ & & & 20.2200 \end{bmatrix}. \quad (88)$$

The empirical trend values for $\mathbf{X}(t)$ are given by

$$E[X_1(t)] = (282.5)\exp(0.033294t), \quad t = 1, 2, \cdots, \quad (89)$$
$$E[X_2(t)] = (192.3)\exp(0.031794t), \quad t = 1, 2, \cdots, \quad (90)$$
$$E[X_3(t)] = (\ 42.4)\exp(0.039844t), \quad t = 1, 2, \cdots, \quad (91)$$
$$E[X_4(t)] = (\ 43.9)\exp(0.051155t), \quad t = 1, 2, \cdots. \quad (92)$$

We observe that the dynamic paths of the variables $\mathbf{X}(t)$ given by Eqs. (89)–(92) depend on the initial state of the economy and the growth rates during the period of our analysis, which are observed as 3.3, 3.2, 4.0, and 5.1 per cent respectively. The trend values of the variables based on these growth rates and the initial values of the variables at time $t = 0$, that is, during the year 1946, are shown in Table 18.

Based on our above model, we try to predict the values of the variables $\mathbf{X}(t)$ for the years 1964 through 1971. These predicted values are shown in Table 19.

TABLE 19

Estimates of National Income $X_1(t)$, Private Consumption $X_2(t)$, Domestic Investment $X_3(t)$, and Government Expenditure $X_4(t)$, for The Years 1964-1971 for The United States

Year	t	$E[X_1(t)]$	$E[X_2(t)]$	$E[X_3(t)]$	$E[X_4(t)]$
1964	18	514.39	340.82	86.86	110.24
1965	19	531.80	351.83	90.39	116.03
1966	20	549.80	363.19	94.07	122.12
1967	21	568.42	374.93	97.89	128.53
1968	22	587.66	38.704	101.87	135.28
1969	23	607.56	399.56	106.01	142.38
1970	24	628.13	412.45	110.32	149.85
1911	25	648.40	425.77	114.81	157.71

The expressions for variances and correlations come out as

$$V[X_1(t)] = (282.5)^2 \exp(0.066588t)[\exp(0.001103t) - 1],$$
$$t = 1, 2, 3, \cdots,$$

$$V[X_2(t)] = (192.3)^2 \exp(0.063588t)[\exp(0.000447t) - 1],$$
$$t = 1, 2, 3, \cdots,$$
$$V[X_3(t)] = (42.4)^2 \exp(0.079688t)[\exp(0.024637t) - 1],$$
$$t = 1, 2, 3, \cdots,$$
$$V[X_4(t)] = (43.9)^2 \exp(0.102310t)[\exp(0.013111t) - 1],$$
$$t = 1, 2, 3, \cdots,$$

$$R[X_i(t), X_j(t)] = \begin{bmatrix} 1.00 & 0.50 & 0.74 & 0.23 \\ & 1.00 & 0.60 & -0.20 \\ & & 1.00 & -0.21 \\ & & & 1.00 \end{bmatrix},$$

$$\text{for} \quad i, j = 1, 2, 3, \cdots.$$

We note that the observed correlation coefficients reflect the nature of the relationship between the increments per unit of time in the variables analyzed. The high correlations 0.74 and 0.60 between the increase in national output and consumption with investment indicate the familiar concepts of multiplier and acceleration principles. Also, the negative correlation -0.21 between the increments in investment and government expenditure show the tendency of maintaining the aggregate expenditure by accelerating the public expenditure whenever investment falls or increases slowly. Further, the approximate confidence ellipsoid of confidence coefficient of 95 per cent for the year 1966 is obtained as

$$\{X_1(20) - 549.80, X_2(20) - 363.19, X_3(20) - 94.07, X_4(20) - 122.12\}$$

$$\cdot 6742.112 \begin{bmatrix} 1378.333 & 4141.114 & & 1186.234 \\ 1184.568 & 1367.944 & -400.426 \\ & & 5634.912 & -839.187 \\ & & & 4471.385 \end{bmatrix}^{-1} \begin{bmatrix} X_1(20) - 549.80 \\ X_2(20) - 363.19 \\ X_3(20) - 94.07 \\ X_4(20) - 122.12 \end{bmatrix}$$

$$\leq 9.49,$$

where 9.49 is the upper 5 per cent point of the Chi-square distribution with four degrees of freedom.

In view of the probabilistic nature of growth in economic activities due to random factors and uncertainty elements, the process of economic growth can be more adequately described by the analysis of stochastic processes. Both temporal dependence in economics and continuity in the evolution of economy suggests continuous stochastic processes of the diffusion type. Moreover, it is appropriate to consider the evolution of an economy in terms of a number of important aggregate economic variables interacting with one another.

Looking to the nature of changes in the economy through the interlinkages of a complex of economic activities and the diffusion amongst them, a hypothesis of proportional change is used to characterize the transition probability of the economic variate $\mathbf{X}(t)$ which specify the state of the economy at any time t, and it turns out that the probability distribution of $\mathbf{X}(t)$ is a multivariate lognormal distribution. The empirical evidence of such skewed distribution is common in economic data. To this end our theory appears reasonable. The mean development path of the variable comes out the well known exponential trend. The exponential growth in the variates are estimated using the postwar U. S. data and the maximum likelihood estimates of the average annual growth rates for the four aggregate economic variables, and the trend projections for each variable are calculated in isolation, with interactions entering through the covariance matrix.

The reasonably well fit of the model to the empirical data suggests the plausibility of the approach of stochastic process to describe and explain the process of economic prowth taking into account the random effects more explicitly. The analysis seems a good first approximation to the complex process of economic development. It may be improved further by introducing the interdependence among the variables more explicitly by studying the conditional trends. It may also be supplemented by incorporating important exogenous variables and *a priori* conditions on the relations between the economic variables, and the theory can be extended to a more complete analysis of the economic development process.

3 STOCHASTIC CONTROL THEORY WITH ECONOMIC APPLICATIONS

One of the most important interesting developments in modern economics is the utilization of ideas which have proved their usefulness in electrical and communications engineering (Tustin, 1953; Allen, 1959). The recent trend in using the theory and methods of optimal control for models of economic growth, planning and economic policy (Fox *et al.*, 1966; Shell, 1967; Kuhn and Szego, 1969) has already proved its important influence on economics, which is very likely to stay. The theory of control has, of course, been used in operations research and industrial engineering in problems of production scheduling, inventory control, and other aspects of enterprise behavior (Sengupta and Fox, 1969; Hanssmann, 1962); but its recent applications in problems of economic growth and capital accumulation have renewed the economists' interest on the pricing system sustaining a growth process, its competitive implications for decentralization and controllability and the various aspects of stability. However, we should emphasize that there is a remarkable prehistory of some of these ideas in the 1930's exemplified by the work of Kalecki (1935), Frisch and Holme (1935) and the books of Evans (1930) and Roos (1934); it seems also connected with the work in macroeconomics resulting from the theory of Keynes (1936).

Methods of stochastic control applied to dynamic economic models are treated in this chapter emphasizing the following aspects connected especially with the theory of quantitative economic policy (Arrow and Kurz, 1970, Sengupta and Fox, 1969): (1) The range of additional restrictions in the decision space due to the stochastic elements incorporated in the deterministic models, for example, in the lack of controllability and observability to different orders, considerations of stability of different types in dynamic portfolio models, and so on. (2) The sensitivity of dynamic decision rules under a stochastic framework, particularly emphasizing the different forms of adaptive control which may be applicable to economic policy models. (3) The interrelationships between stochastic and corresponding deterministic models as regards their characteristics of stability, oscillation, decomposition, and statistical sensitivity of optimal controls.

We believe that our emphasis on the potential and actual applications of stochastic control theory to economic models serves to generalize the framework of deterministic economic theory and policy. Recent advances in the theory of control systems have gone far beyond systems designs which are optimal and adaptive; in particular, distributed parameter systems (i.e., those characterized by partial rather than ordinary differential equations) having mixed difference-differential adjustments are considered in the current literature with very different implications for the concepts of stability, the maximum principle and the decomposition algorithms. We believe that some of these methods would prove operational both statistically and computationally, and hence applicable to economic systems. We have attempted to indicate in Section 3 a selected subset of the current frontiers of research in control theory in its stochastic aspects, in so far as their potentiality of application to economic models is concerned. Although there still exists a number of computational problems and difficulties for solving general models of optimal stochastic control, simplified approximations are available in some cases and in particular the methods of simulation are increasingly resorted to; several economic applications (Fromm and Taubman, 1968, Naylor, Wertz, and Wonnacott, 1968) of simulation methods of control theory are now becoming available.

Our discussion proceeds as follows. The deterministic control models are first reviewed, followed by various extensions to adaptive and stochastic models (Section 2) and then a selected subset of potential applications (Section 3) is illustrated, emphasizing the statistical and computational problems (Sections 3 and 4) of application to realistic multivariate models.

3.1 Deterministic Control Theory and its Economic Applications

The recent economic applications of deterministic control theory may be broadly classified into at least four groups. (a) Economic models of regulation and stabilization policies (Phillips, 1954; Allen, 1959; Fox *et al.*, 1966; Goodwin, 1964; Sengupta, 1965) in a short-run framework, using mostly some form of macrodynamic models of the Keynesian-type. (b) Models of optimum savings and capital accumulation under the neo-classical forms of production functions. These models generalize, on the one hand, the economic problem posed by Ramsay (1928) who characterized the program of capital accumulation which maximizes the integral over time of net utilities of per capita consumption (Phelps, 1966; Cass, 1965; Uzawa, 1966; Koopmans, 1966; Bruno, 1967;

Samuelson, 1967; Radner, 1967). They reveal, on the other hand, the important optimality properties of balanced growth solutions under neo-classical models of economic growth which did not use any social utility functional for optimization (Solow, 1956; Pasinetti, 1965; Uzawa, 1966). (c) Models of investment allocation over time and growth planning on a large-scale framework involving multiple decision-makers and multiple sectors. These models raised the basic questions of computation and attainability of optimum paths through decentralization and decomposition procedures (Kornai and Liptak, 1965; Dantzig, 1963; Sengupta and Fox, 1969). They also led to optimality analysis under realistic restrictions of planning for example, finite horizon, dualism between developed and underdeveloped regions, deviations from competitive framework, problems of specific sectors dealing with external trade (foreign exchange), and skill formation through education, and so on (Shell, 1967; Hamada, 1969; Bardhan, 1967; Chenery and MacEwan, 1966; Alper and Smith, 1967). (d) Decision models of the enterprise or the firm based on alternative frameworks of production, inventory, investment and capacity expansion rules (Sengupta and Fox, 1969; Hanssmann, 1962; Connors and Teichrow, 1967). These models generalize the economic theory of the multi-product firm under conditions of its growth and capacity expansion.

The four groups of economic applications mentioned above have their stochastic counterparts, where stochastic considerations are introduced either through the statistical or extraneous estimation of parameters (e.g., the exponent of capital in a Cobb–Douglas production function) or through random elements in the restrictions of the behavior relations or inequalities of the model (e.g., investment allocation to a sector may be chance-constrained and the aim of planning may be to attain an expected income level with a very low variance). Some of these stochastic considerations and economic applications are treated in great detail in Chapter 4 under the generalized framework of stochastic programming, both static and dynamic. However, some aspects of stochastic control problems over time, which are not illustrated in Chapter 4 in any detail, would be mentioned and analyzed in this chapter. These aspects include, in particular, problems of stability and computation under nonlinear control and of combining the stability and optimality characteristics.

3.1.1 CHARACTERIZATION OF CONTROL: FEEDBACK OPTIMAL, AND ADAPTIVE

From a broad historical viewpoint, the theory of control may be characterized in terms of its three distinct phases based on feedback optimal and adaptive control designs. Considering, for example, a linear controllable system specified by the system of differential equations over time,

$$dx/dt = Ax(t) + Bu(t); \quad \begin{array}{l} x\text{: state vector: n.1} \\ u\text{: control vector: m.1} \\ A\text{: n.n}; \ B\text{: n.m} \end{array} \qquad (1)$$

with constant coefficient matrices A and B, a feedback control design would seek the control vector $u(t)$ as a function of the instantaneous state vector $x(t)$ such that certain properties like stability are attained. The class of feedback controls generally includes the proportional, derivative, and integral controls, for example,

$$u(t) = -C \cdot x(t) - D \cdot (dx/dt) - F \int_0^t x(s)\, ds, \qquad (2)$$

which essentially attempt to determine the unknown coefficients in the matrices C, D, and F. The standard multiplier-accelerator models in macroeconomics usually involve feedback relationships, which have been utilized by Phillips (1954) in discussing the impact of stabilization policies (see in this connection Fox et al., 1966). Proportional controls also arise in buffer-stock models of commodity price stabilization (Sengupta, 1965).

An optimal control design, however, needs a performance criterion, a scalar number which specifies the objective of the controller in terms of system behavior. The performance criteria may represent costs, profits or even stability characteristics measured in a particular way (Oldenburger, 1966). Given a specific performance criterion of system behavior, any control function $u(t)$ which satisfies the system equations in Eq. (1) and other feasibility conditions will be optimal only if it optimizes the performance criterion. Economic applications of such control problems are most frequent in the theory of economic policy (Tinbergen, 1952, 1954, 1956; Theil, 1958, 1964; Fox et al., 1966) and in models of economic growth (Shell, 1967).

If the system equations in Eq. (1) are only approximations to a more complex nonlinear system, or its parameters are only incompletely known, then it may be very difficult to introduce a suitable performance criterion even by observing system behavior in terms of (1). In this case, designing an optimal control may be almost impossible. Adaptive control methods are designed to analyze the various implications of a broad class of admissible controls which may include various sub-optimal controls. Evaluation of alternative learning mechanisms built into the system, comparisons of alternative approximations to the complex model and sequential analysis of system behavior assuming a prior probability distribution of the unknown parameters are some of the common methods used in designing adaptive control. Economic applications of adaptive control are not yet very frequent, although the linear decision rules have been used with great success in problems of dynamic planning for the enterprise (Holt et al., 1960) and for the overall economy (Theil, 1957, 1964; Holt, 1962). Satisfying rather than optimizing and Bayesian methods of incorporating the prior distributions of random elements have also been studied in the context of probabilistic programming (see Section 4.2 in Chapter 4 in this connection), although not specifically in relation to control applications.

It is of great interest to note that the theory of economic policy initiated by Tinbergen (1952, 1956) and Frisch (1955), and extended by Theil (1961, 1964), and others (Fox et al., 1966) has developed along lines parallel to the three distinct phases of control theory mentioned above. Contrary to the usual formulation of economic theory, Tinbergen first emphasized in his theory of economic policy that in a policy model the unknowns are the instrument variables (i.e., control variables) and the target variables (i.e., the state variables) are known. The implications of this result (see Fox et al., 1966) are several. For example, (i) a fixed-target policy model must have policy degrees of freedom defined by the excess of the number of instrument over target variables, and (ii) a flexible-target policy model where a specific welfare function (i.e., performance criterion) is introduced and optimized must have the parameters of the system equations structurally identified. These two aspects are also emphasized in control theory under the names of "controllability" and "observability." For example, any dynamic system specified, for instance, by equations in Eq. (1) is said to be completely state controllable on the time interval $[t_0, t_f]$ if for given t_0 and t_f each initial state vector $x(t_0)$ in its admissible set can be transferred to any final state $x(t_f)$ using some control $u(t)$ over the closed interval $t_0 \leq t \leq t_f$. Complete controllability implies total controllability if the former holds for every arbitrary time interval within $[t_0, t_f]$. The property of observability in control theory is introduced through a distinction between the state vector $x(t)$ and the output vector $y(t)$, where the latter is some function of $x(t)$ or both $x(t)$ and $u(t)$; for example

$$y(t) = G \cdot x(t) + H \cdot u(t), \qquad G = m.n., H = m.m. \qquad (3)$$

(If the vector $y(t)$ contains the target variables in which alone the policy-maker is interested, then the vector $x(t)$ would become an irrelevant variable in Tinbergen's terminology.) Now consider the system equations of Eq. (1) under the condition of no control; that is $u(t) = 0$ for all t. This uncontrolled system is defined to be completely observable on the interval $[t_0, t_f]$, if for given t_0 and t_f every state vector $x(t_0)$ in its admissible domain can be determined from the knowledge of $y(t)$ on the interval $[t_0, t_f]$ where $y(t)$ is given, for example, by Eq. (3) with $u(t) = 0$ for all t. Again, if complete observability holds for every arbitrary interval within $[t_0, t_f]$ then it implies total observability.

These two properties of controllability and observability (Kalman, 1963; Kreindler and Sarachik, 1964) which are very important in feedback control designs are very useful and operational in several respects. First, these properties may not hold in optimal control problems due to various boundary conditions and inequality restrictions, yet they might hold in some interval and in that interval the stability and oscillatory behavior may be analytically characterized; also useful modifications in the weights or forms of the performance criterion

(Aizerman, 1963) may be suggested, particularly when there are random errors in the system equations of Eq. (1). Second, the criteria of complete controllability and observability for linear dynamic systems like Eq. (1) are easy to determine, being essentially dependent on the rank of a certain augmented matrix. Since nonlinear systems can be linearized in a local sense, the above rank criteria can be locally applied in an approximate sense to nonlinear systems also. Third, controllability implies various kinds of decomposability (e.g., block-recursive or vector-causality principles discussed in the theory of economic policy; see Fox et al., 1966) and decentralized procedures. For example, a dynamic system such as in Eq. (1), is said to be strongly controllable if it is controllable by each of the m control variables separately while all others are zero; otherwise, it is weakly controllable. It is apparent that most economic systems would be only weakly controllable, due to the close interaction and interdependence between decisions and state variables.

For the linear system in Eq. (1) the criterion of complete controllability of the state space $x(t)$ in X may be very easily derived by a heuristic argument. Suppose the initial vectors $x(t_0)$, $u(t_0)$ are known and then we define recursively the vectors $x^{(1)}(t_0)$, $x^{(2)}(t_0)$, ..., $x^{(k)}(t_0)$ in Eq. (5) below by using the system of Eq. (1), where

$$\left. \begin{array}{ll} x^{(i)}(t_0) = d^i x(t)/dt^i & \text{at} \quad t = t_0, \\ u^{(i)}(t_0) = d^i u(t)/dt^i & \text{at} \quad t = t_0. \end{array} \right\} \quad \begin{array}{l} \text{derivatives up to} \\ i\text{th order for } i \geq 1. \end{array} \quad (4)$$

Now since the vector $x(t)$ has n components and since any final state $x(t_f)$ is attainable by using some control through the system of Eq. (1), therefore the following recursive system of equations must be independent,

$$\begin{aligned} x^{(1)}(t_0) &= Ax(t_0) + Bu(t_0), \\ x^{(2)}(t_0) &= A^2 x(t_0) + ABu(t_0) + Bu^{(1)}(t_0), \\ &\vdots \\ x^{(n)}(t_0) &= A^n x(t_0) + A^{n-1} Bu(t_0) + \cdots + Bu^{(n-1)}(t_0). \end{aligned} \quad (5)$$

In the language of differential equation theory (Pontryagin, 1961), this is equivalent to saying that the Wronskian matrix comprising the vectors $(x^{(1)}, \ldots, x^{(n)})$ have full rank n. Note that the coefficient matrix relating the matrices $(x^{(1)}(t_0), x^{(2)}(t_0), \ldots, x^{(n)}(t_0))$ and $(u(t_0), u^{(1)}(t_0), \ldots, u^{(n-1)}(t_0))$ is lower-triangular (i.e., recursive in the theory of economic policy), hence, its rank depends on the last row, that is, say

$$[A^{n-1}B, A^{n-2}B, \ldots, AB, B] = P. \quad (6)$$

3.1 Deterministic Control Theory

Therefore, if the rank of matrix P in Eq. (6) is n, the n state system of Eq. (5) can be uniquely solved, starting from the initial position. However, if this holds for $t = t_0$, then by continuity of solutions of differential equations (Struble, 1962) it would hold over a neighborhood of t_0 and hence, over the admissible domain $[t_0, t_f]$. By similar reasoning, it could be shown that the output system $y(t)$ related by Eqs. (3) and (1) is completely observable if the composite matrix

$$Q = [GA^{n-1}, GA^{n-2}, \ldots, GA, G] \tag{7}$$

has rank n.

A number of rigorous mathematical proofs of these results under more general conditions are available in the control literature (Kalman, 1961, 1963; Gilbert, 1963; Wang, 1965). For multivariable economic policy models, these results on controllability and observability have two important specific uses which in our opinion have not been fully explored yet (McFadden, 1969). First, the development of an appropriate measure of controllability for systems less than completely controllable, under convex constraints on the control variables may be usefully explored, especially for near decomposable and block-triangular coefficient matrices (Fisher, 1963; Weil, 1968; Fisher, 1969). Second, for economic models of optimum growth where optimal controls exist, the need for exploring the extent of controllability within the feasible space is very important. This point is reinforced if the random elements are present in the system equations of the model. One fundamental problem of control design is how to combine aspects of feedback control into an optimal control model (Aizerman, 1963).

Now, consider a problem in optimal control in the light of feedback controls. We have the following:

System equations:

$$dx/dt = f(x(t), u(t), t), \tag{8}$$

$x = n.1$, $u = m.1$, $f = n.1$; given initial conditions $x(t_0) = x_0$;

Performance criterion:

$$J = \int_{t_0}^{t_f} L(x(t), u(t)) \, dt \tag{9}$$

(this represents performance costs) $L(\,\cdot\,)$: scalar.

Constraints:

$$\begin{aligned} \text{Terminal:} \quad & x(t_f) \in S, \\ \text{Others:} \quad & u(t) \in U, \quad x(t) \in X, \quad \text{all } t, \end{aligned} \tag{10}$$

where S, U, and X are appropriate closed convex and bounded sets. For instance, S may represent the target region, U the control region, and X the feasible state space.

The optimal control problem is: Given Eq. (8), how to determine the control sequence $u(t)$ as a function of time such that Eq. (9) is minimized subject to Eq. (10)? The feedback control problem is: Given Eq. (8), how to determine a control sequence $u(t)$ as a function of the instantaneous state vector $x(t)$ such that Eq. (10) is satisfied and the time path of $x(t)$ affected by control is more stabilized than before? But the optimal feedback control problem is: Given Eq. (8), how to determine the control sequence $u(t)$ as a function of the instantaneous state vector $x(t)$ such that Eq. (9) is minimized subject to Eq. (10)? In the language of control theory (Ho and Brentani, 1963) the optimal control problem is essentially "open loop," since the optimal solution is defined only in relation to the initial conditions of the system. The solution of the feedback problem, however, defines essentially a "closed loop" operation, which is important in many applications where random disturbances are present. However, obtaining a closed loop form solution in feedback form is very difficult in a general case.

The optimal control problem can be characterized in terms of three related approaches. (a) Maximum principle developed by Pontryagin and his associates (Pontryagin *et al.*, 1962). (b) Optimality principle of dynamic programming developed by Bellman and his associates (Bellman, 1957; Bellman and Kalaba, 1965). (c) The general methods of variational programming which combine the Euler–Lagrange type conditions of variational calculus with Kuhn–Tucker type conditions of nonlinear programming. These methods are conveniently applied by discretizing the continuous model described in Eqs. (8) through (10), for example, by expressing differential equations by difference equation analogs (Sengupta and Fox, 1969; Bellman and Dreyfus, 1962). Of these three approaches the maximum principle of Pontryagin has attracted recent attention due perhaps to the fact that it characterizes the necessary conditions of optimal control in terms of the Hamiltonian function celebrated in mathematical physics and fluid dynamics (Newton, 1969), also developing indirectly different lines of computation algorithms (Balakrishnan and Neustadt, 1964).

It is important to realize, however, that the conditions of Pontryagin's maximum principle are local in nature and in general not sufficient for nonlinear control problems. Hence, in such cases, one must find all the extremal controls and compare the cost functionals $\hat{J} = J(\hat{u}(t))$ associated with each $\hat{u}(t)$ satisfying Pontryagin's conditions and then determine the control which is globally optimal. Define the Hamiltonian function H for the control system of Eqs. (8) through as follows in vector notation.

3.1 Deterministic Control Theory

$$H = L(x(t), u(t)) + \sum_{j=1}^{n} \{f_j(x(t), u(t)) \cdot p_j(t)\}$$
$$= L(x(t), u(t)) + p'(t)f(x(t), u(t)), \quad (11)$$

where $p_j(t)$ are termed adjoint variables or co-state variables. For simplicity, assume the terminal time t_f given and let the functions $L(\cdot)$ and $f(\cdot)$ be differentiable. Then the continuous version of Pontryagin's maximum principle states that, if optimal solution vectors $u^*(t), x^*(t))$ exist, then there must exist co-state vectors $p^*(t)$ satisfying the following three sets of conditions.

Canonical equations:

$$dx_i^*(t)/dt = (\partial H/\partial p_i(t))_*$$
$$\quad (i = 1, \ldots, n) \quad (12)$$
$$dp_i^*(t)/dt = -(\partial H/\partial x_i(t))_*$$

where the asterisk subscript denotes that partial derivatives are evaluated at $(x^*(t), u^*(t), p^*(t))$.

Boundary conditions:

$$x^*(t_0) = x_0; \; x^*(t_f) \in S,$$
$$p^*(t_f)'x^*(t_f) = 0, \quad (13)$$

that is, $p^*(t_f)$ normal to S at $x^*(t_f)$.

Minimization of the Hamiltonian:

$$H(x^*(t), p^*(t), u^*(t)) \leq H(x^*(t), p^*(t), u(t)), \quad (14)$$

for all $t \in [t_0, t_f]$ and all $u(t) \in U$. Note that we are stating a minimum principle rather than a maximum principle. The latter defines the H function as $H = -p_0 L(x(t), u(t)) + p'(t)f(x(t)), u(t))$, where p_0 is normalized to unity.

In case the terminal time t_f is not fixed, we have to add one more condition known as the transversality condition, that is,

$$p'^*(t_f)f(x^*(t_f), u^*(t_f)) = 0. \quad (15)$$

From the viewpoint of applications to economic models several features of Pontryagin's principle are worth emphasizing. First, the above necessary conditions are sufficient if the scalar function $L(x(t), u(t))$ and the vector functions $f(x(t), u(t))$ are convex, for example, $L(\cdot)$ may be a positive definite quadratic form and $f(\cdot)$ may be linear. Second, if the optimal solution time-paths $x^*(t)$,

$u^*(t)$, $p^*(t)$) are in the interior of the feasible space for all $t \in [t_0, t_f]$ and there is no singularity [e.g., H is not linear in $u(t)$], then one can use Pontryagin's maximum principle to derive feedback controls in the function form

$$u^*(t) = h(x^*(t), p^*(t)). \tag{16}$$

Here the notions of complete (or incomplete) controllability and observability can be applied. However, the function $h(\cdot)$ in Eq. (16) may not be unique in the presence of singularity. For instance, if the Hamiltonian H is linear in $u(t)$ (also in other cases of singularity), it may happen that the condition in Eq. (14) holds for $u^*(t)$ and $u(t)$ for a finite interval of time, although $u^*(t)$ is not identical with $u(t)$. These problems involve singularity and special care is needed to treat such problems insofar as the applications of Pontryagin's principle is concerned. For instance, Bruno (1967) has constructed a linear class of discrete capital models with two and more sectors, where the integrand in the performance criterion and the model, that is, $L(\cdot)$ and $f(\cdot)$, are both linear in the control variables. When Pontryagin's principle is applied to determine the optimal control function for this type of problems, the formal solution for the optimal control $u^*(t)$ appears in the form

$$u^*(t) = \begin{cases} a, & \text{if } F(t) > 0, \\ b, & \text{if } F(t) < 0, \end{cases} \tag{17}$$

where a, b are the upper and lower bounds, respectively, on the control function $u(t)$, and $F(t)$ is the switching function which is a part of the Hamiltonian H written in the following way:

$$H(x, p, u) = I(x, p) + u'F(x, p), \tag{18}$$

where $I = I(x, p, t)$ is the collection of terms which are not explicit function of $u(t)$.

$F = F(x, p, t)$ is the collected coefficients of $u(t)$. For the particular case in which the upper and lower bounds on $u(t)$ are equal and opposite in sign that is, $|u(t)| \leq M$ for all t, then the form of the optimal control function in Eq. (17) becomes

$$u^*(t) = M \operatorname{sgn} F(t), \tag{19}$$

where sgn denotes the sign function. For linear problems the switching function $F(t)$ sometimes becomes identically zero over some finite time interval. During this interval, the Hamiltonian function becomes independent of $u(t)$. Hence, Eqs. (17) and (19) do not supply any information on the optimal control function $u^*(t)$. Noting that the optimum value of the Hamiltonian, that is,

$$H^* = H(x, p, u^*) = I + u^{*'}F = 0, \qquad t_0 \leq t \leq t_f$$

is zero when $u(t)$ is replaced by the optimum control function $u^*(t)$. Johnson and Gibson (1963) have derived three types of singular controls as follows:

Type 1. $u^* \equiv 0$ in a finite time interval satisfying Eq. (17) and $I(x, p) = 0$, $F(t) \neq 0$.

Type 2. Any control $u^*(t)$ satisfying Eq. (17) and $I(t) = -u^*(t)F(t)$, where $I(t)$, $F(t) \neq 0$.

Type 3. Any control $u^*(t)$ which will make $I(t) = F(t) = 0$ and satisfy Eq. (17).

Necessary conditions for singular controls and various switching sequences have been analyzed in the control literature (Kelley, 1964; Aizerman, 1963; Johnson and Gibson, 1963). In models of economic growth with heterogeneous capital goods and multiple techniques, problems of technique switching and reswitching (Bruno et al., 1966; Levhari, 1965) have been encountered which would generally involve regions of singular control.

Third, the performance criterion selected in Eq. (9) can easily be generalized in several respects by very minor modifications. If the integrand $L(x(t), u(t))$ is a positive constant for all $x(t)$, $u(t)$, then we get a minimal time problem, That is, given $x(t_0)$ and $x(t_f)$, we want to minimize the time of transition. Such problems have been applied in economic growth models (Kurz, 1965; Stoleru, 1965). Again, if the integrand $L(x(t), u(t), t)$ depends explicitly on time t, it can be taken care of by methods before by simply introducing another state variable, that is, $x_{n+1} = t$ and $dx_{n+1}/dt = 1$ with $x_{n+1}(t_0) = 0$. For finite horizon problems where the planner wants to end up with a desired set of terminal outputs, the performance criterion may have to be slightly modified as

$$J = \lambda(x(t_f)) + \int_t^{t_f} L(x(t), u(t))\, dt,$$

where $\lambda(x(t_f))$ is a scalar function of the vector of terminal outputs $x(t_f)$, which may be interpreted as penalty costs for deviating from the desired terminal output levels. This sort of interpretation may be easily related to turnpike theorems and related planning problems (Morishima, 1964; Radner, 1967).

For problems where the time variable is discrete (e.g., economic data are available quarterly, monthly, or yearly), several versions of discrete maximum principle are now available in the literature (Ho and Brentani, 1963; Rozonoer, 1959). Let $i = 0, 1, \ldots, N - 1$ be a discrete time index.

Consider the following discrete optimal control problem.

System:

$$x_{i+1} - x_i = f_i(x_i, u_i), \tag{20}$$

$x_i = n.1$, $u_i = m.1$, $f_i = n.1$, $x_0 = c$ (initial conditions given).

Performance criterion:

$$W = \sum_{i=0}^{N-1} L_i(x_i, u_i), \tag{21}$$

where $L_i(\cdot)$ is a scalar function (costs) of vectors x_i, u_i.

Constraints:

Terminal: $x_N \in S$,

Others: $u_i \in U$, $x_i \in X$ all $i = 0, 1, \ldots, N-1$ (22)

where the sets S, U, and X are assumed closed bounded and convex. How to determine a sequence of optimal controls $(u_0^*, u_1^*, \ldots, u_{N-1}^*)$ generating the sequence of state vectors $(x_0^*, x_1^*, \ldots, x_N^*)$ such that it minimizes the performance criterion of Eq. (21) subject to Eqs. (20) and (22)? Assuming for simplicity that the functions $f(x_i, u_i)$ and $L_i(x_i, u_i)$ are convex and differentiable and satisfy certain regularity conditions (e.g., directional convexity; see Holtzman and Halkin, 1966; Sengupta and Fox, 1969), one could apply the analogs of the three sets of conditions Eqs. (12) through (14) mentioned for the continuous maximum principle. In terms of the Hamiltonian

$$H = H_i(x_i, p_{i+1}, u_i) = L_i(x_i, u_i) + p'_{i+1} f_i(x_i, u_i), \tag{23}$$

where $i = 0, 1, \ldots, N-1$, prime denotes transpose and f_i is the n-component vector defined by Eq. (20), the necessary conditions for optimal control sequences $(u_0^*, u_1^*, \ldots, u_{N-1}^*)$ are the following.

Canonical equations:

$$\begin{aligned} x_{i+1}^* - x_i^* &= (\partial H/\partial p_{i+1})_*, \\ & \qquad (i = 0, 1, \ldots, N-1) \\ p_{i+1}^* - p_i^* &= -(\partial H/\partial x_i)_*, \end{aligned} \tag{24}$$

where the asterisk subscript indicates evaluation at (x^*, u^*, p^*).

Boundary conditions:

$$x_0^* = c, \quad x_N^* \in S, \quad p_N^{*\prime} x_N^* = 0. \tag{25}$$

Minimization of the Hamiltonian:

$$H_i(x_i^*, p_{i+1}^*, u_i^*) \leq H_i(x_i^*, p_{i+1}^*, u_i) \quad \text{for all} \quad u_i \in U,$$
$$\text{all} \quad i = 0, 1, \ldots, N-1. \tag{26}$$

3.1 Deterministic Control Theory

[Again, note that condition in Eq. (26) can be interpreted as a maximization process by viewing H as $H = -L_i(x_i, u_i) + p'_{i+1} f_i(x_i, u_i)$ and justifying the term maximum principle.]

We have so far discussed feedback and optimal control. An adaptive control system is one where in addition to the system equations, some knowledge of performance characteristics and the constraints, some means are provided to the decision-maker for modifying the control actions in order to make it more acceptable. Adaptive control methods would be discussed briefly in Section 3.2 on stochastic control.

We shall now illustrate a few optimal control methods with the help of a very simple problem in agricultural production in India. Using Indian data for 1951–52 to 1963–64 previously utilized in estimating lognormal diffusion processes (Tintner and Patel, 1966) and denoting by A_t real per capita agricultural production and by E_t real per capita government expenditure on agriculture in India, we derive empirically the following relation

$$A_t = 0.0463 + 0.87325\, A_{t-1} + 1.21524\, E_{t-4}. \qquad (27)$$
$$(0.14723) \phantom{A_{t-1} +} (1.7068)$$

The multiple correlation coefficient is 0.94012, its square $R^2 = 0.883$ shows, that about 88% of the total variance of Indian per capita agricultural real production is explained by the relation. The figures in brackets below the regression coefficients denote standard errors.

This equation is easily integrated and gives the general relation for A_t.

$$\begin{aligned}
A_t = {} & 0.36529 + 1.09681(0.87325)^{t-3} + 1.70681 \\
& [(0.0211)(0.87325)^{t-4} + 0.0278(0.87325)^{t-5} \\
& + 0.0338(0.87325)^{t-6} + 0.0516(0.87325)^{t-7} \\
& + 0.0619(0.87325)^{t-8} + 0.0360(0.87325)^{t-9} \\
& + 0.0379(0.87325)^{t-10} + 0.0405(0.87325)^{t-11} \qquad (28) \\
& + 0.0417(0.87325)^{t-12} + 0.0437(0.87325)^{t-13} \\
& + 0.0452(0.87325)^{t-14} + 0.0529(0.87325)^{t-15} \\
& + 0.0548(0.87325)^{t-16} + E_{13}(0.87325)^{t-17} \\
& + E_{14}(0.87325)^{t-18} + \cdots + E_{t-4}], \quad (t \geq 17).
\end{aligned}$$

Imposing the condition that $A_t \geq A_{16} = 0.8995$ (estimated agricultural production for 1967/68) we derive.

$$E_t \geq 0.557. \qquad (29)$$

The sum:

$$S = A_{17} + A_{18} + A_{19} + A_{20} = 3.0017 + 4.09334\,E_{13}$$
$$+ 3.20314\,E_{14} + 2.27644\,E_{15} + 1.21524\,E_{16}, \quad (30)$$

i.e., the sum of agricultural production 1968/69 to 1971/72 is assumed to be the performance criterion to be maximized. The E_t are the control variables.

Then if a sum $E > 0.2228$ is available, $E - 0.1671 > 0.0557$ should be used for E_{13}, whereas $E_{14} = E_{15} = E_{16} = 0.557$.

A very important idea in deterministic control theory is the special version of Pontryagin's maximum principle due to Rozonoer (1959).

If we use the integral of our difference equation derived empirically from Indian data for food production in India, we have

$$dA(t)/dt = \dot{A}(t) = 0.0463 - 0.135\,A(t) + 1.21524\,E(t-4). \quad (31)$$

We want to maximize $A(20)$ that is, the agricultural production in 1971/72. A dot over a variable is used to denote its time derivative.

We form the Hamiltonian,

$$H = p(t)\,[0.0463 - 0.135\,A(t) + 1.21524\,E(t-4)] \quad (32)$$

From

$$\dot{A}(t) = \partial H/\partial p \quad (33)$$

we get again the above differential Eq. (31) for $A(t)$.

$$p(t) = -\partial H/\partial A = 0.135 \quad (34)$$

and the condition $p(20) = 1$ gives

$$p(t) = -1.7 + 0.135 t. \quad (35)$$

We see from the Hamiltonian, that for a maximum $E(t-4) = \operatorname{sgn} p(t)$, where sgn denotes the sign function.

If we demand that $0 \le E(t) \le E$, we find that $E(t) = 0$ for $t < 16.6$ and $E(t) = E$ for $t > 16.6$.

A discrete version of the maximum principle is also given by Rozonoer (1959) and others mentioned before.

Consider the previous model of agricultural production in India. We derive the relation:

$$A_{t+1} - A_t = \Delta A_t = 0.0463 - 0.12695\,A_t + 1.21524\,E_{t-3}. \quad (36)$$

Now we form the Hamiltonian function

$$H = p_t(0.0463 - 0.12685\, A_t + 1.21524\, E_{t-4}). \tag{37}$$

If we want to maximize A_{20}, that is, agricultural production for 1971/72, we have

$$\Delta A_t = \partial H/\partial p_t, \tag{38}$$

which is the difference equation for A_t. Also

$$\Delta p_{t-1} = -\partial H/\partial A_t = 0.12685 \tag{39}$$

with the initial condition,

$$p(20) = 1. \tag{40}$$

Assume that $0 < E_t < E$. We see from the Hamiltonian that for a maximum

$$E_{t-4} = E \quad \text{if} \quad p_t = 1.537 + 0.12685\, t > 0, \quad \text{for} \quad t > 16.12. \tag{41}$$

Otherwise $E_t = 0$.

This simple illustrative model has a basic similarity with a two-sector planning model developed by Stoleru (1965), who considered his model appropriate for an economy in which the level of unemployment and capital-output ratio are high and the capital goods sector is less capital-intensive than the consumer goods sector. The Indian economy does not, however, have all these characteristics. For example, its capital goods sector is more capital-intensive than the consumer goods sector; yet an experimental application (Tintner and Rao, 1968) of Stoleru-type control model in this situation, which is, of course, very aggregative brings out important and difficult choice problems in development planning. It was found on the basis of an aggregative two-sector model (Tintner and Rao, 1968) very similar to the above that without any change in the structure of industry, it would be possible to obtain full employment and balanced growth by 1972 if the present level of per capita income is allowed to fall about one-half. The optimal policy would then consist of investing in the capital goods sector alone until 1968 and then until 1972 in the consumer goods sector. If consumption is not allowed to fall by more than 10% of the present level, it would be possible to achieve full-employment only by 1982. The optimal policy would then consist of investing in the capital goods sector alone for one year and then in both sectors until 1979. The proportion of investment going into the capital goods sector would increase from 0.72 to 0.93 and after 1979 would be concentrated in the consumer goods sector only.

(For other applications of the maximum principle see: Hardt *et al.*, 1970;

Fox et al., 1966; Montias, 1967; Pugachev, 1963; Tintner and Fels, 1967; Zauberman, 1967; Shell, 1967; Kuhn and Szego, 1969.)

Consider another very simple example based upon the excellent survey article of Dreyfus (1968), which describes the following continuous deterministic control problem. Given an initial time t_0 and a terminal time T, also that

$$x(t_0) = x_0.$$

We want x to get to the origin $x = 0$ (where the origin may be the desired state) and the cost is

$$C = \int_0^T u^2(t)\, dt + x^2(T),$$

where the first term denotes the cost of control, the second the deviation from the origin. The dynamic motion of x is assumed given by

$$dx/dt = ax(t) + bu(t).$$

Introduce a Lagrange multiplier $L(t)$. We have

$$dL/dt = -aL, \qquad L(T) = 2x(T).$$

Optimal control is then easily obtained by applying the maximum principle or variational programming as

$$2u + Lb = 0.$$

Hence, we have:

$$L(t) = 2x(T)e^{a(T-t)}, \qquad u(t) = -x(T)be^{a(T-t)}.$$

Control varies exponentially with time. The ultimate results are

$$x(t) = \frac{e^{a(T-t)}x(t)}{1 - (b^2/2a) + (b^2/2a)e^{2a(T-t)}},$$

$$u(t) = \frac{be^{2a(T-t)}x(t)}{1 - (b^2/2a) + (b^2/2a)e^{2a(T-t)}}.$$

This is the feedback solution. The same result might be obtained by dynamic programming.

(On dynamic programming see also: Aris, 1963; Bellman, 1957, 1961, 1962; Bellman and Dreyfus, 1962; Dantzig, 1963; Dreyfus, 1965; Heady and Candler, 1958; Howard, 1960; Kalaba, 1962; Kromphardt et al., 1962; Nemhauser,

1966; Pallu de la Barriere, 1967; Beckmann, 1968; Bellman and Kalaba, 1965.)

Applying these methods to our example of agricultural production in India, we minimize

$$F = \int_0^T u^2(t)\, dt - A(T)^2,$$

where $u(t) = E(t-4)$. Under the conditions

$$\dot{A}(t) = 0.0463 - 0.135\, A(t) + 1.21524\, u(t),$$

we introduce a Lagrange multiplier $L(t)$ and form the expression

$$G = u^2(t) + L(t)[\dot{A}(t) - 0.0463 + 0.135\, A(t) - 1.21524\, u(t)].$$

The boundary conditions are

$$A(12) = 0.9401, \qquad L(T) = 2A(T).$$

We find the necessary conditions for a minimum

$$\partial G/\partial u = 2u(t) - 1.21524\, L(t) = 0,$$
$$d(\partial G/\partial \dot{A}) = (\partial G/\partial A),$$
$$\dot{L}(t) = 0.135\, L(t).$$

The results are

$$L(t) = 2A(t)e^{-0.135(T-t)},$$
$$u(t) = 1.21524\, A(t)e^{-0.135(T-t)},$$
$$A(t) = 0.342 + [3.050 - 117.214 A(T)e^{-0.135T}]e^{-0.135t}$$
$$\qquad + 4.501 A(T)e^{-0.135T + 0.135t},$$
$$A(t-4) = 0.342 + 5.234 e^{-0.135t} - 201.139\, A(T)e^{-0.135(T+t)}$$
$$\qquad + 2.624 A(T)e^{-0.135T + 0.135t}.$$

Also

$$A(T) = \frac{A(t-4) - 0.342 - 5.234 e^{-0.135t}}{2.624 e^{-0.135T + 0.135t} - 201.139 e^{-0.135(T+t)}}.$$

Finally, since

$$u(t+4) = E(t) = 2.085\, A(T)e^{-0.135(T-t)}$$

we have for the control variable

$$E(t) = \frac{2.085 A(t) - 0.713 e^{0.135t} - 10.913}{2.624 - 201.139 e^{-0.135t}}.$$

This feedback relation gives the control variable $E(t)$ (i.e., real per capita government expenditure on agriculture) as a function of the contemporary state variable $A(t)$ (i.e., real per capita agricultural production.).

3.1.2 Economic Applications of Deterministic Optimal Control

Of the recent upsurge in substantive applications of deterministic optimal control theory to various economic problems, we would mention a few selected applications which are intended to illustrate various aspects of control policies in macroeconomic models. For example, stability, controllability, and decentralization property. Specifically, the models are selected from the following fields: (A) theory of optimum growth, (B) fiscal and monetary policy, (C) problems of borrowing in international capital markets, and (D) investment allocation problems in underdeveloped countries.

(A) *Optimal Growth Modeles*

In the recent theory of economic growth, the Harrod–Domar type model with its associated knife-edge instability due to limitational production functions has been generalized in several respects. Starting with a neo-classical production function (with no technical change and no depreciation of capital) homogeneous of first degree (i.e., constant returns to scale), and the equilibrium condition that savings equal investment. Solow (1956) derived the following fundamental equation in the capital-labor ratio $k = K/L$ for characterizing the stability of equilibrium or balanced growth,

$$\dot{k} = s f(k) - nk, \tag{42}$$

where dot over k denotes time derivative, n is the constant proportional growth rate of labor L assumed exogenous, s is the constant savings propensity, and $f(k)$ is a neo-classical production function with positive but diminishing marginal productivity that is,

$$f(k)_k = \partial f/\partial k > 0, \quad f(k)_{kk} = \partial^2 f/\partial k^2 < 0. \tag{43}$$

It is apparent from the differential equation of Eq. (42) that if there exists a particular value k^* of k such that

$$f(k^*) = (nk^*/s) \tag{44}$$

3.1 Deterministic Control Theory

then at that value $dk^*/dt = \dot{k}^* = 0$. If this capital-labor ratio k^* is economically meaningful and ever established by a competitive market process (i.e., through factor price flexibility), then it will be maintained and capital and labor will grow therefrom in constant proportion. Also by constant returns to scale, real output will also grow at the same relative rate n and the output-labor ratio will be constant. This particular balanced growth solution k^* would be stable provided the following relations hold

$$nk > sf(k), \quad \text{if} \quad k > k^*,$$
$$nk < sf(k), \quad \text{if} \quad k < k^*, \qquad (45)$$

along with certain regularity and continuity conditions.

Now consider the following growth process in optimum growth theory.

$$\dot{k} = f(k) - nk - c, \quad c = \text{per capita consumption} \qquad (46)$$

which maximizes a utility functional of the form,

$$\int_0^\infty \exp(-\delta t) \cdot U(c)\, dt, \qquad \delta \geqq 0, \qquad (47)$$

where $\exp(x)$ denotes e^x. Equation (47) is subject to the constraint of Eq. (46) and possibly the given initial stock $k(0)$. Defining the Hamiltonian H as

$$H = \exp(-\delta t)[U(c) + p(t)\{f(k) - nk - c\}] \qquad (48)$$

and applying Pontryagin's maximum principle, we obtain the following necessary conditions [which are also sufficient if $U(c)$ is strictly concave] characterizing the optimal time paths

$$\dot{p}(t) = (\delta + n - f_k(k))p(t), \qquad (49)$$
$$U_c(c) = p(t), \qquad (50)$$
$$\lim \exp(-\delta t) \cdot p(t) = 0, \quad \text{(transversality condition}[1]) \qquad (51)$$

---

[1] Note that the condition of Eq. (51) specifies here a special case of the generalized transversality condition, which needs be specified whenever the endpoints of the integral in Eq. (47) are free (i.e., belonging either to specified sets or to improper integrals). For the terminal point T with its associated capital stock (i.e., the state vector) $k(T)$, let S be the tangent plane passing through $k(T)$; then the adjoint variable (i.e., the adjoint vector) $p(T)$ is said to satisfy the transversality condition, if it is orthogonal to S for every vector parallel to S. In our simple case this means

$$\lim_{T \to \infty} p(T) \exp(-\delta T) k(T) = 0$$

but since, feasibility implies

where dot over a variable denotes its time derivative and subscripts denote partial derivatives (e.g., $U_c = \partial U/\partial c$). Again suppose there exist a pair (p^*, k^*) and the associated variables $(c^*, f(k^*) - c^*)$ such that $\dot{p}^* = 0$, $\dot{k}^* = 0$. Then these would define solutions of the equations

$$f_{k^*} = \delta + n; \quad f(k^*) = nk^* + c^*; \quad U_{c^*} = p^* \qquad (52)$$

provided they are on the optimal paths. Again the point (p^*, k^*) would be stable provided the following equation holds, along with certain continuity and regularity properties

$$\begin{cases} nk^* + c^* > f(k^*), & \text{if } k > k^* \quad \text{and} \quad nk^* + c^* < f(k^*), \\ & \qquad\qquad\qquad\qquad\qquad\qquad \text{if } k < k^*, \\ f_{k^*} > \delta + n, & \text{if } p > p^* \quad \text{and} \quad f_{k^*} < \delta + n, \\ & \qquad\qquad\qquad\qquad\qquad\qquad \text{if } p < p^*. \end{cases} \qquad (53)$$

The balanced growth solution (k^*, p^*) is also termed quasi-stationary (optimum) path or quasi-golden rule path. Three characteristics of this path are worth emphasizing. First, note that if the initial conditions happen to be such that $k(0) = k^*$ and $p(0) = p^*$, then the balanced path $(k^*, c^*, f(k^*) - c^*)$ would be voluntarily maintained forever as optimum; also from Eq. (52) it is evident that k^* and c^* do not depend on the form of the social utility function $U(c)$ and if the rate of discount $\delta \to 0$, then the conditions of the golden rule path (Phelps, 1966; Koopmans, 1967; Samuelson, 1965; Newman, 1968) are reached [e.g., $f_{k^*} = n$, $c^* = f(k^*) - nk^*$]. Second, the optimal path $(k^*(t), p^*(t))$ is a saddle point in the class of all pairs of functions $(k(t), p(t))$ which satisfy the necessary conditions of Pontryagin's maximum principle that is, Eqs. (46), (49)–(51). This may be easily noted by linearizing the nonlinear terms in

---

$$0 < \lim_{T \to \infty} k(T) < \infty,$$

the 'interior,' endpoint condition is that given by Eq. (51). Allowing for a corner condition at the terminal time T [e.g., $k(T) \leq k_T$] one may write the full transversality condition as

$$\lim_{T \to \infty} p(T) \exp(-\delta T)[k(T) - k_T] = 0$$

and

$$\lim_{T \to \infty} p(T) \exp(-\delta T) \geq 0.$$

In words, the transversality condition expresses the point, as shown very explicitly and clearly by Shell (1969) that on the optimal trajectory capital is not to be "left over" beyond its terminal requirements $p(T)k_T$. A similar economic interpretation has been offered by Arrow and Kurz (1970).

Eqs. (46) and (49) by Taylor series expansions around the point $(k^*(t), p^*(t))$ and then computing the characteristic roots of the resulting nonlinear system, which in this case turn out to be real and opposite in sign. Also

$$\left(\frac{dp}{dk}\right)_{\substack{\dot{k}=0 \\ k^*, p^*}} \gtreqless \text{zero, according as } k(t) \gtreqless k^*(t).$$

This saddle point property which is an instability property (Kurz, 1968) implies that, for any set of initial conditions different from $k^*(0)$ and $p^*(0)$, the resulting paths become either infeasible (in finite time or asymptotically) or infinitely inferior. Third, noting from Eq. (52) that c^* and k^* do not depend on the specific form of the utility function $U(c)$, one could compare Eqs. (42) and (46) under the condition $\dot{k} = 0 = \dot{k}^*$. This implies that the balanced growth solution in Eq. (44) of the Solow-model [Eq. (42)] may under certain conditions be imputed an optimal path interpretation. Furthermore, the golden rule path [Eq. (52)] with $\delta = 0$ can be easily generalized to include technical changes of different types, for example, labor-augmenting (Harrod), capital-augmenting (Solow, Fei-Ranis), neutral (Hicks) technical progress (Allen, 1967). Thus Phelps (1966) has shown, for example, that if we have a labor-augmenting technical progress in the production function $F(K, AL)$ where A is a function of t satisfying $\dot{L}/L + \dot{A}/A > 0$, then in the zero depreciation case, total consumption C is

$$C = F(K, AL) - \dot{K}.$$

If the production function $F(\cdot)$ is homogeneous of degree one in K and AL, then per head of augmented labor, consumption (c) can be expressed as

$$c = C/(AL) = f(k) - \left(\frac{\dot{L}}{L} + \frac{\dot{A}}{A}\right)k - \dot{k},$$

where $f(k) = F(K/AL, 1)$, $k = K/AL$.

But since for quasi-golden rule path $\dot{k}^* = 0$, therefore, we get

$$f(k^*) = c^* + \left(\frac{\dot{L}}{L} + \frac{\dot{A}}{A}\right)k^*. \tag{54}$$

Also the quasi-golden rule growth path characterized by Eq. (54) has a local optimality property within a class of parallel paths. Following Phelps (1966) define any perturbed path around $k^* = k^*(t)$ by

$$k(t) = k^*(t) + \epsilon, \quad \epsilon \text{ a small constant.} \tag{55}$$

3. Stochastic Control Theory

Since $\dot{k}(t) = \dot{k}^*(t)$ for all t, then from Eq. (54) we get the perturbed consumption relation as

$$c(t) = f(k^*(t) + \epsilon) - \left(\frac{\dot{L}}{L} + \frac{\dot{A}}{A}\right)(k^*(t) + \epsilon) - \dot{k}^*(t).$$

Taking derivative of $c(t)$ with respect to ϵ and equating it to zero, we get,

$$f_{k^*(t)+\epsilon}(k^*(t) + \epsilon) = \frac{\dot{L}}{L} + \frac{\dot{A}}{A}, \quad \text{also} \quad \partial f_{k^*+\epsilon}/\partial(k^* + \epsilon) < 0$$

and in particular for $\epsilon \to 0$ we get back the condition in Eq. (52) with $\delta = 0$.

Now consider a generalization of the optimum growth model of Eq. (46) and (47) in a multisector framework, where there are n heterogenous capital goods and m consumption goods (i.e., $n + m$ sectors). Then as Kurz (1968) has shown, the general model could be specified in the following simple form, using first sector's investment as numeraire in the production transformation frontier (Samuelson, 1959)

$$\text{maximize } J = \int_0^\infty R(t)U(c_{n+1}, c_{n+2}, \ldots, c_m, t) \, dt, \tag{56}$$

subject to

$$\dot{k} = g(v, k, c) - nk, \tag{57}$$

$$k(0) > 0, \text{ given initial conditions,} \tag{58}$$

where, $R(t)$ is the exogenous discount function satisfying the conditions $R(0) = 1$, $R(\infty) = 0$, and $R(t)$ monotonically declining for increasing t and $\dot{k}$ is a vector with a typical element

$$\dot{k}_i = g^i(v_2, v_3, \ldots, v_n; k_1, k_2, \ldots, k_n; c_{n+1}, \ldots, c_m) - nk_i,$$
$$i = 1, 2, \ldots, n, \tag{59}$$

where the superscript i in $g^i(\cdot)$ only identifies the ith sector's transformation frontier.

It is assumed that $U(c, t)$ is concave and differentiable and that the functions in Eq. (59) are well defined and allow partial derivatives, and so on. Also v_i is defined as $\dot{k}_i/L$, where L is the exogenously determined labor force. We assume that an optimal growth program (v^*, c^*) of the above multisectoral model exists in the sense that there exists a pair of vectors $(p^*(t), k^*(t))$ defining the unique maximum of the associated Hamiltonian function (where $U(c, t)$ is used for $U(c_{n+1}, \ldots, c_m, t)$),

3.1 Deterministic Control Theory

$$H = R(t)[U(c, t) + p'(t)\{g(v, k, c) - nk\}]. \quad (60)$$

We would discuss the local asymptotic stability of the optimal growth solution for $k(0) \neq k^*(0)$ and $p(0) \neq p^*(0)$. In the current literature on optimum growth in a multisectoral framework, two aspects of instability have been particularly emphasized, thus raising doubts about the stability properties of neo-classical growth theory. These are

(a) the saddle point property (Kurz, 1968; Hahn, 1966);
(b) the lack of determinacy in the presence of heterogenous capital goods[1] (Hahn, 1966; Samuelson, 1967).

However, both these properties have been discussed in a local sense. Our purpose here is to characterize these in a slightly broader framework. Broadly speaking, it will be shown for the system

$$\dot{z}(t) = F(t, z) \quad \text{and} \quad F(t, 0) = 0 \quad (61)$$

all $t \geq 0$ associated with the optimal H in Eq. (60) that

(i) all interesting solutions are unstable (stable) if the scalar function $V = z'F(t, z)$ is a strictly convex (concave) function of z,

(ii) oscillations in terms of the presence of complex roots cannot be avoided if the discount function $R(t)$ has the property that

$$\dot{r}(t) \to -\infty \quad \text{as} \quad t \to \infty, \quad \text{where} \quad r(t) = \dot{R}(t)/R(t),$$

(iii) the policy space of the associated control problem is unlikely to have "complete controllability" and "complete observability" in the sense described in Section 3.1.1.

---

[1] Recently Shell and Stiglitz (1967) have examined more closely the two propositions of the Hahn model that an economy with several heterogenous capital goods faces two crucial problems. (1) The growth path may be indeterminate and (2) not all paths will converge to balanced growth. Using a simple one-sector model with two capital goods, Shell and Stiglitz (1967) have shown that under reasonable behavioral assumptions (e.g., short-run profit maximization with short-run perfect foresight about capital gains, etc.) and given initial endowment of resources there exists one and only one assignment of initial prices that is consistent with long-run balanced growth. Furthermore, in this dynamical system of the form; $\dot{x} \in f(x)$, where $f(\cdot)$ is an upper semi-continuous correspondence there exists a unique (long-run) equilibrium point which is also a saddlepoint. This implies that, whenever momentary equilibrium is not unique, one and only one allocation of investment exists which is consistent with long-run equilibrium (or balanced) growth. The crucial decision problem is whether such an allocation of investment can actually be secured in a capitalistic economy through appropriate institutions, particularly when future markets for all captial goods and also perfectly competitive conditions do not exist.

We consider the optimal Hamiltonian H^* defined in Eq. (60) and expand the canonical equations (i. e., necessary conditions) of Pontryagin's maximum principle by Taylor series around $k^*(t)$ and $p^*(t)$. Thus, we obtain the following linearized system

$$\dot{g} = H^*_{pk}(t) \cdot g(t) + H^*_{pp}(t) \cdot h(t),$$
$$\dot{h} = -H^*_{kk}(t) \cdot g(t) - (H^*_{kp}(t) + r(t)I) \cdot h(t), \qquad (62)$$

or,

$$\dot{z} = B(t) \cdot z,$$

where

$$z = (g, h)', \qquad g = k(t) - k^*(t), \qquad k(0) \text{ given},$$
$$h = p(t) - p^*(t), \qquad p(0) \text{ given},$$

and

$$B(t) = \begin{bmatrix} H^*_{pk}(t) & H^*_{pp}(t) \\ -H^*_{kk}(t) & -(H^*_{kp}(t) + r(t)I) \end{bmatrix},$$

where I is identity matrix of order n and $r(t) = \dot{R}(t)/R(t)$.

The local instability properties of systems like Eq. (62) have been analyzed in details by Kurz (1968) for $r(t)$ fixed and independent of time. Also we consider a more general system of Eq. (61) which is nonlinear and which contains Eq. (62). Our result may be stated as theorems (see Massera, 1956; Mangasarian, 1963; Rosen, 1962).

Theorem 1 Let the function $F(t, z)$ in Eq. (61) be continuous in $z(t)$ at $z(t) = 0$ and $F(t, 0) = 0$ for $0 \leq t < \infty$ and let $M^*(z)$ be defined as $[\liminf_{t \to \infty} z'F(t, z)]$. Then $z(t) = 0$ is an unstable equilibrium point in the Liapunov sense, if the scalar function $V = z'F(t, z)$ is a strictly convex function of z for $0 \leq t < \infty$ and if either $M^*(z) > 0$ for $z \neq 0$ or $M^*(z)$ is strictly convex in z.

Proof: Consider points $z = z(t) \neq 0$ and $z^0 = 0$. By the assumed strict convexity of $V = V(t, z)$, $(1 - \lambda)V(t, z^0) + \lambda V(t, z) > V[t, (1 - \lambda)z^0 + \lambda z]$ for $0 < \lambda < 1$ (scalar).

Since $z^0 = 0$ and $V(t, z^0) = 0$, we have from above

$$V(t, z) > V(t, \lambda z)/\lambda, \qquad \text{for} \quad 0 \leq t < \infty, 0 < \lambda < 1, z \neq 0$$

also,

3.1 Deterministic Control Theory

$$V(t, z) > z'F(t, \lambda z) \geq \lim_{\lambda \to 0} z'F(t, \lambda z),$$

for

$$0 \leq t < \infty, \quad 0 < \lambda < 1, \quad z \neq 0.$$

Since $F(t, z)$ is continuous in z at $z = 0$ for $0 \leq t < \infty$, it follows that $V(t, z) > 0$ for $0 \leq t \infty$ and $z \neq 0$. Now, if the function $M^*(z) > 0$ for $z \neq 0$, then the function $V(t, z)$ is positive definite for $0 \leq t < \infty$, $z \neq 0$. Alternatively, if $M^*(z)$ is strictly convex in z, then for $0 < \lambda < 1$, $z \neq 0$

$$M^*(z) > M^*(\lambda z)/\lambda,$$

but $M^*(z) \geq 0$ for $z \neq 0$, since $V(t, z) > 0$ for all $z \neq 0$, $0 \leq t < \infty$. Hence, $M^*(z) > 0$ for $z \neq 0$ implying $V(t, z)$ to be positive definite.

Now consider a function $W(t, z) = z'z$ for testing the Liapunov stability. This function satisfies the properties of a (Euclidean) distance function. For example, $W(t, z) > 0$ for $z \neq 0$, $W(t, 0) = 0$ and it has first-order continuity (i.e., existence of first derivatives). But since

$$dW(t, z)/dt = \dot{W} = 2z'\dot{z} = 2z'F(t, z) > 0,$$

Hence, it follows from Liapunov's instability theorem that $z(t) = 0$ is an unstable equilibrium point.

Corollary For autonomous systems, where $F(t, z) = F(z)$ does not involve time t explicitly, it is sufficient for $z = 0$ to be an unstable point that $F(z)$ be continuous at $z = 0$ and $V = z'F(z)$ be strictly convex in $z = z(t)$.

By following the argument above, the case when $V = z'F(t, z)$ is concave in z can be easily covered.

Theorem 2 If $F(t, z)$ is continuous in z at $z = 0$ and $F(t, 0) = 0$ for $0 \leq t < \infty$ and if $V = z'F(t, z)$ is a concave function of z for $0 \leq z < \infty$, then the point $z = z(t) = 0$ is a stable equilibrium point of the system in Eq. (61).

The above two theorems suggest that stability in the space of $z(t) = (k(t) = k^*(t), p(t) - p^*(t))'$ depends essentially on a certain type of concavity of the function V and the linearizing approximations in Eq. (62) may sometimes conceal instability due to failure of concavity of V.

Also, it is not necessarily true that oscillations in terms of the presence of complex characteristic roots will be absent in an aggregative one-sector model. For instance, consider the optimal control problem

$$\text{maximize } J = \int_0^\infty R(t)U(c)\, dt$$

subject to

$$\dot{k} = f(k) - nk - c; \quad k(0) > 0 \text{ given,}$$

here c, k are scalars and w_1, w_2 are constant positive weights. Assume for simplicity: $U(c) = w_1 c - \frac{1}{2} w_2 c^2$; $f(k) = bk$. By Pontryagin's maximum principle which is both necessary and sufficient in this case, the optimal path must satisfy

$$\ddot{k} + r(t)\dot{k} - \{(b-n)^2 + r(t)(b-n)\}k = (w_1/w_2)(n - b - r(t)),$$

where $r(t) = \dot{R}(t)/R(t)$. Let

$$k = z \exp\left(-\frac{1}{2}\int_0^t r(s)\, ds\right)$$

then we get

$$\ddot{z} = d\dot{z}/dt = -\alpha(t)z + \beta(t), \tag{63}$$

where

$$-\alpha(t) = \frac{1}{2}\dot{r}(t) + \left(b - n + \frac{r(t)}{2}\right)^2,$$

$$\beta(t) = (w_1/w_2)(n - b - r(t)) \exp\left\{\frac{1}{2}\int_0^t r(s)\, ds\right\}.$$

Then the following theorem can be derived from the results of Nehari (1957) and others (Hartman and Winter, 1954).

Theorem 3 The second-order differential equation in Eq. (63) is strongly oscillatory if

$$\limsup_{\tau \to \infty} \tau \int_\tau^\infty \alpha(t)\, dt = \infty,$$

and mildly oscillatory if

$$\int_0^\infty \alpha(t)\, dt = \infty$$

Note that if $r(t) = -\delta$ is constant independent of time t, then $-\alpha(t)$ is always positive. Hence, the conditions of strong or mild oscillation cannot be fulfilled. However, one should not infer that if $\alpha(t)$ is positive in Eq. (63), the solutions $z(t)$ of Eq. (63) would be necessarily oscillatory. The following Euler equation is a counterexample

$$z + \gamma(t+1)^{-2}z = 0; \quad (\gamma = \text{constant}),$$

which has the general solution

$$z(t) = \begin{cases} A_1(1+t)^s + A_2(1+t)^{1-s}; s = \tfrac{1}{2} + \tfrac{1}{2}(1-4\gamma)^{1/2}, \gamma \neq \tfrac{1}{4}, \\ (1+t)^{1/2}[A_1 + A_2 \log_e(1+t)]; \gamma = \tfrac{1}{4}. \end{cases}$$

Here A_1, A_2 are constants of integration, which is nonoscillatory if the constant $\gamma \leq 1/4$ and oscillatory if $\gamma > 1/4$.

Note also that the matrices H^*_{kk}, $H^*_{kp} + r(t)I$, etc., in Eq. (62) may not have full rank, so that the rank condition of complete controllability may not be satisfied; to that extent the nonlinear system in Eq. (61) of which Eq. (62) is only a linearized part may have zones of less than complete controllability in the policy space. Also we would have problems of singular control, if the Hamiltonian H in Eq. (60) involves the control variables k, v linearly.

(B) Fiscal and Monetary Policy

From the viewpoint of economic policy, an important question is whether appropriate fiscal and monetary policies can be designed to attain optimal growth under neo-classical conditions (Tobin, 1965; Uzawa, 1966; Foley et al., 1969). Uzawa (1966) considers an economic system with two sectors, private (comprising business firms and households) and public. Goods produced by the public sector are not accumulated (i. e., only consumed) and capital accumulation occurs only in the private sector. At any given time t the volume of capital $K(t)$ and labor $L(t)$ is allocated between the two sectors to produce outputs $Y_c(t)$ (paivate goods) and $Y_v(t)$ (public goods). The model contains the following relationships

$$Y_c(t) = F_c(K_c(t), L_c(t)), \quad F_c(\cdot): \text{neo-classical} \tag{64}$$
(private sector production function)

$$Y_v(t) = F_v(K_v(t), L_v(t)), \quad F_v(\cdot): \text{neo-classical} \tag{65}$$
(public sector production function)

$$Y(t) = \text{real gross national income}$$
$$= r(t)K(t) + w(t)L(t), \tag{66}$$

where real wage $w(t)$ and rental rates $r(t)$ satisfy

$$w(t) = \partial F_c/\partial L_c, \, r(t) = \partial F_c/\partial K_c$$

and

$$\frac{w(t)}{r(t)} = \frac{\partial F_v/\partial L_v}{\partial F_v/\partial K_v},$$

$\tau(t) \cdot Y(t) = $ total revenue in public sector.

$$= \tau(t) \cdot Y(t), \quad \tau(t): \text{tax rate.} \tag{67}$$

$r(t)K_v(t) + w(t)L_v(t) = $ total expenditure in public sector. (68)

$\theta(t) \cdot M(t)/p(t) = $ increase in real money supply to meet budget deficit,

$$= [r(t)K_v(t) + w(t)L_v(t)] - \tau(t) \cdot Y(t), \tag{69}$$

where $p(t) = $ market price of private goods, and $\theta(t) = $ rate at which money increases

$$\begin{aligned} K_c(t) + K_v(t) &= K(t): \quad \text{capital allocation,} \\ L_c(t) + L_v(t) &= L(t): \quad \text{labor allocation.} \end{aligned} \tag{70}$$

$$Y_c(t) = C(t) + Z(t), \quad \text{(demand for } Y_c) \tag{71}$$

where $C(t) = $ total consumption, $Z(t) = $ total investment.

$$C(t) = C(\pi(t), Y^d(t), A(t)), \quad \text{(consumption function)} \tag{72}$$

where $\pi(t) = $ rate of interest, $Y^d(t) = $ disposable income, $A(t) = $ total assets held by households.
with

$$Y^d(t) = (1 - \tau(t)) \cdot Y(t) \quad \text{(disposable income)}$$

$$A(t) = K(t) + M(t)/p(t) = \text{physical and monetary assets}$$

$\pi(t)$ determined by $\dfrac{M(t)/p(t)}{K(t)} = \lambda\left(\pi(t), \dfrac{Y^d(t)}{K(t)}\right)$ (demand for money)

$$\frac{Z(t)}{K(t)} = Z\left(\pi(t), \frac{Y(t)}{K(t)}\right): \quad \text{investment demand function.} \tag{73}$$

$$\dot{K}(t) = dK(t)/dt = Z(t) - \mu \cdot K(t): \quad \text{capital accumulation,} \tag{74}$$

where μ is the rate of depreciation.

$$\dot{L}(t) = nL(t): \quad \text{exogenous growth of labor.} \tag{75}$$

Note that the dynamic fiscal policy is here specfied in terms of the time paths of income tax rates $\tau(t)$ and growth rates of money supply $\theta(t)$. If these are predetermined through time, then the differential equation in Eq. (74) specifies

3.1 Deterministic Control Theory

the time path of capital $K(t)$ through time for a given initial stock $K(0)$; also the equations in Eqs. (64) through (75) determine other variables of the system, $K_c(t)$, $K_v(t)$, $L_c(t)$, $L_v(t)$, $Y_c(t)$, $Y_v(t)$, $C(t)$, $Z(t)$ and the various prices $w(t)$, $r(t)$, $p(t)$, and $\pi(t)$. The dynamic system could be uniquely solved under certain conditions on the forms of the above functions, certain predetermined fiscal policy and given initial conditions. Also quasi-stationary solution may exist under certain conditions on the parameters and it may be globally stable. However, since in principle the public sector is free to choose any fiscal policy, it desires in terms of varying the instruments, that is, tax rates $\tau(t)$ and rate of money supply $\theta(t)$, one could impute a social welfare function J as the discounted sum of instantaneous utilities through time. That is,

$$J = \int_0^\infty U\left[\frac{C(t)}{L(t)}, \frac{X(t)}{L(t)}\right] \exp(-\delta t)\, dt, \qquad (76)$$

where $X(t)$ is the consumption of public sector's output. A dynamic fiscal policy may then be defined to be optimal, if it maximizes the social utility function (76) subject to feasible capital accumulation paths constrained by Eqs. (64) through (74).

However, this may be a very complicated nonlinear control problem to solve. Uzawa (1966) suggests an ingenious interpretation of an alternative solution which is based on a long-run optimal control model involving no prices, although the Lagrange multipliers (i. e., co-state variables) are interpreted as imputed prices (the stability of which may then be related to the money market conditions introduced in the original system). The alternative solution is characterized by an optimal time path of $(K_c(t), K_v(t), L_c(t), L_v(t), C(t), Z(t), X(t))$ which maximizes the social welfare function J

$$J \text{ defined in Eq. (76)} \qquad (77)$$

subject to given initial capital $K(0)$ and labor $L(0)$, nonnegativity of variables, and

$$C(t) + Z(t) \leq F_c(K_c(t), L_c(t)), \qquad (78)$$

$$X(t) \leq F_v(K_v(t), L_v(t)), \qquad (79)$$

$$K_c(t) + K_v(t) \leq K(t), \qquad (80)$$

$$L_c(t) + L_v(t) \leq L(t), \qquad (81)$$

$$\dot{K}(t) = Z(t) - \mu K(t), \qquad (82)$$

$$\dot{L}(t) = nL(t). \qquad (83)$$

If the integrand U in Eq. (76) is strictly concave and twice differentiable in its arguments, then this problem may have optimal solutions characterized by Pontryagin's principle. Such optimal paths would be associated with optimal imputed prices (i. e., co-state variables). That is, prices of two goods $q_c(t)$, $q_v(t)$ for Eqs. (78) and (79), price of investment $q(t)$ for Eq. (82), rental rate $r(t)$ for Eq. (80), wage rate $w(t)$ for Eq. (81) (all measured in terms of private goods). The market price $p(t)$ of private goods is then equated to the ratio $q_v(t)/q_c(t)$. Once these prices are available at the optimal solution, we would also have the optimal division of output between consumption and investment and the optimum relative share of private output in gross national product. With given velocity of money λ and given propensity to save out of disposable income, we might ask what type of dynamic fiscal policy $(\tau(t), \theta(t))$ would attain the above conditions of the optimal (growth) solution? If such a dynamic fiscal policy in terms of piecewise continuous functions $\tau(t)$, $\theta(t)$ exists and is economically meaningful, then this is to be interpreted as an optimal fiscal policy.

Consideration of monetary policy has also been introduced in similar framework by Foley *et al.* (1969) in terms of the government (i. e., the public sector) determining the composition of its outstanding debt through open-market purchases and sales.

Two brief comments may be added here. First, although it is remarkable that imputed prices (i. e., co-state variables) allow a decomposition of the overall control model into two subsystems (i. e., the primal and the dual in some sense), yet there remain great problems of "identification" of these imputed prices as ordinary market prices (or price indices). Because the former are "internal prices" dependent not only on the specific way the constraints are set up in the model but also on the computational convergence, which may, however, be interpreted as a competitive tatonnement process (Sengupta and Fox, 1969). What would happen if there is a non-tatonnement process or a deviation from the competitive framework? Second, so long as specific forms of utility functions $U(\cdot)$ are not introduced, certain types of stability could be associated only with the quasi-stationary point in optimal growth. With a nonlinear framework more investigation is needed for ensuring stability in a wider zone of the policy space. In particular, sub-optimal solutions (or satisfying solutions) having wider stability zones (Sengupta and Fox, 1969) may have to be preferred sometimes.

(C) Optimum Borrowing in Capital Markets

A number of models of optimum foreign borrowing in the context of intertemporal optimization has been considered in recent times (Bardhan, 1967, Hamada, 1969) based on neo-classical production functions. We consider here a model due to Hamada (1969) which discusses the problem of optimal borrowing and accumulation policy of an economy facing an imperfectly competitive capital market. This model has two interesting features; for example, imperfect

3.1 Deterministic Control Theory

capital market and the possibility of switch-over from one phase to another in following the optimum borrowing policy.

Under neo-classical conditions of aggregative production function with two factors, labor and capital (with no depreciation), a country is assumed to have capital borrowing (lending), when it increases its capital more (less) than its domestic savings. The per capita net (real) income (y) of this country is defined as

$$y = f(k) + g(z), \qquad (84)$$

where k, z are per capital domestic capital and capital invested abroad $f(k)$ satisfies neo-classical conditions of production, (i. e., $f_k > 0$, $f_{kk} < 0$) and $g(z)$ is the return function satisfying,

$$\partial g(z)/\partial z = g_z > 0, \qquad \partial g_z/\partial z = g_{zz} < 0, \qquad g(0) = 0, \qquad g_z(0) = i(0), \qquad (85)$$

where $i(0)$ is the value of the function $i(b_t)$ at $b_t = 0$, b_t being the per capita indebtedness of the country (interest rate i is assumed an increasing function of b_t). Note that in Eq. (85) a negative (positive) value of z implies a net borrowing, that is, $b_t > 0$ (net lending, i.e., $b_t < 0$) position. Per capita consumption c_t is defined as

$$c_t = (1 - s_1 - s_2)(f(k) + g(z)), \qquad (86)$$

where s_1, s_2 are the control variables denoting the proportions of net income (y) invested domestically and abroad respectively (note a negative s_2 implies borrowing from abroad), which are subject to the following control region

(a) $s_1 \geqq 0$,
(b) $s_2 \geqq -\beta$, ($\beta > 0$ fixed; upper limit on foreign borrowing),   (87)
(c) $1 - s_1 - s_2 \geqq 0$.

The dynamic equations of the system are ($n = \dot{L}/L$ = growth rate of labor);

$$\dot{k} = dk/dt = s_1(f(k) + g(z)) - nk, \qquad (88)$$
$$\dot{z} = dz/dt = s_2(f(k) + g(z)) - nk, \qquad (89)$$

subject to the initial conditions

$$k(0) = k_0, \qquad z(0) = z_0, \qquad \text{assuming} \quad f(k_0) = g(z_0) > 0. \qquad (90)$$

The social welfare function is assumed to be

$$J = \int_0^\infty \exp(-\delta t)u(c_t)dt$$

$$= \int_0^\infty u[(1 - s_1 - s_2)(f(k) + g(z))]\exp(-\delta t)\, dt, \quad (91)$$

where

$$\partial u(c)/\partial c = u_c > 0, \quad \partial u_c/\partial_c = u_{cc} < 0 \quad \text{for all} \quad c > 0,$$
$$\lim_{c \to 0} u(c) = -\infty, \quad \lim_{c \to 0} u_c = +\infty, \quad \delta > 0: \text{social rate of discount}.$$

Defining the Hamiltonian H as

$$H = \exp(-\delta t)[u\{(1 - s_1 - s_2)(f(k) + g(z))\} + p_1(t)\{s_1(f(k) + g(z)) - nk\} + p_2(t)\{s_2(f(k) + g(z)) - nk\}]$$

and applying the necessary conditions (i. e., canonical equations) of Pontryagin's maximum principle, we may characterize the following four interesting phases in the optimal control region according as the inequalities of Eq. (89) are binding or not {here q is $(1 - s_1 - s_2)u_c + p_1s_1 + p_2s_2$].

Phase I: $\quad s_1 = 0, \quad s_2 > -\beta, \quad 1 - s_1 - s_2 > 0; \quad p_1 \leq p_2 = u_c,$
$\quad\quad\quad q = u_c.$

Phase II: $\quad s_1 > 0, \quad s_2 = -\beta, \quad 1 - s_1 - s_2 > 0; \quad p_2 \leq p_1 = u_c,$
$\quad\quad\quad q = (1 + \beta)p_1 - \beta p_2 \geq u_c.$

Phase III: $\quad s_1 = 0, \quad s_2 = -\beta; \quad p_1 \leq u_c \quad p_2 \leq u_c, \quad q = (1 + \beta)u_c > u_c.$

Phase IV: $\quad s_1 > 0, \quad s_2 > -\beta, \quad 1 - s_1 - s_2 > 0; \quad p_1 = p_2 = u_c = q.$

Note that the transversality conditions, that is,

$$\lim_{t \to 0} p_1 \exp(-\delta t) = 0, \quad \lim_{t \to \infty} p_2 \exp(-\delta t) = 0$$

are satisfied only in Phase IV (i. e., the stationary solutions in this phase will be on the optimal path), where the canonical equations are

$$\begin{aligned}
\dot{k} &= s_1(f(k) + g(z)) - nk, \\
\dot{z} &= s_2(f(k) + g(z)) - nk, \\
\dot{p}_1 &= (\delta + n)p_1 - p_1 f_k(k), \\
\dot{p}_2 &= (\delta + n)p_2 - p_2 g_z(z),
\end{aligned} \quad (92)$$

with $p_1 = p_2 = u_c$. The stationary solution (k^*, z^*) and the associated prices $p_1^* = p_2^* = p^*$ are easily obtained from the system in Eq. (92) by equating to zero all time derivatives. Hence,

$$f_k^* = g_z^* = n + \delta. \tag{93}$$

But this gives feasible values of the control variables s_1, s_2 if

$$\beta > -nz^*/(f(k^*) + g(z^*)). \tag{94}$$

As before the quasi-golden rule optimal solution (k^*, z^*, p^*) has the interpretation that once reached it would be maintained forever. But if we start with initial values $k(0) \neq k^*$, $z(0) \neq z^*$, $p(0) \neq p^*$ then switch-over may occur from one phase to another in the direction of Phase IV stationary solution. For instance, it is necessary if the optimal path switches from Phase I (i. e., $p_1 \leq p_2$) to Phase IV (i. e., $p_1 = p_2 = p$) that the following condition

$$\dot{p}_1/p_1 \geq \dot{p}_2/p_2 \tag{95}$$

holds in the neighborhood of the transition point. Thus, if the initial conditions $(k_0, z_0, p_1(0), p_2(0))$ in Phase I are such that they result in condition Eq. (95), then the optimal control switches from Phase I to Phase IV. Thus, depending on the initial conditions and the indebtedness position $i(0) = g_z(0)$ the different phases of optimum borrowing strategy in an open economy may be laid down.

Note that this type of model could be generalized very easily by incorporating an external trade sector with balance of payments adjustments and foreign borrowing. However, we should emphasize that the long-run stationary (or quasi-golden rule) solution may be very unrealistic from the viewpoint of a short-run horizon.

(D) Investment Allocation in Underdeveloped Economies

Several planning models for characterizing optimal investment allocation between sectors have recently been developed for underdeveloped countries (Fox et al., 1966; Johansen, 1967; Dixit, 1968; Marglin, 1967; Bose, 1968). The role of optimal control theory is explicitly recognized in the recent work on the subject. Some of these models of investment planning and their empirical applications are considered in Chapter 4 in both deterministic and stochastic programming framework.

In this section, we refer to an important formulation due to Bruno (1967), which has great potentiality of application in our view to investment allocation problems in underdeveloped countries. This formulation considers in the general case an economy producing $n + 1$ goods (i. e., one composite consumption good and n depreciable capital goods with constant depreciation rates μ_i, $i = 1, \ldots, n$), where each good may be producible by m_i alternative activities ($i = 0, 1, \ldots, n$). The control problem is to

3. Stochastic Control Theory

$$\text{maximize } J = \int_0^\infty \exp(-\delta t) \cdot c \, dt \tag{96}$$

Subject to the constraints

$$\sum_{j=1}^{m_0} a_r^j c_j + \sum_{i=1}^{n} a_{ri}^j z_i^j \leq k_r, \qquad r = 0, 1, \ldots, n,$$

$$(n+1) \text{ constraints}, \tag{97}$$

$$\dot{k}_i = -\lambda_i k_i + z_i, \qquad \lambda_i = n + \mu_i, \qquad n = \text{growth rate of labor}$$
$$i = 1, 2, \ldots, n; \; (n \text{ constraints}) \tag{98}$$

$$k_i(0) \text{ given } (i = 1, \ldots, n) \text{ and all } c_j, z_i^j, k_r \geq 0. \tag{99}$$

Here $k_0 = 1$, $\sum_{j=1}^{m_j} z_i^j = z_i =$ per capita gross investment in capital type i, $\sum_{j=1}^{m_0} c_j$ $= c =$ consumption per capita and a_{ri}^j denotes the input coefficient of the rth factor in the jth activity for the production of the ith good (where a_r^j is the same for the single consumption good). Note that there is only one type of labor growing exogenously at the rate n. The above problem is decomposed into primal and dual problems by means of the Hamiltonian, for example,

Primal:

$$\text{maximize } H = \exp(-\delta t) \cdot [c + \sum_{i=1}^{n} p_i(z_i - \lambda_i k_i)] \tag{100}$$

subject to Eqs. (97), (98), and (99),

Dual:

$$\text{maximize } D = \exp(-\delta t) \cdot [w + \sum_{r=1}^{n} s_r k_r] \tag{101}$$

subject to

$$w a_0^j + \sum_{r=1}^{n} s_r a_r^j \geq 1; \qquad j = 1, \ldots, m_0,$$

$$w a_{0i}^j + \sum_{r=1}^{n} s_r a_{ji}^r \geq p_i; \qquad i = 1, \ldots, n,$$

$$j = 1, \ldots, m_i,$$

$$\dot{p}_i = (\lambda_i + \delta) p_i - s_i; \qquad i = 1, \ldots, n,$$

$p_i(0)$ given and all $s_i, w, p_i \geq 0$.

To complete the specification of the optimal path of $k_i(t)$, $p_i(t)$ and the rest of the system, we have to satisfy for each capital good

3.1 Deterministic Control Theory

the transversality condition[1]: $\lim_{t \to \infty} \exp(-\delta t) \cdot p_i(t) = 0$, $i = 1, \ldots, n$

the jump condition at corners:[2] $\lim_{\theta \to 0} [p_i(\tau + \theta) - p_i(\tau - \theta)] = 0$,

at $t = \tau$, all $i = 1, \ldots, n$.

Note that the imputed prices s_i may be interpreted as the rental rates for capital of type $i = 1, \ldots, n$ and w as the real wage rate.

Three interesting fertures of this model may be emphasized. First, as in linear programming models, the inequalities necessitate switchover from one set of boundary conditions to another. Second, the primal problem in Eq.(100) which characterizes the production subsystem exhibits under certain conditions a contrasting behavior (associated with the saddle point) to the dual problem in Eq. (101) which characterizes the price subsystem. For example if the primal system is unstable around the steady state $k_i = k_i^*$ all i, then the dual system is stable (Morishima, 1964; Jorgenson, 1960). Third, the model shows that under certain cases oscillations would be present in the approach to the equilibrium point which satisfies the quasi-golden rule properties.

Two comments may be added. First, from the operational viewpoint of applications to investment planning in underdeveloped countries, the optimal problem with a shorter planning horizon $t_f < \infty$ and especially the sensitivity of the quasi-stationary (optimal) solution to variations in the end-point of the planning horizon becomes very important. Some work in this line is available (Goodwin, 1961; Sengupta, 1968; Chakravarty, 1966; Sengupta and Fox, 1969), although not in a multisector framework. Second, the uncertainty of estimates and variability of the parameters characterizing the input coefficients (e. g., a_{ji}^r) would raise very fundamental problems in empirical application of such models. Considerations of stochastic programming and stability of optimal stochastic control become very important here.

---

[1] For a generalized specification of the transversality condition see the footnote on page 111. Mathematically, the adjoint vector $p(T)$ is said to satisfy the transversality condition, if it is orthogonal to the tangent plane S passing through the terminal vector point $k(T)$ for every vector belonging to (or parallel to) S, where T is the terminal point ($T \to \infty$).

[2] This specification of the jump condition due to Bruno (1967) appears to be more restrictive than is required. A careful reading of the proofs of Theorems 22-24 in Pontryagin et al (1962) shows that we merely require the existence of only one of the functions $p^+(t) = (p_1^+(t), \ldots, p_n^+(t))$ and $p^-(t) = (p_1^-(t), \ldots, p_n^-(t))$, which must be continuous, nonzero and satisfy the usual necessary conditions of the maximal principle. The authors are indebted to Professor Karl Shell for pointing out that one should examine this condition very carefully. Note that the existence of a continuous nonzero function, either $p^+(t)$ or $p^-(t)$ does not preclude switch over of optimal control from one phase to another. We have discussed this before in connection with singular controls.

3.2 Stochatic Control: Methods and Applications

Problems of stochastic control are usually characterized by methods of (a) dynamic programming, (b) adaptive control, (c) certainty equivalents, and (d) stochastic differential games. Methods of dynamic stochastic programming and its various economic applications are discussed in detail in Chapter 4. Here we consider some of the other methods.

We consider first methods of stochastic control which are rather simple and in some sense straightforward extensions of deterministic control models which do not involve inequalities. A possible generalization of deterministic control theory consists in the assumption that all parameters involved in the model are random variables with a known joint probability distribution. The methods using this assumption do not always lead to very realistic and in some sense efficient models, but the latter models although at times very approximate, point to the great need of taking the random influeuces in economic development into account as much as policy decisions are concerned. Abandoning the somewhat unreal but convenient assumption of single valued anticipations, we have a case of risk. Here we obtain not a unique value of anticipated conditions (prices, etc.) but probability distributions. However, the assumption is that at least these probability distributions are known with certainty. Since we are dealing here with dynamic control problems, stochastic processes would be naturally involved. In the very general case of a multivariable optimal stochastic control, there remain a great many problems still unsolved. Computational difficulties are quite great in this area.

In order to analyze some operational results in the theory of stochastic control, we proceed first with simple models which are intended to provide some motivation for our later disucussion. It is assumed here that we have a system characterized by simple linear difference or differential equations connecting certain inputs and outputs. There is a control variable which should enable us to produce some desired output. The output is the controlled variable. Open loop control exists where the output has no effect on the input. Feedback control provides the means of feeding back the output in order to enable a comparison with the desired output. Closed loop control means a system where output has effects on the input in order to obtain the desired result. Also, adaptive control systems take change in the environment into account, especially stochastic variations.

We describe first a simple application of the lognormal diffusion process (which was discussed in Chapter 2) to a feedback control problem.

Consider a stochastic variable such as real agricultural production in India per capita. Assume that real agricultural production per capita in India follows a lognormal diffusion process. The logarithm of real agricultural production in India is normally distributed, and the mathematical expectation of the logarithm and its variance are linear functions of time. Define a transition probability density as follows.

$$f(st\ x;\ t,\ y) = \frac{\exp\{-[\log y - \log x - b(t - s)]^2/2c(t - s)\}}{y\sqrt{2\pi c(t - s)}}$$

This is the probability that real national agricultural production will be y at time t if it was x at time s ($< t$). We assume now that our random variable has with probability one the value x_0 at the point in time $t = 0$. Then the mathematical expectation of the random variable is

$$EY(t) = x_0 \exp(b + 1/2c)t$$

and its variance is given by

$$\sigma^2(t) = [Ey(t)]^2[\exp(ct) - 1].$$

Hence, the mathematical expectation and variance of our random variable are both exponential funtions of time.

We present Table 1 with our predictions for real per capita agricultural production in India.

TABLE 1

Year	t	$X_t =$ Real per capita agricultural produbtion
1964–65	13	0.9517
1965–66	14	0.9456
1966–67	15	0.9396
1967–68	16	0.9335
1968–69	17	0.9274
1969–70	18	0.9215

We see from this table that predicted real per capita agricultural production in India has a decreasing trend. It is also possible to compute confidence limits, for example, for $t = 18$, (1969–70) we predict a real per capita agricultural production of only 0.9215. The 95% confidence or fiducial limits are 0.708 and 1.113.

Now we consider also the influence of an exogenous factor on real per capita agricultural production in India, namely, real per capita government expenditure on agriculture (G_t). This is our feedback control variable. Define

$$H_t = G_0 + G_1 + \cdots + G_t,$$

the cumulative real per capita government expenditure in India after t years. Assume that the effect of expenditure will only be felt after p years. The transition probability density becomes now

$$f(s, x; t, y) = \frac{\exp\{-[\log y - \log x - b_0(t - s) - b_1(H_{t-p} - H_{s-p})]^2/2c(t - s)\}}{y[2\pi c(t - s)]^{1/2}}.$$

There is now the conditional probability that our random variable will have the value y at time t if it had the vabue x at time $s(< t)$ given the government expenditure between $s - p$ and $t - p$.

Now we consider the forecasts of the real per capita agricultural production in India under a variety of hypotheses for $t = 18$ (1969–70): (a) Same real per capita government expenditure in the intervening years as in 1963–64. (b) An increase of expenditure by one-half. (c) Doubling of expenditure. (d) Two and a half times the 1964–65 expenditure. (e) Three times the expenditure of 1963–64. The resulting forecasts and their 95% confidence or fiducial limits are given in Table 2.

TABLE 2

Hypothesis	Estimated Real Per Capita Agricultural Production 1969–70		95% Limits
(a)	0.8541	0.5952	1,1130
(b)	0.9107	0.6347	1.1867
(c)	0.9712	0.6768	1.2656
(d)	1.0350	0.7212	1.3488
(e)	1.1040	0.7694	1.4386

This table shows that, in the intervening years between 1964–65 and 1969–70, real per capita government expenditure on agriculture must be almost doubled [Hypothesis (c)] in order to keep real per capita agricultural production at least on the level of 1964–65.

Methods which are more closely related to electro-technical control theory are due to Holt et al. (1960), (see also Whittle, 1963). Consider, for example, planning for an enterprise and use the following notation. P_t production, W_t work force, I_t end of the period inventory, S_t orders during the period. The total cost function to be minimized is given by a quadratic function,

$$F = E[A_1(W_t - W_{t-1})^2 + A_2(P_t - W_t)^2 + A_3(I_t - B)^2],$$

where E is the mathematical expectation, A_1, A_2, A_3, and B are constants. The first term represents hiring and laying off costs caused by variation in the work force. The second term represents overtime or idle time costs incurred if the size of the work force is not appropriate for current production. The third term represents storage or inventory cost if the stock is too large or too small.

All orders are satisfied, a negative inventory corresponds to supply of goods from another source at extra cost. We have

$$P_t - S_t = I_t - I_{t-1}.$$

Assuming now that orders follow a stationary stochastic process, we derive the optimum transfer functions as

$$W/S = (1 + Hf^2 + Gf^4)^{-1} = B^*,$$
$$I/S = -(A_1/A_2)(1 - z^{-1})^2(1 - z) \cdot B = iGf^3B/z^{1/2} = C^*$$

where

$$H = A_1/A_2, \qquad G = A_1/A_3,$$
$$f = (z^{1/2} - z^{-1/2})/i = 2\sin(w/2)z = e^{-iw},$$
$$P/S = 1 + (1 - z)C^* = (1 + Hf^2)B^*.$$

This shows that response in the work force will fall off rather quickly with increasing frequency. Inventory is relatively sensitive to higher frequencies, that is, sudden changes in orders.

The moduli of the actual components of the cost function are

$$|I/S| = G|f|^3 B^*,$$
$$|(W_t - W_{t-1})/S| = |f|B^*,$$
$$|(P - W)/S| = H|f|^2 B^*.$$

Inventory costs are sensitive to rapid variation of orders. Costs of changes in the work force are sensitive to slow variation. Costs due to incorrect size of the work force are intermediate in response. Some empirical applications of this model under chance-constrained and reliability programming are reported in Chapter 4. Another useful method in the theory of stochastic optimal control is the idea of certainty equivalents (Simon, 1956; Theil, 1957). Least squares estimates are substituted for all variables in the criterion function which have not yet been observed and the quadratic form is minimized under such conditions (see Fox *et al.* 1966).

(See also: Aoki, 1967; Balakrishan and Neustadt, 1967; Bellman, 1961; Box and Jenkins, 1962; D'Azzo and Houpis, 1966; Howard, 1960; Kalaba, 1962; Karreman, 1968; Kushner, 1967; Lee and Markus, 1967; Leitman, 1962; 1967; Oldenburger, 1966; Solodnikov, 1963; Whittle, 1963; Wiener, 1949.)

Consider now a continuous stochastic problem (Dreyfus, 1968). Let x be determined by

$$x(t + \Delta t) = x(t) + [ax(t) + bu(t)]\Delta t + \xi(t)(\Delta t),$$

where $\xi(t)$ is stochastic process with

$$E\xi(\Delta t) = 0,$$
$$E\xi^2(\Delta t) = \sigma^2 \Delta t,$$
$$E\xi^n(\Delta t) = o(\Delta t), \quad \text{for} \quad n > 2,$$

$\xi(\Delta t_1), \ldots, \xi(\Delta t_n)$ are independent in not overlapping intervals $\Delta t_1, \ldots, \Delta t_n$. E is mathematical expectation. If $z(t)$ is a Brownian motion process

$$\xi(\Delta t) = \Delta z_t = z(t + \Delta t) - z(t).$$

We wish to minimize

$$E\left\{\int_{t_0}^{T} u^2(t)\, dt + x^2(T)\right\} = \int_{t_0}^{T} u^2 \, dt + [Ex(T)]^2 + \text{a constant}.$$

One solution of this problem might be obtained by the certainty equivalent. We replace the stochastic process $\xi(\Delta t)$ by its mean value, zero (Theil, 1957; Simon, 1956; Sengupta, 1965). Let $f(x, t)$ be the minimum value of the objective function

$$f(x, t) = \min_u E_\xi[u^2 \Delta t + f(x + (ax + bu)\Delta t + \xi, t + \Delta t)],$$
$$0 = \min_u [u^2 + (\partial f/\partial x)(ax + bu) + (\sigma^2/2)(\partial^2 f/\partial x^2) + (\partial f/\partial t)],$$
$$u = -b(\partial f/\partial x)/2.$$

The result is again the same as in the deterministic case.

Now we assume that the variance of $\xi(t)$ depends upon the control decision, with no randomness in the evolution of x if no control is exerted. Now,

$$E\xi(\Delta t)^2 = u^2 \sigma^2 \Delta t,$$

where σ^2 is a constant. The criterion function is simply

$$\min E[x(T)^2],$$

where

3.2 Stochastic Control

$$x(t + \Delta t) = x(t) + [bu(t)]\Delta t + \xi(t)\Delta t.$$

Open loop control implies that the variance of $x(T)$ is

$$\int_{t_0}^{T} u^2(t)\sigma^2 \, dt$$

and the criterion function is

$$[Ex(T)]^2 + \int_{t_0}^{T} u^2\sigma^2 \, dt.$$

Now the Lagrange multiplier is a constant, $L = 2Ex(T)$. Optimal control is given by

$$u(t) = -E[x(T)]b/\sigma^2.$$

We find empirically in our example for Indian agricultural production $b = 1.21524$ and $\sigma^2 = 0.049246$. Hence, the optimal control, if we minimize $\int_{t_0}^{T} u^2(t) \, dt - [Ex(T)]^2$, is given by

$$E(t - 4) = -Ex(T) \, (24.677).$$

This is the same as

$$u(t) = \frac{-x(t)}{b[T - t + (\sigma^2/b^2)]}.$$

In our example, this is

$$E(t - 4) = \frac{-A(t)}{1.21524(T - t) + 0.0405},$$

$$E(t) = \frac{A(t - 4)}{1.21524(T - t - 4) + 0.0405}.$$

The expected terminal value of x is given by

$$E[x(T)] = \frac{\sigma^2 x(t_0)}{b^2[T - t_0 + (\sigma^2/b^2)]}.$$

In our example

$$EA(T) = \frac{0.033 A(t_0)}{t - t_0 + 0.033}.$$

The variance of the random variable $x(T)$ is

$$\text{var } x(T) = \sigma^2(x(T)) = \frac{\sigma^2 x^2(t_0)(T - t_0)}{b^2[T - t_0 + (\sigma^2/b^2)]^2}.$$

This is in our example

$$\text{var } A(T) = \sigma^2(A(T)) = \frac{0.033 A^2(t_0)(T - t_0)}{(T - t_0 + 0.033)^2}.$$

The value of the criterion function is

$$Ex^2(T) = \frac{\sigma^2 x^2(t_0)}{b^2[T - t_0 + (\sigma^2/b^2)]}$$

This is in our example

$$EA(T) = \frac{0.033 A^2(t_0)}{T - t_0 + 0.033}.$$

Open loop optimal feedback proceeds as follows. The equation of motion is

$$x(t + \Delta t) = x(t) - \frac{x(t)\Delta t}{T - t + (\sigma^2/b^2)} + \xi(\Delta t).$$

Let $f(x, t) = E[x(T)]^2$. Dynamic programming gives

$$f(x, t) = E_\xi f\left[x - \frac{x\Delta t}{T - t + (\sigma^2/b^2)} + \xi, t + \Delta t\right].$$

Letting $\Delta t \to 0$

$$0 = -x(\partial f/\partial x)/[T - t + (\sigma^2/b^2)]$$
$$+ x^2\sigma^2\{(\partial^2 f/\partial x^2)/2b^2[(T - t + (\sigma^2/b^2)]^2\} + (\partial f/\partial t).$$

Hence

$$f(x, t) = x^2 \exp\{1 - \sigma^2/b^2[(T - t + (\sigma^2/b^2)] + 2\log(\sigma^2/b^2)$$
$$- 2\log[T - t + (\sigma^2/b^2)]\},$$
$$E[x(T)] = \sigma^2 x(t_0)/b^2[T - t_0 + (\sigma^2/b^2)].$$

This is again the same result as for the pure open loop process.

For feedback, we use the optimality principle of dynamic programming. Let now $f(x, t)$ be the value of the criterion if we start with x at time t, $t_0 < t < T$.

3.2 Stochastic Control

$$f(x, t) = \min_u E_\xi[f(x + (bu)\Delta t + \xi, t + \Delta t],$$

that is,

$$0 = \min_u [bu(\partial f/\partial x) + (1/2)u^2\sigma^2(\partial^2 f/\partial x^2) + (\partial f/\partial t)].$$

This yields after some transformations

$$f(x, t) = [\exp\{-(b^2/\sigma^2)(T - t)\}]x^2,$$
$$u = -bx/\sigma^2.$$

In our example, using the original difference equation

$$E(t) = 0.003825 + 0.017025 A(t),$$
$$f(x, t) = [\exp\{-0.033(T - t)\} \cdot A^2(t)],$$
$$u(t) = -0.025 A(t).$$

Let now $h(x, t)$ be the expected terminal value of x using this control. Dynamic programming gives

$$h(x, t) = E_\xi h[x - (b^2 x/\sigma^2)\Delta t + \xi, \quad t + \Delta t]$$

with the boundary condition

$$h(x, T) = x.$$

The solution is

$$h(x, t) = x \cdot \exp[-\{(b^2/\sigma^2)(T - t)\}].$$

This gives for our example, dealing with agricultural production in India,

$$h(x, t) = e^{-0.033(T-t)} \cdot A(t)$$

for the optimal terminal value $A(T)$.

Next, we consider a discrete open loop control. Let i be the stage ($i = 0, 1, \ldots, N - 1$) and $j = 1, 2, \ldots, M$ the state. Let u_i be the control chosen at stage i. The probability of transition from state j in stage i to state k in stage $i + 1$ is $p(j, k, i, u_i)$ if control u_i is applied to stage i. The cost is $L(j, k, i, u_i)$. The process terminates at state N and if the final stage is j, an additional cost $F(j)$ is incurred. The control sequence $u_i(i = 0, i, \ldots, N - 1)$ is sought given the probability distribution at stage zero, which minimizes the total expected cost of the N stage dynamical process. It is open loop control, since it depends on the stage, not the state.

Let u_m^0 be the optimal control sequence; $p^0(j, i)$ the resulting probability of the process being in state j at stage i and $f^0(j, i)$ the expected value of the criterion function for a process starting at state j at time i when control u_m^0 ($m = i, i + 1, \cdots, N - 1$) is used. The equation

$$J(i) = \sum_j p^0(j, i) f^0(j, i)$$

is the minimum attainable expected cost of the process starting from i, given $p^0(j, i)$ the state probability at the time i.

The new expected cost is the expected cost of the remaining process using u_m^0 ($m = i + 1, \ldots, N - 1$) with updated initial probabilities at time $i + 1$. Alternatively, it is the average over all states j with weights $p^0(j, i)$, of the expected cost of the first step plus the expected cost of the remaining process

$$\begin{aligned}\sum_j p^0(j, i) f^0(j, i) &= \min_u \{\sum_j p^0(j, i) \sum_k p(j, k, i, u) L(j, k, i, u) \\ &\quad + \sum_j \sum_k p^0(k, i) p(k, j, i, u) f^0(j, i + 1)\} \\ &= \min_u \{\sum_j p^0(j, i) [\sum_k p(j, k, i, u) \\ &\quad \cdot (L(j, k, i, u) + f^0(k, i + 1))]\}.\end{aligned}$$

These two formulations are equivalent.

Given control u_m^0, f^0 is given by

$$f^0(j, i) = \sum_k p(j, k, i, u_i^0)[L(j, k, i, u_i^0) + f^0(k, i + 1)],$$
$$f^0(j, N) = F(j).$$

Let

$$x(t + \Delta t) = x(t) + g(x, t, u)\Delta t + \xi(\Delta t)$$

and $t \to 0$, where $\xi(\Delta t)$ is a stochastic process,

$$\begin{aligned}E\xi(\Delta t) &= 0 \\ E\xi^2(\Delta t) &= \sigma^2(x, t, u), \\ &\vdots \\ E\xi^n(\Delta t) &= o(\Delta t), \qquad n > 2.\end{aligned}$$

The criterion is taken to be the minimization of the expected value of the integral of $L(x, t, u)$ plus the terminal contribution $F(x)$

$$\min\left[E\int_0^T L(x, t, u)\, dt + F(x)\right]$$

which leads to

$$\int p^0(x, t) f^0(x, t)\, dx = \min_u \left\{ \int p^0(x, t) L(x, t, u) \Delta t\, dx \right.$$
$$\left. + \int (p^0(x, t) + (\partial p^0/\partial t)\Delta t) f^0(x, t + \Delta t)\, dx \right\}.$$

Expanding and going to the limit we have

$$\partial p^0(x, t, u)/\partial t = -(\partial g p^0/\partial x) + (1/2)(\partial^2 \sigma^2 p^0/\partial x^2)$$

and

$$L(x, t, u) + (\partial f^0/\partial x)g + (1/2)(\partial^2 f^0/\partial x^2)\sigma^2 + (\partial f^0/\partial t) = 0.$$

Also, $p^0(x, 0)$ is the initial and $f^0(x, T) = F(x)$ the terminal information.

Various applications of such problems have been considered in the literature (Howard, 1960; Sengupta and Fox, 1969). Two aspects about the various control schemes presented above may be emphasized. First, the convergence speed and characteristics of the asymptotic criterion function, that is, $E[x^2(T)]$ for $T \to \infty$ may be different in feedback, open-loop and open-loop optimal feedback controls. Second, the problems of estimating the parameters (e. g., variance) of the underlying stochastic processes have been completely ignored.

3.2.1 STOCHASTIC OPTIMAL CONTROL: GENERAL CHARACTERIZATION

Consider now a dynamic system characterized by n stochastic differential equations, that is,

$$dx(s, t) = f(x(s, t), u(s, t))\, dt + dz(s, t), \tag{1}$$

where the vectors $x(s, t)$, $f(x, u)$, and $z(s, t)$ have n-components, $u(s, t)$ is the control vector with m elements and $z(s, t)$ is a measurable stochastic process [e. g., $z(s, t)$ may be the Brownian motion process which is the zero mean Gaussian process with stationary independent increments; here the scalar component s indicates the sample space of points $P(s)$ and t denotes the space of time points]. The control vector $u(s, t)$ is to be selected such that it satisfies the system in Eq. (1) and a set U of constraints,

$$u(t) \in U \tag{2}$$

and also minimizes a risk function in the form

$$Ec'x(s, t), \tag{3}$$

where the end period T is fixed, c is a known vector with n elements, and E is expectation over the sample space. If such a control exists then it would be defined as one possible form of optimal stochastic control, which could be considered a straightforward stochastic analog of the deterministic problem of optimal control. Note that the linear form of the performance criterion in Eq. (3) is really no restriction on generality, since as Rozonoer (1959) has shown that any nonlinear objective function can be reduced to this form by merely introducing an additional coordinate into the state space, for example,

$$x_{n+1}(T) = \int_0^T L(x(t), u(t))\, dt$$

and

$$\dot{x}_{n+1} = 1, \qquad x_{n+1}(0) = 0.$$

Since the stochastic process $z(s, t)$ in the system of Eq. (1) could be very general, we need some conditions for characterizing the class from which the optimal stochastic control for this problem is sought. These conditions have been derived rigorously in the stochastic control literature (Kushner 1963, 1965; Ho and Newbold, 1967; Florentin, 1961). Very broadly, these conditions define the class of admissible controls by a stochastic process such that for each s, the control $u(s, t)$ is Lebesgue integrable and

$$\int_0^T \|u(t)\|\, dt < \infty, \quad \text{where} \quad \|u(t)\| = \text{norm as } \sum_{i=1}^m |u_i(t)|, \tag{4}$$

$$\int_0^T E\,\|u(t)\|\, dt < \infty,$$

and $u(s, t)$ is measurable over a field $S_u(t)$. From now on, we will drop the sample points in the notations, that is, $u(t)$, $p(t)$, $x(t)$, and so on. Also, we need boundedness and continuity conditions on the vector functions $f(\cdot)$ and $z(\cdot)$, they are the following.

There is a finite non-random K such that

$$|f_i(x, u)| \leq K(1 + \|x\| + \|u\|). \tag{5}$$

Each f_i satisfies the uniform Lipschitz condition

$$|f_i(x + \delta x, u + \delta u) - f_i(x, u)| \leq K(\|\delta x\| + \|\delta u\|). \tag{6}$$

$z(s, t)$ is a measurable stochastic process such that

$$\int_0^T E\,\|z(t)\|\, dt < \infty, \tag{7}$$

3.2 Stochastic Control

The partial derivative of each f_i with respect to x_j satisfies the uniform Lipschitz condition. (8)

Under these regularity conditions, Eqs. (4)-(8), if there exists an optimal stochastic control in the class of admissible controls, then there must exist a vector of stochastic variables $p(t)$ (uniformly bounded in s and t) satisfying the canonical equations

$$dp(t) = -f'_x(x(t), u(t))p(t)\,dt; \qquad f_x \equiv \partial f/\partial x$$
$$dp_i: \text{integrable} \qquad (9)$$
$$i = 1, \ldots, n$$

and the boundary conditions

$$p(T) = c, \qquad u(t) \in U \qquad (10)$$

and minimizing the conditional expected value of the Hamiltonian; i. e.,

$$E[H(x(t), u(t), p(t)) \mid S_u(t)]$$
$$\leqq E[H(x(t), \hat{u}(t), p(t)) \mid S_u(t)], \qquad (11)$$

where $E(a \mid b)$ denotes the conditional expectation of a given b and H defines the Hamiltonian, which is well defined, integrable, and $\int_0^T E|H|\,dt$ is finite,

$$H(x(t), u(t), p(t)) = p'(t)f(x(t), u(t)), \qquad (12)$$

and $\hat{u}(t)$ is the set of admissible controls satisfying Eq. (4).

Intuitively speaking, the above necessary conditions state that the optimal stochastic control vector $u(t)$ must minimize the conditional expectation of the Hamiltonian given the information available at t over the class of admissible controls. (These necessary conditions would be sufficient only through additional concavity restrictions on the functions.)

Two points about the above stochastic analog of Pontryagin's maximum principle may be noted. First, the canonical equations are random differential equations and, in particular, if the set $P(s)$ of sample points s contains a finite number of points, say N, then the stochastic control problem could be immediately reduced to an equivalent deterministic problem, since the functions $p(s, t)$, $x(s, t)$, and $u(s, t)$ may then be obtained explicitly as functions of s and t from the following $2N$ differential equations (i. e., for $i = 1, \ldots, N$),

$$dp(s_i, t) = -f'_x(x(s_i, t), u(s_i, t))p(s_i, t)\,dt,$$
$$p(s_i, T) = c,$$

$$dx(s_i, t) = f(x(s_i, t), u(s_i, t)) \, dt + dz(s_i, t),$$

and $u(s_i, t)$ is given by

$$\min E[p'(s, t) f(x(s, t), u(s, t)) \mid S_u(t)], \quad u \in U.$$

Second, when the stochastic process $z(s, t)$ is of Brownian motion (i. e., a zero mean Gaussian process with stationary independent increments) and the available information at time t is $x(t)$ [i. e, $S_u(t) = x(t)$, case of optimal feedback control], then the canonical equations could be utilized to derive the behavior of a scalar function $V(x(t), t)$ defined as

$$V(x(t), t) = E[c'x(T) \mid x(t)],$$
$$V(x(T), T) = c'x(T),$$
$$\operatorname{grad} V(x(T), T) = c,$$

where grad V is the gradient vector. The following theorem has been derived by Kushner (1965).

Theorem 1 If all admissible controls $\hat{u}(s, t)$ are functions of $x(t)$ and t and satisfy for small ϵ, the uniform Lipschitz condition,

$$\|\hat{u}(x(t) + \epsilon, t) - u(x(t), t)\| \leq K\|\epsilon\|. \tag{13}$$

then for the optimal control $u(s, t)$ and the associated optimal process $x(t)$ with the initial condition $x(0)$, it holds

$$\operatorname{grad} V(x(t), t) = E(p(t) \mid x(t)).$$

Also let $\delta x(0)$ be a perturbation in $x(0)$ and $\hat{u}(\cdot)$ be the associated admissible control

$$\hat{u}(x(t) + \delta x(t), t) \equiv u(x(t), t) + \delta u(x(t), t, \delta x(t)).$$

If $\delta u(x(t), \delta x(t))$ is abbreviated as $\delta u(\delta x(t))$ and it satisfies $\delta u(0) = 0$ and the uniform Lipschitz condition, Eq. (13), then

$$E p'(t) \delta x(t) = C + o(\delta x(0)),$$

with probability one, where $\lim_{\epsilon \to 0} o(\epsilon)/\epsilon = 0$ and C is a scalar independent of t and the symbol $o(a)$ means terms of lower order of smallness than a.

Proof For perturbations around the original system, we have

$$\delta \dot{x} = f(x + \delta x, u + \delta u) - f(x, u).$$

3.2 Stochastic Control

By conditions (8) on f and (13) on u we may linearize the right-hand side by Taylor expansion (f_x denotes the partials $\partial f/\partial x$)

$$\delta \dot{x} = f_x(x, u)\delta x + f_u(x, u)\delta u + o(\delta x).$$

This linearized system has the solution

$$\delta x(t) = W(t, 0)\delta x(0) + \int_0^t W(t, \tau)f_u(x, u)\delta u\, d\tau + o(\delta x(0)), \tag{15}$$

where $W(T, t)$ is the fundamental matrix solution of the linear system; also for almost all sample points $s \in P(s)$,

$$\dot{p} = -f_x'(x, u)p, \quad p(T) = c, \quad p(t) = W'(T, t)p(T). \tag{16}$$

Pre-multiplying Eq. (15) by $p'(t)$ and noting from the third equation of Eq. (7) that for $t \geq r$, we get

$$p'(t)W(t, r) = p'(T)W(T, r) = p'(r).$$

Since

$$W(T, t)W(t, r) = W(T, r)$$

we get

$$p'(t)\,\delta x(t) = p'(0)\,\delta x(0) + \int_0^t p'(r)f_u(x, u)\,\delta u\, dr + o(\delta x(0)). \tag{17}$$

But since δu is unconstrained in magnitude, the minimization of

$$E[p'(t)f(x, u) \mid x(t)] \text{ yields } E[p'(t) \mid x(t)] \cdot f_u(x, u) = 0.$$

Hence, the expectation of the integral on the right-hand side of Eq. (15), given $x(t)$ is zero. Also $E(p'(0)\,\delta x(0))$ is a constant, hence Eq. (14) holds. Again,

$$Ec'\,\delta x(T) = E[p'(0)|\delta x(0)] \cdot \delta x(0) + o(\delta x(0))$$

using Eqs. (15) and (16). But since time zero is arbitrary, the condition

$$\operatorname{grad} V(x(t), t) = E[p(t) \mid x(t)]$$

holds.

Note that in the deterministic case, these results hold without expectation. In a sense, this generalizes the certainty equivalence result (Theil, 1957; Simon, 1956; Sengupta, 1970). However, the numerical computation of optimal feedback controls when there are no restrictions on their magnitudes may be still difficult. For instance, consider the following multiplier-accelerator model due to Phillips (see Allen, 1959) which is used to discuss stabilization policies through varying government expenditure

$$\dot{x}_1 = x_2 \quad (x_1 \text{ denotes aggregate output}),$$
$$\dot{x}_2 = -b_1 x_2 - b_2 x_1 - (m/k)u_1 - m u_2, \quad \dot{x}_i \equiv dx_i/dt,$$
$$u_2 = \dot{u}_1 \quad (u_1 \text{ denotes government expenditure plus autonomous constant demand}), \text{ given } x_1(0), x_2(0);$$

Here the control variables are u_1, u_2. If a quadratic performance criterion is assumed for this system, that is,

$$\text{minimize } J_T = \int_0^T (w_1 x_1^2 + w_2 u_1^2)\, dt, \quad T \text{ fixed}; w_1, w_2: \text{ positive weights,}$$

then with no restriction on the magnitude of the control variables, this leads to the following optimal stabilization policy (u^*, x^*) which is also feedback

$$u^*(t) = -(1/w_2)B'N(t)x^*(t), \quad u^* = \begin{pmatrix} u_1^*(t) \\ u_2^*(t) \end{pmatrix}, \quad x^*(t) = \begin{pmatrix} x_1^*(t) \\ x_2^*(t) \end{pmatrix},$$

where

$$B' = \begin{bmatrix} 0 & m/k \\ 0 & m \end{bmatrix}$$

(prime on B denotes transpose), and the elements $n_{ij}(t)$ of the second-order symmetric matrix $N(t)$ are the unique solution of the following matrix Riccati equation subject to the boundary condition $N(T) = 0$,

$$\dot{N}(t) = -N(t)A - A'N(t) - w_1 I, \tag{18}$$

where

$$A = \begin{bmatrix} 0 & 1 \\ -b_2 & -b_1 \end{bmatrix}, \quad I = \begin{bmatrix} 1 & 0 \\ 0 & 1 \end{bmatrix}.$$

The optimal feedback policy in terms of $u_1^*(t)$ is

3.2 Stochastic Control

$$u_1^*(t) = -(m/kw_2)\{n_{12}(t)x_1^*(t) + n_{22}(t)\dot{x}_1^*(t)\}, \tag{19}$$

where

$$n_{22}(t) = n_{22}(0) \exp(-w_1 t),$$
$$n_{12}(t) = n_{12}(0) \exp(-b_1 t) \int_0^t \{\exp(b_1 s - w_1 s)(n_{11}(0) + b_2 n_{22}(0))\}\,ds,$$
$$n_{ij}(0) = \text{initial values.}$$

Note that the matrix $N(t)$ is required to be positive definite if the optimal feedback policy in Eq. (19) has to be stabilizing and this is possible by appropriate choice of the initial values $n_{ij}(0)$. For instance, if $n_{12}(0)$ and $n_{22}(0)$ are positive, then a substitution of Eq. (19) into the system equations shows that the speed of convergence, that is, the stabilizing tendency to a stationary value is improved for the system $x(t)$. As a matter of fact, it has a damping tendency much more than a no-control policy, since the characteristic roots have all negative real parts. Comparisons of other types of stabilization policies under aggregative models have been discussed in the literature (Sengupta, 1965, 1970; Naylor et al., 1969).

For applied economic models and several models in operations research (e.g., inventory control problems over time), time enters in a discrete fashion and hence, we have to characterize the dynamic system in terms of difference rather than differential equations. Maximum principles have also been developed for such discrete stochastic systems (Fan and Wang, 1964; Rozonoer, 1959, Kushner, 1965).

Consider the following characterization of the discrete system in terms of the control (u_i) and state (x_i) vectors and random ξ_i:
System:

$$x_{i+1} = f(x_i, u_i, \xi_i), \quad i = 0, 1, \ldots, m-1, \tag{20}$$

where x_0 is given and known (in a general case, x_0 can be random), x_i, u_i, ξ_i are vectors, with ξ_i the random vector.
Performance criterion (scalar):

$$\text{minimize } Eg(x_m), \tag{21}$$

where E denotes expectation, and the scalar function $g(x_m)$ of vector x_m has differentiability, concavity and other required regularity properties.
Constraints:

$$h(y(\xi), \xi) = 0$$

where h is a vector with components h_j and y, ξ are generic expressions for any or all of $(x_0, \ldots, x_m, u_0, u_i, \ldots, u_m)$ and $\xi_0, \xi_1, \ldots, \xi_{m-1}$, respectively.

Note that expressing the performance criterion in terms of the terminal period variables x_m is no restriction of generality. However, as in the continuous problem, it is very important to specify the distribution characteristics of the random vector and the way they enter into the system. The following characterization of the conditions to be satisfied by a stochastic optimal control satisfying Eqs. (20), (22) and minimizing Eq. (21) is due to Kushner (1965). It is assumed that

(a) The multivariate distribution of the random vector ξ has a continuous probability density.

(b) The functions $h_j(y(\xi), \xi)$, $g(y(\xi), \xi)$ and their derivatives with respect to each y_j are continuous with respect to each ξ_j and y_k.

(c) For the vector function $\bar{y}(\xi)$, [which makes $Eg(y(\xi), \xi)$ stationary with respect to perturbations $\epsilon\phi(\xi)$, where ϵ is small and the perturbation vector $\phi(\xi)$ of the same dimension as $y(\xi)$, has components $\phi_i(\xi)$ such that $E|\phi_i(\xi)|^2 < \infty$] the following conditions hold:

$$E|g(\bar{y}(\xi) + \epsilon\,\phi(\xi), \xi)| < \infty,$$

$$E|g_{y_i}(\bar{y}(\xi) + \epsilon\,\phi(\xi), \xi)|^2 < \infty, \qquad g_{y_i} \equiv \partial g/\partial y_i,$$

$$E\left|\frac{\partial h_j}{\partial y_k}(\bar{y}(\xi) + \epsilon\,\phi(\xi), \xi)\right|^4 < \infty.$$

Now we consider two simple forms in which the random vector ξ in the system of Eq. (20) may enter into the optimal decision.

Case I. All $u_0, u_1, \ldots, u_{m-1}$ are to be determined at time $i = 0$. Here x_i is a function of $\xi_1, \xi_2, \ldots, \xi_{i-1}$, but u_i does not depend on any ξ_i. But there are constraints on u_i in the relation $h(u_0, \ldots, u_{m-1}) = 0$, where the function h is non-stochastic, since u_i are constants.

Case II. Here u_i like x_i may be functions of $\xi_0, \ldots, \xi_{i-k}$, for $k > 0$; we denote this by $u_i(\xi^{i-1})$ and $x_i(\xi^{i-1})$. When decisions one time point ahead are considered, the variables $\xi_0, \xi_1, \ldots, \xi_{i-1}$ are often known at time i through observing the sequence $x_0, x_1, \ldots, x_i$. In these cases we may have the constraints as $h(u_i, \xi) = 0$ which are stochastic. The following theorem is proved in Kushner (1965), where $\xi = (\xi_0, \ldots, \xi_{m-1})$, $\xi^i = (\xi_0, \ldots, \xi_i)$.

Theorem 2 If there exists for Case I type problems a stochastic optimal control which satisfies the conditions (a), (b), and (c), the system of Eq. (20) and the constraints of Eq. (22) and also minimizes the performance criterion Eq. (21), then there must exist piecewise continuous random Lagrange multipliers (or co-state vectors) $\alpha_i(\xi)$ and $\beta(\xi)$ associated with Eq. (20) and (22) with

$E|\alpha_i(\xi)|^2 < \infty$ and $E|\beta_j(\xi)|^2 < \infty$ satisfying in terms of function

$$G = Eg(x_m(\xi)) + E \sum_{i=1}^{m} \alpha'_i(\xi) \cdot [f(x_{i-1}(\xi^{i-2}), u_{i-1}, \xi_{i-1}) - x_i(\xi^{i-1})] + E\beta(\xi)' \cdot h(u),$$

the following necessary conditions:

$$E[f_{u_i}(x_i(\xi^{i-1}), u_i, \xi^i) + h_{u_i}(u)\beta(\xi)] = 0, \tag{23}$$

$$\alpha_i(\xi) = f_{x_i}(x_i(\xi^{i-1}), u_i, \xi_i) \cdot \alpha_{i+1}(\xi), \quad i \leq m-1, \tag{24}$$

$$\alpha_m(\xi) = g_{x_m}(x_m(\xi)); \quad \text{here } f_a, g_b \text{ are } \partial f/\partial a, \partial g/\partial b \tag{25}$$

Note that these conditions can be alternatively written in terms of the expected value of the Hamiltonian H

$$EH = Eg(x_m(\xi)) + E \sum_{i=1}^{m} \alpha_i(\xi).$$

Noting that Eq. (23) expresses the condition of minimization of the modified Hamiltonian EH, Eq. (24) expresses one of the canonical equations and Eq. (25) notes the boundary condition. On noting these points one can usually extend these necessary conditions to include the case of inequalities on the constraints by means of the Kuhn–Tucker-type conditions (see Sengupta, 1970).

For Case-II-type problems with stochastic constraints in the form $h(u_i(\xi^{i-2})) = 0$, both $u_i(\xi^{i-1})$ and $x_i(\xi^{i-1})$ are stochastic and must satisfy the necessary conditions

$$E[f_{u_i}(x_i(\xi^{i-1}), \xi_i) \cdot \alpha_{i+1}(\xi) + h_{u_i}(u_i(\xi^{i-1})) \cdot \beta_i(\xi) \mid \xi^{i-1}] = 0, \tag{26}$$

$$\alpha_i(\xi) = f_{x_i}(x_i(\xi^{i-1}), u_i(\xi^{i-1}), \xi_i) \cdot \alpha_{i+1}(\xi), \quad n \leq m-1, \tag{27}$$

$$\alpha_m(\xi) = g_{x_m}(x_m(\xi)). \tag{28}$$

Note that the expectation used in Eq. (26) is conditioned on the random vector sequence $\xi^{i-1} = (\xi_0, \ldots, \xi_{i-1})$ already observed.

Optimal discrete stochastic control problems have been used in prediction and communication theory (Box and Jenkins, 1962; Kalaba, 1962) and problems of inventory control. An illustrative model based on Naslund's dynamic investment model (Naslund, 1967) is applied in Chapter 4 to compare selection of random and nonrandom decision rules (like Case I and Case II problems) when the constraints are in the form of probabilistic inequalities.

3.2.2 ADAPTIVE CONTROL METHODS

We characterized so far stochastic cases of feedback and optimal control. Now we discuss a few cases of adaptive control, which has great potentiality of application in economic policy and growth models. An adaptive control system is one where, in addition to the state variables and the performance characteristics of the dynamic system, some means are provided to the monitor or the decision-maker for modifying the control action in order to make it more acceptable. Various types of adaptive control methods are possible due first to the fact that the system equations, the performance criterion or the constraints may be incompletely or approximately specified and secondly, to the fact that there is uncertainty in the model in the sense that the probability distributions of random elements in the system are unknown, or incompletely known.

Some examples of adaptive control problems of the first type are the following.

(A) *Parameter Adjustment Through a Reference Model*

In this case, (McGrath et al., 1961) we have a reference model with its dynamic equations which specify the desired relation between outputs and inputs (control), so far as the parameters of the equations are concerned. The system parameters of the given model are then adjusted to conform with those of the reference model by appropriate choice of control. For instance, assume the following second-order differential equation in $x(t)$ which may represent the reduced form equation

$$\ddot{x}(t) + a_1 \dot{x}(t) + a_0 x(t) = r(t), \tag{29}$$

(dot = time derivative) of a model of economic regulation discussed by Phillips (Phillips, 1954; Allen, 1959; Sengupta, 1970) based on multiplier and acceleration principles, where $r(t)$ is a certain function of the control variable $u(t)$ and its time derivatives. In the Phillips model, the parameters a_1, a_0 are functions of the saving coefficient s, the acceleration coefficients v and k and the coefficient of adjustment m as follows (see Sengupta, 1970):

$$a_1 = ms + (1/k) - mv/k, \qquad a_0 = ms/k.$$

Now assume the desired or the reference model expressing the desired path of $x(t)$ by

$$\ddot{x}^*(t) + a_1^* \dot{x}^*(t) + a_0^* x^*(t) = r(t), \tag{30}$$

where the asterisk denotes desired values. Define the deviations or errors as

$$e(t) = x(t) - x^*(t), \tag{31}$$

and consider varying the parameters a_1, a_0 of the given model of Eq. (29) (e.g., through decomposing the saving coefficient $s = s_1 + s_2$, where s_2 may be con-

3.2 Stochastic Control

trolled by the government through tax rates) so as to conform to those of the reference model of Eq. (30) such that a performance criterion in terms of the errors,

$$J(e) = \tfrac{1}{2}(w_0 e(t) + w_1 \dot{e}(t) + w_2 \ddot{e}(t))^2, \quad (w_i\text{'s: suitable constant weights}) \quad (32)$$

is minimized. Choices of various appropriate performance measures more general than Eq. (32) are discussed in the control literature (Aizerman, 1963; Feldbaum, 1960). Note that these parameter adjustment techniques are useful only when the deviations $a_1 - a_1^*$, $a_0 - a_0^*$ are not too large and there are some empirical basis (e.g., experiment) of choosing the weights w_i in the performance criterion.

(B) Adjustment Through Initial Conditions and the Planning Horizon

As an example, assume the following linear system

$$\dot{x}(t) = b(t)x(t) + c(t)u(t), \quad x(t), \; u(t)\text{: scalar}$$
$$x(0) = a, \text{ constant}, \quad u(t) = \text{control}.$$

The complete solution of this system is

$$x(t) = aw(t) + \int_0^t w(t)w^{-1}(s)c(s)u(s)\, ds,$$

where $w(t)$ is the homogeneous solution satisfying

$$\dot{w}(t) = b(t)w(t), \quad w(0) = 1$$

and $w^{-1}(t)$ is the inverse of $w(t)$. Now, if instead of starting at time zero, the system starts at arbitrary time h with the arbitrary (initial) condition $x(h)$, then the output $x(h + \tau)$ at future time $h + \tau$ can be described dy using the complete solution,

$$x(h + \tau) = w(h + \tau)w^{-1}(h)x(h) + \int_h^{h+\tau} w(h + \tau)w^{-1}(s)c(s)u(s)\, ds.$$

We now choose a suitable performance critirion $J(u)$ based on the desired input $u^*(t)$ and output $x^*(t)$. For example, the quadratic scalar function

$$J(u) = \int_h^{t+\tau} [\alpha(s)\{x^*(s) - x(s)\}^2 + \beta(s)\{u^*(s) - u(s)\}^2]\, ds, \quad (33)$$

which for specified weighting functions $\alpha(s)$, $\beta(s)$ denotes the average weighted squared error during the planning horizon $t + \tau - h$ and has been frequently used in control literature (Merriam, 1960; Wonham and Johnson, 1964). The control problem is to select $u(t)$ which minimizes the error form $J(u)$ in Eq. (33) subject to the model and the variables $x(h)$ and h. If for given $x(h)$ and h the function $J(u)$ in Eq. (33) is minimized, a further minimization could be performed in some cases by varying $x(h)$ and h optimally. Adaptive problems of this type under multivariable outputs and controls are receiving utmost attention in current research on control theory (Balakrishnan and Neustadt, 1967; Leitman, 1967).

(C) Sensitivity Analysis of Optimal Control

Suppose we derive optimal controls $u^*(t)$, the associated state vector $x^*(t)$ and the associated co-state vector $p^*(t)$ by applying Pontryagin's maximum principle and other methods. Apart from the initial and terminal boundary conditions and the constraints, these vectors $u^*(t)$, $x^*(t)$, $p^*(t)$ would be (say) functions of the parameters of the system denoted by $\theta(t)$. For example, if the system equations are $\dot{x}(t) = A(t)x(t) + B(t)u(t)$, then $\theta(t)$ comprises $A(t)$, $B(t)$. How sensitive are the optimal vectors u^*, x^*, p^* to variations in these parameter $\theta(t)$? If the variations in the parameters are random with known or incompletely known probability distributions and the optimal controls $u^*(t)$, $x^*(t)$, $p^*(t)$ are computed on the basis of expected values of all random variables, then what is the stability of these optimal control variables in terms of their variances under a fixed horizon model? Several types of sensitivity and perturbation analysis in this framework are available in the control literature (Tomovic, 1963; Balakrishnan and Neustadt, 1967; Sengupta and Fox, 1969).

Some examples of adaptive control problems of the second type may now be presented in a simplified framework.

(E) Discrete Bayesian Control

Assume a discrete system in terms of the state vector x_i, the control vector u_i, the random vector ξ_i, and the vector of parameters θ

$$x_{i+1} = f(x_i, u_i, \xi_i, \theta),$$

where ξ_i is a sequence of independent random variables. At any time k the system could be observed in terms of a vector $y_k = g(x_k, u_k, \xi_k, \theta)$, which contains noise elements represented by the random vector ξ_k. The optimal control problem at any time h is assumed to be one of minimizing a function $J(u, \theta)$ denoting the expected value of future costs, that is,

$$J(u,\theta) = E[\sum_{i=h}^{N} L(x_i, u_i, \xi_i, \theta)]; \quad E = \text{expectation},$$

subject to a class of controls in the form

3.2 Stochastic Control

$$u_i = u_i(y_0, y_1, \ldots, y_i;\qquad u_0, u_1, \ldots, u_{i-1}).$$

This means that at time k, the control u_k is selected as a function of available information at that time in terms of the observations $y_0, y_1, \ldots, y_k$ and the controls $u_0, u_1, \ldots, u_{k-1}$. The sequence $(u_0, u_1, \ldots, u_N)$ is denoted by u in $J(u, \theta)$. Now assume that for each vector θ tentatively fixed, an optimal control vector $u = u(\theta)$ can be found. However, θ is not fixed but has a prior probability distribution $P(\theta)$. Hence, $J(u, \theta)$ would vary due to θ. However, if the prior distribution of θ is known, a modified performance criterion based on $J(u, \theta)$ can be set up as

$$\hat{J} = \int_R J(u, \theta) dP(\theta), \qquad R = \text{range of } \theta \text{ in } P(\theta).$$

Now for a given prior distribution $P(\theta)$ of θ, the stochastic control problem is to determine a set of control vectors $u = u_0, u_1, \ldots, u_k$ which minimizes $\hat{J}$. If such a control exists, it would be called a Bayes optimal control using the prior distribution $P(\theta)$. The determination of Bayes optimal controls has two distinct phases. At the first phase we have to estimate the conditional distribution or the posterior distribution (for $k = 0, \ldots, N$)

$$P(x_k, \theta \mid z_k),$$

where $z_k = z_k(y_0, y_1, \ldots, y_k;\quad u_0, \ldots, u_{k-1})$, the information available up to k.
(This can be recursively carried till we reach the initial distribution of x_0). At the second phase, we have to determine the control as a function of the information, that is,

$$u_k = u_k(z_k).$$

Dynamic programming methods are most suitable for computation of this sort of Bayes optimal policies. Various control problems related to this framework are discussed in the theory of discrete stochastic control, specially the dual control theory of Feldbaum (1960) and the control problem as stochastic approximation in the sample space (Schultz, 1965; Chernoff, 1968).

(F) Games Against Nature

This is a case where the unknown vector θ is such that its probability distribution is not known but the decision-maker (or the statistician) has a set of strategies $u \in U$ against nature's set of strategies $v \in V$ and the loss function of the decision-maker $L(U, V)$ depends on nature's strategies. Note that $U(\theta)$, $V(\theta)$ are functions of θ and one type of problem is to find control functions

$u^*(t)$, $v^*(t)$, $t \in [0, T]$ such that it is feasible, satisfies $\dot{x}(t) = f(x(t), u(t), v(t))$ and that

$$J(u^*, v^*) = \min_{u \in U} \max_{v \in V} L(U, V).$$

Various types of decision rules can now be conceived. For example,
Minimax Rule:

$$\min_{u \in U} \{\max_{v \in V} L(U, V)\},$$

where $L(U, V)$ may be

$$\int_0^T f_0(x(t), u(t), v(t)) \, dt.$$

Mixed Minimax Rule:

$$\min_{u \in U} \{w \min_{v \in V} L(U, V) + (1 - w) \max_{v \in V} L(U, V)\},$$

where w = pessimism-optimism index.
Min-expectation Rule:

$$\min_{u \in U} \{\text{expectation of } L(U, V) \text{ over the prior distribution of } \theta\}.$$

Minimax Randomized Rule:

$$\min_{p \in P(\theta)} \{\max_{v \in V} \int L(U, V) \, dP(u(\theta))\},$$

where the strategy U is played with probability distribution $P(u(\theta))$.

Introduction of these decision rules into a control model leads to the problem of differential games and several new sorts of questions arise. For example information availability, truncation and duration of play, and the problems of computation (Isaacs, 1965; Ho, 1965; Pontryagin, 1965; Rosenfeld, 1964). This is a new research field and the computational aspects of this method are yet to be investigated specially for large-scale systems. The experiences of application of these techniques to simple decomposition problems (Kornai and Liptak, 1965) are not very encouraging so far.

3.3 Economic Applications of Stochastic Control

From the viewpoint of operational applications of stochastic control methods to economic models (e.g., growth and planning models), there are three basic characterizations by stochastic control which are greatly useful in deterministic models. First, in economic models, the dynamic system equations (e.g., the behavior equations) are intended to be good approximations to the real economic phenomena. However, due to aggregation difficulties, they contain all kinds of specification errors. Hence, it is all the more necessary in discussions of optimal growth under aggregative deterministic models that the effects of variations in parameters like the capital-output coefficient on optimal control systems are evaluated. Second, the social welfare functional assumed in optimal growth models may also be intended as an approximation. Since a number of utility functions with positive and diminishing marginal untility can be built up and since the convergence characteristics of optimal paths under different utility functions would differ, particularly due to nonlinearities in the model, it is sometimes more important to consider sub-optimal policies (i.e., policies which are better than a standard) rather than the optimal one, provided the former are more stabilizing than the latter. As in control theory, feedback controls which impart a stabilizing tendency in the presence of random errors or noise may have to be combined with optimal controls, in the design of optimal growth and stabilization policies. However, in the stochastic plane, mere stability of the deterministic part of the model is not enough. It is known in control theory that differential equation models which are unstable (i.e., having characteristic roots with positive real parts) in their deterministic form may be quite stable in their stochastic framework. Again, for effective realization of an optimal economic policy one should analyze the aspects of stochastiic stability of the optimal growth paths discussed in Ramsay-type models. Third, most economic problems of optimal investment allocation in a dynamic multi-sectoral framework involve multiple levels of decision-making and a tremendous amount of computation. Methods of decentralization and decomposition of economic decisions have been greatly emphasized in recent times as a technique for getting the final optimal solution through a series of sub-opimal solutions linked under shadow price-guided rules. Stochastic implications of such decentralized sub-optimal solutions are particularly important in view of the risks associated with inadequate information channels, mistakes in the specification of the sector and overall constraints, and so on.

Our various economic applications in this section are selected with a view to illustrate the three basic aspects of stochastic control mentioned above. Also, we should mention that a number of empirical economic models of dynamic stochastic programming are presented in Chapter 4, which also discusses the problems of uncertainty in economic models of competitive equilibrium.

3.3.1 OPTIMAL CONTROL: SENSITIVITY AND SPECIFICATION PROBLEMS

In models of Ramsay-type optimal economic growth, we considered before the deterministic control problem in the following simple form:

$$\text{maximize} \int_0^T U(f(k) - z)\, dt$$

subject to

$$\dot{k} = dk/dt = z - nk, \qquad k(0) = k_0,$$

where z (the per capita gross investment) is the single control variable, k (the per capita capital) is the state variable, $U(\cdot)$ is the instantaneous utility function with per capita consumption as the argument, $f(k)$ is the production function satisfying neo-classical conditions, and n is the constant rate of growth of labor. In actual applications of such a model to problems of economic planning, the end point T of the planning horizon must be finite, since the parameters θ, say, [in the production function $f(k)$ and n] would definitely change over periods. (Goodwin, 1961, Bruno et al., 1966). What is the sensitivity of optimal growth paths under variations of the parameters θ both deterministically and stochastically? Again, if the integrand $U(\cdot)$ is approximated by a quadratic function in terms of a Taylor series expansion, although it is much more nonlinear, then what is the sensitivity of a sub-optimal path computed on the basis of a quadratically approximated performance criterion? It is apparent that for finite horizon ($T < \infty$) problems of investment planning, quasi-stationary (or quasi-golden rule) optimal paths are of very little practical relevance; although in the long run, we are not all dead, the parameters of production and technology may change quite significantly, as also the tastes and social preferences. We would discuss these two aspects of sensitivity in a multi-sectoral framework, where $x(t)$ is the state vector with n elements, $u(t)$ is the control vecter with m elements, and θ is the parameter vector with k elements. It is assumed that optimal control $u^*(t)$ exists and is a piecewise continuous vector function of time t. This optimal control satisfies:

(a) The system: $\dot{x} = dx/dt = f(x, u, \theta, t), \qquad f = n \cdot 1$ vector,

(b) The initial condition: $x(0) = x_0$

and minimizes the performance criterion

(c) $J = \int_0^T f_0(x, u)\, dt, \qquad T$ fixed, $\quad f_0 =$ scalar function.

(1)

For each fixed parameter vector θ_0 belonging to a set S, it is assumed that there is a unique solution $x(\theta_0, t)$ of the system in Eq. (a) satisfying Eq. (b). Now if the true (or actual) parameter vector θ is not identical with the nominal (or observed) vector θ_0, then the actual optimal trajectory will differ from the nominal one and this results in an error in the terminal state and a change in costs measured by the performance criterion. It is assumed, however, that the devia-

3.3 Economic Applications of Stochastic Control

tions of θ from θ_0 is sufficiently small such that the solution of the linearized version of the system in Eq. (a) provides a sufficiently accurate estimate of changes in optimal trajectory $x^*(t)$. This assumption is made for simplicity and the fact that linearized system equations are very frequently used in economic models. Now since the optimal control $u^*(t)$ exists and is a known piecewise continuous function, we substitute it in Eq. (a) to get

$$\dot{x}^* = h(x^*, \theta, t), \qquad x^*(0) = x_0. \tag{2}$$

Expanding the function $h(\cdot)$ in a Taylor series around the point $(x^*(\theta_0, t), \theta_0, t)$ and retaining only linear terms we obtain the variations along the optimal trajectory

$$\begin{align}
\text{(a)} \quad & \Delta \dot{x}^* = F\Delta x^* + G\Delta \theta, \quad F = [\partial h/\partial x] \quad \text{at} \quad (x^*(\theta_0, t), \theta_0, t), \\
\text{(b)} \quad & \Delta x^*(0) = 0, \quad G = [\partial h/\partial \theta] \quad \text{at} \quad (x^*(\theta_0, t), \theta_0, t),
\end{align} \tag{3}$$

where F and G are Jacobian matrices with partial derivatives of h evaluated at the point $(x^*(\theta_0, t), \theta_0, t)$. Using the transition matrix $W(t, \tau)$ the solution of the linear system in Eq. (a) can be explicitly computed as

$$\Delta x^*(\theta_0, t) = \int_0^t W(t, \tau) \cdot G(\tau) \cdot \Delta \theta \, d\tau, \quad \text{for } T \geq t \geq 0. \tag{4}$$

Similarly the change in performance criterion can be evaluated along the optimal trajectory as

$$\Delta J^* = \int_0^T \sum_{j=1}^n (\partial f_0/\partial x_j^*) \cdot \Delta x_j^*(\theta_0, t) \, dt. \tag{5}$$

Now if the sensitivity of the optimal trajectory measured by Eq. (4) and (5) is large compared to an expected standard (e. g., one standard may be that $\Delta x_i^*/x_i^* \leq 0.05$), then we might say that the optimal trajectory is not robust. In particular, the optimal trajectory may be more sensitive to a few parameters characterizing technology and productivity in economic models, say, and special attention is to be given to estimate it more efficiently or with more precise data or information.

Now suppose the variations in the parameter vector are due to random disturbances $\xi(\tau)$ which satisfy the conditions of a Markov process which is described completely by the conditional probability distribution $P(t, x \mid \tau, y)$ that at time τ the random vector $\xi(\tau)$ would belong to $[y, y + dy]$ under the condition that at time t, $\xi(\tau)$ takes the value x. Still the random variations are assumed to be small enough such that the linearized version of the system in Eq. (a) still provides a sufficiently accurate estimate of changes in optimal trajectory. Then following the same method as before, we may derive

(a) $\dot{z} = Fz + G\alpha$,
(b) $z(0) = z_0$ (6)

where z is now the vector of variations along the optimal trajectory, F and G are the same as in Eq. (a), and α is the random vector with elements α_i which are assumed to satisfy the following conditions.

(a) Each component α_i is a Gaussian white noise (i.e., zero mean with constant spectral density)
(b) $E[\alpha_i(t_1)\alpha_j(t_2)] = 0$, if $i \neq j$, [i.e., $\alpha_i(t_1)$, $\alpha_j(t_2)$ uncorrelated].

If the initial probability distribution $P(0, z_0)$ is known, then by the Markov process property, the probability distribution $P(t, z)$ can be computed from the integral

$$P(t, z) = \int_{\text{over } R_n} P(0, z_0) P(0, z_0 \mid t, z)\, dz_0, \quad (7)$$

R_n: entire range of variation.

Hence, the average variations along the optimal trajectory can be explicitly computed as

(a) $E(z_i(t)) = \int_{\text{over } R_n} z_i(t) P(t, z)\, dz, \quad i = 1, \ldots, n,$

(b) $E\Delta J = \int_0^T \sum_{j=1}^n \frac{\partial f_0}{\partial x_j^*} \cdot E(z_j(t))\, dt.$ (8)

Note that these sensitivity indicators could be interpreted as expected marginal variations along the optimal trajectory and hence, related to the static concepts of sensitivity measures for programming problems (Sengupta and Fox, 1969) which have been applied in a number of practical problems. However, the above sensitivity measures although computable in principle may be difficult to estimate numerically except for problems of lower dimensions.

Next consider a problem of determining sub-optimal controls when the nonlinear performance criterion and the nonlinear model is suitably approximated. Now we consider the system differential equations (in vectors) in the Ito-form (Ito, 1951) as

$$dx = f(x, u)\, dt + C\, dw, \quad x: n.1$$
$$u: m.1 \quad (9)$$
$$C: \text{constant matrix}$$

where w is a vector of components w_i, each following a zero mean Gaussian

3.3 Economic Applications of Stochastic Control

process with stationary independent increments, that is, independent Wiener processes for Brownian motion. We seek a feedback control function

$$u = K(x) \tag{10}$$

among some class U of admissible controls, which minimizes a performance criterion

$$J(x) = E(L(x, u)) = \int_{\text{over } R_n} L(x, K(x)) p_F(x) \, dx, \tag{11}$$

where E is expectation defined over the sample space R_n, $L(\cdot)$ is a suitable positive definite function of x and p_F is the probability density of the random process X_F determined by Eqs. (9) and (10). The class U of admissible controls is defined over the process X_F satisfying certain regularity conditions [e.g., Exx' is bounded and the stochastic process X_F determined in Eqs. (9) and (10) are stationary and Gaussian].

It is known in the theory of deterministic control (Athans and Falb, 1966) that if $f(x, u) = Ax + Bu$ and $L(x, u) = x'Mx + u'Nu$, where A, B, M, and N are constant matrices and M, N positive definite, then an optimal linear feedback exists in the form

$$u^* = -N^{-1}B'Dx(t),$$

where the matrix D satisfies the equation

$$A'D + DA - DBN^{-1}B'D + M = 0, \qquad \text{prime} = \text{transpose}.$$

Since this is easily computable one may ask whether for suitable nonlinear functions $L(\cdot)$ and $f(\cdot)$ the above could be assumed as approximations and then optimal controls based on the approximations derived. Such approximately optimal controls are called sub-optimal controls and since they are of feedback form, they are likely to be more stabilizing. This gives the motivation for seeking sub-optimal stochastic controls. This type of problem has been investigated in control literature by several authors (Athans and Falb, 1966; Ho and Brentani, 1963).

The following formulation due to Smith and Man (1969) assumes the stochastic process X_u determined through Eq. (9) by some control $u(t)$ to be stationary and "nearly" Gaussian with mean vector $\bar{x}$ and positive definite covariance matrix Q. Only linear feedback controls are considered, so that the control vector u also is a nearly Gaussian process. Define

$$y = x - \bar{x}, \qquad v = u - \bar{u}, \qquad \bar{y} \equiv Ey \equiv 0 \equiv Ev \equiv \bar{v}. \tag{12}$$

3. Stochastic Control Theory

By assumption, the probability density $p(y, Q)$ of the stochastic process X_u is nearly equal to

$$P(y, Q) \doteq (2\pi)^{-n/2} |Q|^{-1/2} \exp\left\{-\frac{1}{2} y' Q^{-1} y\right\}. \tag{13}$$

Now we have to choose a control $\bar{u}$ and a linear incremental feedback control

$$v = -K \cdot y$$

such that the performance criterion is minimized. For this purpose, the original system equations (9) and (10) are replaced by their linear equivalents. That is,

$$dy = Ay\,dt + Bv\,dt + C\,dw, \quad v = -Ky, \tag{14}$$
$$E\{f_i(\bar{x} + y, \bar{u} + v)\} = 0, \quad \text{for all} \quad i = 1, \cdots, n.$$

The row vectors a'_i and b'_i of A and B are found from the requirement that minimum mean square exists between the outputs of the nonlinear and the linear model, that is,

$$\int_{R_{n+m}} [f_i(\bar{z} + s) - \alpha'_i s]^2 p(s, Q_a)\,ds = \text{minimum}, \tag{15}$$

where

$$\bar{z}' = (\bar{x}', \bar{u}'), \quad s' = (y', v'), \quad \alpha'_i = (a'_i, b'_i),$$

$$Q_a = \begin{bmatrix} Q & -QK' \\ -KQ & KQK' \end{bmatrix},$$

from which $\alpha_i = \begin{pmatrix} a_i \\ b_i \end{pmatrix}$ could be solved if Q_a^{-1} exists. Again, the objective function is replaced by its equivalent

$$J(\bar{x} + y, \bar{u} + v) = E[L(\bar{x} + y, \bar{u} + v)]. \tag{16}$$

However, since y and v are Gaussian random vectors with zero mean, the expectation of the function L may be expanded in a convergent power series and hence expressed as a weighted sum of the elements of the matrix Q_a. Hence Eq. (16) becomes

$$J = L(\bar{x}, \bar{u}) + E[s' R(\bar{x}, \bar{u}) s], \tag{17}$$

R: the weight matrix of order $n + m$.

3.3 Economic Applications of Stochastic Control

The optimization problem is now decomposed into formally separate problems, one static the other dynamic.

Static problem: For fixed Q_a we have to choose $\bar{u}$ so as to

$$\text{minimize } J(\bar{x}, \bar{u}) = L(\bar{x}, \bar{u}) + E\{s'R(\bar{x}, \bar{u})s\}$$
$$\text{subject to } E\{f_i(\bar{x} + y, \bar{u} + v)\} = g_i(\bar{x}, \bar{u}, Q_a) = 0.$$

Dynamic problem: For fixed $\bar{x}$ and $\bar{u}$ we have to choose the control matrix K and thus the quantities s and Q_a to minmize

$$J(y, v) = E[s'R(\bar{x}, \bar{u})s]$$

subject to the first two equations of (14).

The final solution is obtained only when we find a set $(\bar{x}, \bar{u}, K, Q_a)$ which simultaneously solves both problems. Several types of iterative algorithms are discussed in the literature (Smith and Man, 1969).

The above analytical characterization of stochastic optimal feedback controls is presented to illustrate two points. First, even under very simple types of stochastic assumptions the problems of numerical calculation of such sub-optimal controls are quite significant. Second, the stability aspects of such control are effective not simply in the deterministic but also in the stochastic framework.

In order to understand the meaning of stochastic stability imparted by a feedback sub-optimal control, consider a scalar stochastic linear differential equation

$$dx = \alpha_0 x \, dt + \alpha_1 x \, dw(t), \tag{18}$$

where x_0 is given in which α_0, α_1 are constants and the component $w(t)$ of $dw(t)$ denotes a Brownian motion process (i.e., zero mean Gaussian process with stationary independent increments). Suppose α_0 is positive so that the deterministic system derived from Eq. (18) is not stable in the Lyapunov sense, since $x(t)$ tends to infinity as $t \to \infty$. However, the stochastic model in Eq. (18) has the solution

$$x(t) = x_0 \exp[(\alpha_0 - \alpha_1^2/2)t + \alpha_1 w(t)], \tag{19}$$

which is stable in the Lyapunov sense if $\alpha_0 < \alpha_1^2/2$. Also the solution in Eq. (19) when interpreted in the sample space has stability of the kth moment, that is,

$$E[x^k(t)] = x_0^k \exp[(\alpha_0 - \alpha_1^2/2)kt + \alpha_1^2 k^2 t/2] < \infty,$$

as $t \to \infty$ if $\alpha_0 < \alpha_1^2(k - k^2)/2$, $k = 1, 2, \ldots$. Hence, for $k = 2$, the condition $\alpha_0 < -\alpha_1^2$ is sufficient. A word about the economic model where such a stochastic differential equation as Eq. (18) may arise. In Section 3.1.2 we noted Solow's contribution to growth, where the stability of the following scalar differential equation

$$\dot{k} = dk/dt = sf(k) - nk \qquad (20)$$

was discussed, where k is per capita capital stock, $f(k)$ is the neoclassical production function, n is the constant rate of growth of labor, and s is the constant marginal propensity to save. Note, however, that the production function $f(k)$ is highly aggregative and to that extent the estimate of variable k may be regarded stochastic. That is, $k + \xi$, where ξ is a Gaussian (i.e., normal) random variable with zero mean. Assuming $f(k + \xi)$ to be of the form $b_1 k + b_2 k \xi$, with b_1, b_2 constants and defining $dw(t) = \xi \, dt$ where $w(t)$ is a Brownian motion stochastic process, we get the differential equation in Eq. (18) by equating $x = k$, $\alpha_0 = sb_1 - n$, $\alpha_1 = b_2$.

The reasons why stochastic differential equations may have stochastic stability (or stability in its different moments) in spite of the deterministic system being unstable are basically due first to the fact that stochastic differentials are subject to special rules called Ito-rules (Kailath and Frost, 1968) and secondly to the nonlinearity of the stochastic process involved (e.g., the so-called logistic process shows a divergence of deterministic and stochastic stability; see Sengupta and Tintner, 1964).

In multivariable systems, there has been considerable amount of work on the stochastic stability of stochastic differential equation systems (Kushner, 1965, 1967; Bucy, 1965; Kozin, 1965, Samuels, 1960). Consider a general system of differential equations

$$dx/dt = f(x, t, y(t)), \qquad (21)$$

where x, f are vectors with n components, f is continuous and satisfies $f(0, t, y) = 0$ and a Lipschitz condition and the $y(t)$ has appropriate interpretation as a stochastic process [i.e., as $dy(t)$] possessing well behaved sample functions so that the sample equations can be treated like ordinary differential equations. The following result has been proven by Bertram and Sarachik (1959) for the system in Eq. (21).

Theorem 1 If there exists a Lyapunov function $V(x, t)$ defined over the state space $x(t)$ which satisfies the conditions

(a) $V(0, t) = 0$.

(b) $V(x, t)$ is continuous in x and t and the first derivatives with respect to x and t exist.

(c) $V(x, t) \geq \alpha \cdot \|x\|$, for some constant $\alpha > 0$, and $\|x\|$ = absolute value norm of x.

(d) $E[dV(x(t), t)/dt] < -g(\|x\|)$, where $g(0) = 0$ and $g(\|x\|)$ is an increasing function of the norm of x.

Then the equilibrium solution possesses asymptotic stability of the first mean in the large.

Note that these results are merely analogs of the deterministic theory of stability by Lyapunov methods, the applications of which in various economic models (Uzawa, 1961; Morishima, 1964; Fox et al., 1966) are well known. However, in the stochastic plane the following concepts are useful.

(a) Lyapunov stability in probability: The equilibrium solution $x(t; x_0, t_0)$ having norm $\|x\|$ as $\sum_i |x_i|$ is said to possess stability in probability if given ϵ, $\bar{\epsilon} > 0$, there exists a positive $\delta = \delta(\epsilon, \bar{\epsilon}, t_0)$ such that $\|x_0\| < \delta$ implies

$$\text{prob}\, [\|x(t; x_0, t_0)\| > \bar{\epsilon}] < \epsilon.$$

(b) Lyapunov stability of the mth mean: The equilibrium solution is said to possess stability of the mth mean, if given $\epsilon > 0$, there exists a positive $\delta(\epsilon, t_0)$ such that $\|x_0\| < \delta$ implies

$$E[\|x(t, x_0, t_0)\|_m] < \epsilon,$$

where E is expectation and $\|y\|_m$ denotes $\sum_{i=1}^n |y_i|^m$.

(c) Exponential stability of the mth mean: The equilibrium solution is said to have exponential stability of the mth mean if there exists a positive δ and constants α, β such that $\|x_0\| < \delta$ implies for all $t \geq t_0$

$$E[\|x(t, x_0, t_0)\|_m] \leq \beta \|x_0\|_m \cdot \exp[-\alpha(t - t_0)].$$

In the above definitions x_0 denotes the initial state of x at the initial time t_0 and the equilibrium solution may refer to quasi-stationary solutions as described in Section 3.1.2 in relation to optimal growth theory.

Note, however, that if the y-process in Eq. (21) is a stationary Markov process with a finite number of states $(y_1, \ldots, y_r)$ such that the probability of transition from state y_i to state y_j in time Δt is

$$\text{prob}\,(y_i \text{ to } y_j) = \alpha_{ij} \Delta t + o(\Delta t), \quad \alpha_{ij}\colon \text{constant},\ i, j = 1, \ldots, r.$$

Then the $\{x, y\}$ process becomes a Markov process and the Lyapunov function can then be defined as a stochastic quantity $V(x, t, y)$ and one could determine specific conditions on the quantity $dE[V(x, t, y)]/dt$ or the conditional quantity $dE[V(x, t, y) \mid x(t) = x, y(t) = y_i, t]/dt$ to guarantee stochastic stability. Techniques for using these considerations of stochastic stability in the design of adaptive control methods are available in control literature (Aizerman, 1963; Rekasius, 1961).

As an application of the above idea of stability in the mean, consider the variable $x(t)$ as per capita capital stock, subject to a stochastic birth and death process discussed before in Chapter 2. For simplicity $x(t)$ is considered a discrete variable and the following assumptions are made of the transition probability $p_x(t) = \text{prob}\,[x(t) = x]$ for $x = 0, 1, \ldots$.

(i) Changes in value of $x(t)$ during a small interval between t and $t + \Delta t$ are subject to the assumption of stationary independent increments. That is the transition from x to $x + 1$ is given by $\lambda_x \Delta t + o(\Delta t)$, from x to $x - 1$ by $\mu_x \Delta t + o(\Delta t)$.

(ii) The probability of no transition to a neighboring value is $1 - (\lambda_x + \mu_x)\Delta t + o(\Delta t)$.

(iii) The probability of transition to a value other than the neighboring value is $o(\Delta t)$.

(iv) The transition from the initial state $x = 0$ is not possible, that is, the state $x = 0$ is an absorbing state in the language of Markov chains.

Here $o(\Delta t)$ denotes terms which become negligible in the limit as $\Delta t \to 0$. Note also that the initial state could also be the equilibrium or the stationary state, since it is arbitrary. Under the above assumptions a recurrence relation for the transition probability can be derived as follows.

$$p_x(t + \Delta t) = \lambda_{x-1} p_{x-1}(t) + [1 - (\lambda_x + \mu_x \Delta t] p_x(t) + \mu_{x+1} p_{x+1}(t) + o(\Delta t).$$

Taking the limit $\Delta t \to 0$

$$\frac{dp_x(t)}{dt} = \lambda_{x-1} p_{x-1}(t) - (\lambda_x + \mu_x) p_x(t) + \mu_{x+1} p_{x+1}(t), \tag{22}$$

where the initial conditions are

$$p_x(0) = \delta_{xx_0} = \begin{cases} 1 & \text{for } x = x_0 \text{ where } x_0 \text{ is the value of } x(t) \text{ at } t = 0, \\ 0 & \text{otherwise.} \end{cases}$$

As we noted before (see Chapter 2) the parameter λ_x, μ_x which depend on the state x are called the birth and death rate parameters, since the former may lead to positive growth and the latter to decay. Now assume

$$\begin{aligned} \lambda_x &= \lambda x, & \lambda &= a(h_2 - x) \\ \mu_x &= \mu x, & \mu &= b(x - h_1), \quad h_1 < h_2 \end{aligned} \tag{23}$$

3.3 Economic Applications of Stochastic Control

where a, b, h_1, h_2 are constants such that x_0 lies in the closed interval $[h_1, h_2]$.

The assumptions (23) imply that birth rates may fall and death rates may rise. Note that in the vicinity of the stationary solution k^* of the Solow model (20) of balanced growth, this sort of assumptions are needed for stability of k^*. For example, it is required for stability around k^* that for $k > k^*$ we have $nk > sf(k)$ (e.g., death rate higher) and for $k > k^*$, we have $sf(k) > nk$ (e.g., birth rate increases faster). Now, if we combine Eqs. (23) with (22) we get a system of difference-differential equations, the explicit solutions of which in terms of a probability function $p_x(t)$ is yet unknown. However, in this case it has been shown (D. Kendall, 1949; Takashima, 1956) that the mean value function $m(t) = Ex(t)$ satisfies the following differential equation

$$dm(t)/dt = \left\{(a+b)\left[\frac{ah_2 + bh_1}{a+b} \cdot m(t) - m^2(t)\right] - (a+b)v(t)\right\} \quad (24)$$

where $v(t)$ is the variance function of the stochastic process $x(t)$. Note that if $x(t)$ would have been treated deterministically [e.g., the logistic growth model considered by Haavelmo (1954)] then we would get

$$dm(t)/dt = (a+b)\left[\frac{ah_2 + bh_1}{a+b} \cdot x(t) - x^2(t)\right]. \quad (25)$$

It is apparent that $m(t)$ is not equal to $x(t)$, so long as the variance function is other than zero. In fact $m(t) < x(t)$ as $v(t) > 0$. This shows that the method of replacing stochastic variables by their expectations may not always lead to a deterministic system having identical steady-state characteristics.

Another important use of the formulation of Eq. (24) is to utilize the means and variances in a portfolio allocation type model, particularly when the parameters a, b, h_1, h_2 of the birth and death processes can be influenced by other control variables (e.g., public investment, changes in tax rates, etc). Such models have been considered in the economic literature by Samuelson (1969) and Merton (1969) for lifetime portfolio allocation problems of an individual's wealth. A more interesting use of this formulation lies in testing the sensitivity of optimal growth paths in the simple aggregative framework. For instance, denote $m(t) = Ex(t)$ by $k(t)$ the capital per head, gross investment per head as $z = z(t)$ and the per capita consumption by c, then consider the optimum growth model as

$$\text{maximize } J = \int_0^\infty \exp(-\delta t) \cdot U(c)\, dt$$

subject to

$$\dot{k} = g_1 k - g_2 k^2 - g_3(t) - nk,$$

$k(0)$ given where $c = f(k) - z$ and the terms g_1, g_2, g_3 are from Eq. (24). That is, $g_1 = ah_2 + bh_1$, $g_2 = a + b$, $g_3(t) = (a + b)v(t)$ and the constant proportional rate of growth of labor is n. Also the variable $z = \dot{k} + nk = f(k) - c$, where $f(k)$ is related to the neoclassical production function. The optimal trajectory in this case would depend on the assumptions made about the term $g_3(t)$ which reflects the variance of the $x(t)$-process. The cases with $g_3(t) > 0$ may be compared with those assuming $g_3(t) = 0$ and there is considerable scope for applying simulation techniques to evaluate the sensitivity of the approximate optimal trajectory. Note, however, that since $g_3(t)$ is likely to be positive, it would tend to stabilize the dynamic movement of $k(t)$, for instance, $g_3(t)$ may be a positive linear function of $k(t)$.

3.3.2 APPLICATIONS OF ADAPTIVE CONTLOL

We have already referred to the technique of parameter adjustment as a method of adaptive control. We consider now a very general method of deriving and applying this technique. Assume that we have a differential equation system in linear form,

$$\dot{x}_i = \sum_{j=1}^{n} a_{ij} x_j, \quad i = 1, \ldots, n \quad (\text{i.e., } \dot{x} = Ax), \tag{26}$$

where there is no constant inhomogenous term, as it is absorbed in the definition of each x_i measured from some standard value. This system model of Eq. (26) may arise in two ways. First, if we have a feedback control

$$u(t) = \text{control} = -Kx, \quad K = \text{matrix}$$

applied to the system

$$\dot{x} = Bx + Cu(t), \quad B, C = \text{appropriate matrices}, \tag{27}$$

we would obtain Eq. (26) with $A = B - CK$. However, the problem is that, although the elements of the matrices B and C of the system model in Eq. (27) are known, the elements of matrix k are usually not known. The decision problem is how to select the unknown elements of the matrix K when there is no explicit performance criterion given. Second, consider an optimal control model with a quadratic performance criterion with no inequality constraints on the state and the control vector, and suppose there exists an optimal control $u^*(t)$ and the associated state and co-state vectors $x^*(t)$ and $p^*(t)$. Now, if there are errors in the system due to the presence of random noise or otherwise (i.e., deviations between the desired and the actual trajectory), the optimal trajectory may have perturbations; for instance, if it had oscillations before the noise, the oscillations may be augmented due to noise. One way to improve the control design in this case is to introduce a separate criterion of stability of system behavior and then consider variations of the control or its parameters in order to attain stability as defined. Such adjustments in the strategy of optimal control are usually necessary when the sample data are used to estimate the (population) parameters of the system model.

(A) Adaptive System Analysis

Various criteria of stable system behavior could be introduced (Schultz and Rideout, 1961; Rekasius, 1961) as a measure of stability. However, we consider here a simple quadratic integral estimate due to Aizerman (1963), that is,

$$J = \int_0^\infty [x^2(t) + k^2 \cdot \dot{x}^2(t)]\, dt, \qquad (28)$$

where $\dot{x} = dx/dt$, $x(t) =$ scalar, $k =$ constant,

because it avoids obtaining strongly oscillatory processes and it is closely related to the Lyapunov theory of stability. The system parameters [e.g., a_{ij} in Eq. (26)] must be so selected that the integral estimate J in Eq. (28) is minimized. Also, it is assumed that the system to which Eq. (28) is applied is stable in the sense that $x(\infty) = 0$ (i.e., the origin could be interpreted as a steady state value). Now express Eq. (28) as the difference of two integrals,

$$J = \int_0^\infty [x(t) + k\dot{x}(t)]^2 - 2k \int_0^\infty x(t)\dot{x}(t)\, dt,$$

where the last integral can be reduced to $-\tfrac{1}{2}x^2(0)$ using integration by parts and the condition $x(\infty) = 0$; and the first integral can be minimum if

$$x(t) + k\dot{x}(t) = 0, \qquad (29)$$

since the integrand is always positive. Hence, the minimum value of J denoted by J^{**} is attained as

$$J^{**} = kx^2(0), \qquad (30)$$

if the system parameters can be so chosen or varied that $J = J^{**}$. From Eq. (29) it follows that if such a process of varying system parameters is possible, then this limit process will be defined by $x(t) = x(0) \exp(-t/k)$. The value of k may also be chosen so that the exponential $x(t) = x(0) \exp(-t/k)$ will satisfy the stability behavior with a known tolerance. In particular, the value $k = 0$ defines a simple quadratic integral estimate in Eq. (28) which could be related to the certainty equivalence theorem (Theil, 1957, 1964; Simon, 1956) under some assumption of the error generation process. Note that we have indicated the minimum value of J in Eq. (30) by double asterisks. This is deliberately done to differentiate it from the minimum value of the same integral J^* computed on the basis of nominal values of, say, the parameters θ. The integral estimate J^{**} is such that the transient process of the system coincides with the exponential $x^{**} = C_0 \exp(-t/k)$, C_0 is an appropriate constant. If the difference between these two integral estimates is denoted by $\epsilon = J^* - J^{**}$, where x^* and x^{**} denote the coordinates associated with J^* and J^{**}, respectively, then it can be easily shown (Aizerman, 1963) that

$$\epsilon = J^* - J^{**} = \int_0^\infty \left[\Delta x^2 + k^2 \cdot \left(\frac{d\Delta x}{dt}\right)^2\right] dt, \tag{31}$$

where $x^* - x^{**} = \Delta x$ and

$$\Delta x^2 \leq \frac{\epsilon}{k}, \qquad \|\Delta x\| \leq (\epsilon/k)^{1/2}. \tag{32}$$

The second relation in Eq. (32) implies that the deviation of the nominal from the extremal processes defined by Δx lies within a tolerance level of $(\epsilon/k)^{1/2}$.

Now consider a general multivariable case, where we have a n-variable dynamic system of Eq. (26) and a general integral quadratic estimate

$$J = \int_0^T V \, dt, \qquad \text{where} \quad V = \sum_{i=1}^n B_i x_i^2. \tag{33}$$

[Note that the case of Eq. (28) is obtained by $x_1 = x$, $x_2 = \dot{x}$, $B_1 = 1$, $B_2 = k^2$, and $B_i = 0$ for $i \geq 3$.] Choose another integral R such that

$$dR/dt = -V, \qquad V \, dt = -dR \tag{34}$$

and then

$$J = -[R(\infty) - R(0)] = R(0) \tag{35}$$

if we assume that the system is stable, that is, $x_i(\infty) = 0$, $i = 1, \ldots, n$. For determining $R(t)$ we seek it in the quadratic class, that is,

$$R = \sum_{i,j=1}^n M_{ij} x_i x_j = x'Mx, \tag{36}$$

where the constants M_{ij} of matrix M are so determined that Eq. (34), namely,

3.3 Economic Applications of Stochastic Control

$$\frac{dR}{dt} = \sum_{i=1}^{n} \frac{\partial R}{\partial x_1} \dot{x}_i = -V. \tag{37}$$

is satisfied. For instance, from Eqs. (36), (37), and (33)

$$\sum_{i=1}^{n} \frac{\partial R}{\partial x_i} \dot{x}_i = \sum_{i=1}^{n} [\sum_{j=1}^{n} M_{ij} x_j] \dot{x}_i = -\sum_{i=1}^{n} B_i x_i^2.$$

Equating coefficients on both sides we obtain a system of linear algebraic equations for determining all the unknown coefficients M_{ij} of R in Eq. (36) as a function of the system parameters. Hence, we can express $R(0)$ defining J in Eq. (35) as a function of the system parameters θ say. Then θ could be varied in order to minimize J.

As an application of the above type of adaptive control, consider the following per capita version of the aggregative model of economic regulation discussed by Phillips (see Sengupta, 1970a):

$$\dot{x}_1 = -b_1 x_1 + m x_3 + m u,$$
$$\dot{x}_1 = x_2,$$
$$\dot{x}_3 = b_2 x_2 - b_3 u + b_4 x_1,,$$

where x_1 = real income per capita, x_3 = per capita gross investment, u = per capita government expenditure and the parameters (assumed constant) are $b_1 = ms + n$, $b_2 = v/k$, $b_3 = n + (1/k)$, $b_4 = vn/k$ where s = saving coefficient, n = proportional rate of growth of labor exogenously given, v, k = parameters of the induced investment function and m = coefficient of output adjustment. Assuming for simplicity that controls are proportional, that is, $u = gx$ where g is an unknown constant, we obtain from above the system equation

$$\ddot{x}_1 + a_1 \dot{x}_1 + a_2 x_1 = 0, \tag{38}$$

where $a_1 = b_1 - mb_2 - mg$, $a_2 = mb_3 g - mb_4$.

We replace this single second-order equation by an equivalent two-equation system

$$\dot{x}_1 = x_2 \quad \text{and} \quad \dot{x}_2 = -a_1 x_2 - a_2 x_1.$$

Let

$$V = x_1^2 + k^2 x_2^2 \quad \text{and} \quad R = M_1 x_1^2 + M_{12} x_1 x_2 + M_2 x_2^2.$$

Applying Eq. (37) we get

$$\sum_{i=1}^{2} \frac{\partial R}{\partial x_i} \dot{x}_i = (2M_1 x_1 + M_{12} x_2) x_2 + (2M_2 x_2 + M_{12} x_1)(-a_1 x_2 - a_2 x_1)$$
$$= -(x_1^2 + k^2 x_2^2).$$

Equating coefficients on the left and right, we obtain

$$-a_2 M_{12} = -1, \quad M_{12} - 2a_1 M_2 = -k^2, \quad 2M_1 - a_1 M_{12} - 2M_2 a_2 = 0,$$

the solution of which is

$$M_{12} = 1/a_2, \quad M_2 = \left(\frac{1}{a_2} + k^2\right) \cdot \frac{1}{2a_1}, \quad M_1 = \frac{1}{2a_1 a_2}(a_1^2 + a_2^2 k^2 + a_2),$$

hence,

$$J = \int_0^\infty (x_1^2 + k^2 \dot{x}_1^2)\, dt = R(0)$$
$$= M_1 x_1^2(0) + M_{12} x_1(0) x_2(0) + M_2 x_2^2(0), \tag{39}$$

where $x_1(0)$, $x_2(0)$ denote the initial values of x_1 and $\dot{x}_1$ at $t = 0$.

Note that the measure of system stability J in Eq. (39) is a function of the control variable g and the index k which may be thought of as weights. In particular, if there is a value g^{**} of g which is feasible (i.e., $0 < g < 1$ for economic realism) and which minimizes J in Eq. (39), then this value denoted by J^{**} can be compared with any other nominal value g^* which is feasible and which leads to J^*. Also, if the time series of income x_1 has cyclical fluctuations or oscillations over quarters, the weight coefficient k could be so selected as to give more importance to reduction of those undesired fluctuations; after a proper value of k is chosen, an optimal value of the proportion g of the feedback control can be determined by minimizing the instability performance criterion in Eq. (39). Note that the above criterion could be applied in the short-run framework also for $T < \infty$, although the minimum value of J would be slightly more complicated in this case. Also if there are random elements in the parameters but with small magnitudes, then under certain regularity conditions, the performance criterion J could be replaced by its expected value and then we would have problems of designing controls with stochastic stability of the mean.

The above type of adaptive control methods may be easily applied to problems of optimal growth under two or more capital goods. For instance, consider the following three-sector single-technique model due to Bruno (1967):

$$g_{00}c + g_{01}z_1 + g_{02}z_2 \leq 1 \quad \text{(labor constraint)}, \tag{40}$$
$$g_{10}c + g_{11}z_1 + g_{12}z_2 \leq k_1 \quad \text{(first sector capital constraint)}, \tag{41}$$

$$g_{20}c + g_{21}z_1 + g_{22}z_2 \leq k_2 \quad \text{(second sector capital constraint)}, \quad (42)$$

$$\dot{k}_1 = -\lambda_1 k_1 + z_1 \quad \text{(first sector capital accumulation path)}, \quad (43)$$

$$\dot{k}_2 = -\lambda_2 k_2 + z_2 \quad \text{(second sector capital accumulation path)}, \quad (44)$$

where the variables c, z_i, k_i represent in per capita terms, consumption good, two capital goods ($i = 1, 2$) and capital allocation in sector $i = 1, 2$ respectively; $\lambda_i = n + \mu_i$, where $n = $ proportional rate of growth of labor, $\mu_i = $ depreciation rate for capital good of type $i = 1, 2$ and g_{ij} are the allocation ratios of labor and capital. Assuming g_{ij} to be given, one could determine the optimal paths of k_i, z_i and the associated co-state vectors p_i if they exist, in terms of the Hamiltonian function,

$$H = \exp(-\delta t)\left[c + \sum_{i=1}^{2} p_i(z_i - \lambda_i k_i)\right], \quad \delta = \text{exogenous discount rate},$$

assumed, subject to the conditions in Eqs. (40) through (42). Bruno (1967) has discussed various cases where the optimal trajectory in terms of $k_1(t)$ and $k_2(t)$ may have around its equilibrium (i.e., stationary or singular) point k_i^* the following types of behavior: stable node, unstable node, saddle point, stable, or unstable focus (i.e., spiral approach due to complex characteristic roots). Some of the latter characteristics, specially the spiral and unstable behavior, could be modified if not eliminated through considering two sets of values g_{ij}^*, g_{ij}^{**} of the coefficients g_{ij} ($i, j = 0, 1, 2$), where g_{ij}^* is the nominal set and g_{ij}^{**} is the optimal set which minimizes the performance criterion of stability defined before in Eq. (33). Also the deviation $\epsilon = J^*(g_{ij}^*) - J^{**}(g_{ij}^{**})$ of the two values of the performance criterion may be analyzed in terms of the tolerance measures built into the system.

Two remarks may be added about the criterion function of Eq. (28) for the adaptive behavior assumed as a quadratic integral estimate. First, if we have a desired (or ideal) system response (e.g, in terms of either the feedback or the optimal trajectories satisfying the necessary and sufficient conditions of Pontryagin's maximum principle), which is different from the actual system response, one could introduce the system error e as a function of this difference and introduce a generalized performance index proposed by Aizerman (1963),

$$J = \int_0^\infty \left[e^2 + \sum_{i=1}^{n} k_i^{2i}\left(\frac{d^i e}{dt^i}\right)^2\right] dt, \quad (45)$$

where $d^i e/dt^i$ denotes the ith-time derivative of error e and the constants k_i are to be determined by the differential equation system (containing n coordinates) of the ideal system response. Several other generalized forms alternative to Eq. (45) have been proposed in the control literature (Ruubel, 1959; Dostupov,

1957; Rekasius, 1961). The following due to Rekasius (1961) is computationally very useful:

$$J = \int_0^\infty \left[x + \sum_{i=1}^m k_i \left(\frac{d^i x}{dt^i}\right) \right]^2 dt, \qquad m \leq n, \qquad (46)$$

where $x = x(t)$ is a scalar representing the state of the system appropriately measured (e.g., a single nth order linear differential equation computed from n first-order linear differential equations). The absolute minimum of this performance index in Eq. (46) and hence the optimum system is obtained in this case when

$$x + k_1(dx/dt) + \ldots + k_m(d^m x/dt^m) = 0, \qquad m \leq n. \qquad (47)$$

But since Eq. (47) represents the characteristic equation of the ideal (or desired) system of linear differential equations, the weight indices $k_1, \ldots, k_m$ may be easily analyzed in terms of the system parameters. Also the performance criterion of Eq. (46) would select as optimum any system of differential equations which satisfy Eq. (47). That is, the need for adjustment of the characteristic roots for securing system stability becomes readily apparent. An economic application which readily comes to mind is in the field of economic planning models based on modified Leontief-type input-output relations (Lange, 1957; Johansen, 1960; Frisch, 1960; Fox et al. 1966). The contribution by Lange (1957, 1960) introduced an investment allocation matrix I_{ij} into the usual Leontief-type static input–output model

$$x_i = \sum_{j=1}^n a_{ij} x_j + C_i + I_i, \qquad I_i = \sum_{j=1}^n I_{ij}, \qquad \Delta x_j = b_{ij} I_{ij}, \qquad (48)$$

where for the ith sector $x_i = $ gross output, $C_i = $ consumption, $I_i = $ investment demand, $a_{ij} = $ current input-output coefficients, $b_{ij} = $ investment coefficients which indicate the amount by which capital stock produced in the ith sector and installed in (or transferred to) the jth sector must increase in order that in the latter sector output should increase by one unit per year. Given a set of initial conditions $x_i(0)$ and a given time path for consumption $C_i = C_i^*(t)$), the differential equation system derived from Eq. (48), that is,

$$-\sum_{j=1}^n \beta_{ij} \dot{x}_j + x_i - \sum_{j=1}^n a_{ij} x_j = C_i^*(t) \qquad (49)$$

where $\beta_{ij} = 1/b_{ij}$, $\Delta x_j = \dot{x}_j$, $i, j = 1, \cdots, n$), would completely determine the time path of $x_i(t)$. Suppose the desired time path of consumption which the central planning agency intends to follow is given by

3.3 Economic Applications of Stochastic Control

$$C_i^{**}(t) = C_i(0) \exp(h_i r t), \tag{50}$$

where $C_i^{**} = C_T^{hi}$, $\dot{C}_T/C_T = r$, $C_T =$ aggregate consumption, and $C_i(0) =$ initial value.

The deviations $[C_i^*(t) - C_i^{**}(t)]$ of nominal from desired time paths of consumption cannot be reduced or modified unless some of the coefficients or parameters in Eq. (49) are flexible. Indeed, a closer look reveals that b_{ij} (or its reciprocal) is not entirely technologically determined, since we could write

$$\Delta x_j = \sum_{i=1}^{n} b_{ij} g_{ij} I_i, \quad g_{ij} \geq 0, \quad \sum_{j=1}^{n} g_{ij} = 1, \quad I_j = \sum_{i=1}^{n} g_{ij} I_i, \tag{51}$$

where g_{ij} indicates the proportion of investment in sector i allocated or transferred to sector j for augmenting output in sector j. The allocated amount $(g_{ij} I_i)$ has the gross productivity of b_{ij} per unit. Hence, the system differential equations become

$$-G^{-1} B^{-1} \dot{X} + (I - A) X = C^{**}(t),$$
$$G = (g_{ij}), \quad B = (b_{ij}), \quad A = (a_{ij}), \quad C^{**}(t) = (C_i^{**}(t)).$$

Here, however, the allocation ratios g_{ij} in matrix G provide the decision variables, which could be adjusted to minimize a performance measure like Eq. (46), where the system error $e = \sum_i w_i(C_i^*(t) - C_i^{**}(t))$, w_i being suitable nonnegative can be easily defined. This type of parameter adjustment technique can also be applied to various types of macrodynamic policy models available in the economic literature (Theil, 1964, 1967; Stone and Brown, 1962; Chenery and Bruno, 1962; Chenery and MacEwan, 1966).

As a second remark, we should point out that this type of adaptive control design is not limited to either linear models like Eq. (26) or to the integrand of the performance criterion in Eq. (28) which does not involve the control variables. The following formulation due to Letov (1961) discusses a very general problem of finding a class of optimal continuous scalar functions $f(\sigma)$, where σ is a scalar function of $(x_1, x_2, \ldots, x_n$ and u, where u is the single control variable and x_i is the ith state variable), that is,

$$\text{minimize } J = J(f(\sigma)) = \int_0^\sigma [\sum_{i=1}^n k_i x_i^2 + k_0 u^2 + \dot{u}^2] \, dt$$

subject to

$$\dot{x}_i = \sum_{j=1}^{n} b_{ij} x_j + c_i u, \quad \dot{u} = f(\sigma),$$
$$x_i(0) \text{ given, } i = 1, \ldots, n, \quad u(0) \text{ given,}$$

$$x_i(\infty) = 0, \text{ all } i = 1, \cdots, n,$$

and the class of $f(\sigma)$ satisfying

$$\sigma f(\sigma) > 0, \quad \sigma \neq 0, \quad \text{and} \quad (df/d\sigma)\underset{\sigma=0}{>} 0.$$

Here k_i, k_0 are suitable weights that can be adjusted and b_{ij}, c_i are fixed coefficients. Defining the Lagrangian as

$$L = \sum_{i=1}^{n} k_i x_i^2 + k_0 u^2 + f^2(\sigma) + \lambda_0\{\dot{u} - f(\sigma)\} + \sum_{i=1}^{n} \lambda_i \left\{ x_i - \sum_{j=1}^{n} b_{ij} x_j - c_i u \right\},$$

where λ_0, λ_i are Lagrange multipliers and applying the Euler–Lagrange necessary conditions for minimum, we obtain

$$\dot{x}_i = \sum_j b_{ij} x_j + c_i u, \quad \dot{u} = f(\sigma),$$
$$\dot{\lambda}_i = -\sum_j b_{ij} \lambda_j + 2k_i x_i,$$
$$\dot{\lambda}_0 = 2k_0 u - \sum_j \lambda_j c_j, \quad i, j = 1, \ldots, n,$$
$$0 = [2f(\sigma) - \lambda_0] \cdot (\partial f/\partial \sigma).$$

There are quite interesting problems of stability of this system for the two phases $2f(\sigma) - \lambda_0 = 0$ and $(\partial f/\partial \sigma) = 0$. Also, if there exists an optimal trajectory system satisfying the above conditions and it is asymptotically stable in the large in the Lyapunov sense (La Salle, 1962; Brockett, 1966), then the optimal system is called "absolutely stable" or "structurally stable," since these systems are then stable for a whole class of functions $f(\sigma)$ and they possess a strong stability property relative to perturbations of f.

We feel that the above design of optimal feedback controls is directly applicable to a large number of macrodynamic policy models in economics where, like government expenditure, there is only one control variable and, often for simplicity, a linearized system of differential equations is used as a substitute. Monte Carlo methods using alternative forms of the function $f(\sigma)$ have great scope of application here; also the various types of stability and how to incorporate these in an adaptive performance criterion offer a challenging field of research.

(B) Adaptive Allocation Process

The dynamics of economic growth through capital accumulation may be viewed in terms of resource allocation, transfer, and substitutions (e.g., postponement of current consumption for augmenting capital accumulation). In models of aggregative growth used for economic planning, the allocation of investment

3.3 Economic Applications of Stochastic Control

between sectors has attracted considerable interest in the economic literature. In its very simplest form the problem of optimal sectoral allocation of total investment (I) at a given time period t may be represented in terms of a discrete version of a Harrod–Domar model. That is,

$$y_t = \left(1 + s \sum_{i=1}^{n} \lambda_i \beta_i - n\right) y_{t-1}, \quad 0 \leq \lambda_i \leq 1, \quad \sum_i \lambda_i = 1, \tag{52}$$

where y is per capita real income, s is marginal saving-income ratio, n is the proportional rate of growth of labor exogenously given, β_i is the incremental output-investment ratio for ith sector output, and λ_i is the proportion of total investment allocated to sector i (i.e., $I_i/I = \lambda_i$). A large number of investment allocation models for economic growth and planning have been built along the lines indicated in Eq. (52) in a more generalized framework, for example, using the case of open economy and foreign borrowing, nonlinear production functions, introduction of exogenous performance criteria, or social welfare functions over a planning horizon, and so on. (Mahalanobis, 1955; Radner, 1966; Fox et al., 1966). However, all these generalizations mostly assume that the coefficients β_i are technically given and known with certainty like the other parameters s and n. However, in real life, like portfolio allocation problems involving a number of assets of varying profitability, these coefficients β_i have significant variability due to several reasons. For example, each sector itself is highly aggregative, information process is not perfectly distributed and the estimation process specially in less-developed countries may have a high degree of subjective and non-market imputation. The empirical studies (Chenery, 1963; Eckaus and Parikh, 1966) also indicate the variability of these output-investment coefficients. If there are two sectors in an aggregative model, agriculture and the rest, including industry, services, and so on, then it is easy to see that the output-investment ratio for agriculture would have higher variability year to year due to random variations in weather, irrigation and uneven diffusion of technical knowledge. Similarly if the two sectors are consumption goods and capital goods, the output investment ratio would be more stable for the latter (i.e., would have less variance). There exists empirical evidence (Sengupta, 1963) to support these inferences. How should an optimal decision be made in the face of such random variability of coefficients like β_i? A number of approaches may be suggested.

(a) A stochastic process analogue of the deterministic model in Eq. (52) could be developed and utilized to specify the time paths of expected value and variance functions of the y_t process; the methods of stochastic control are relevant here.

(b) A stochastic programming approach could be utilized if the probability distribution of each β_i could be estimated or approximated and a performance criterion like the expected value of y_T where T is the end year of the planning horizon is imputed. Several empirical applications of such models are discussed in Chapter 4.

(c) One could also utilize an adaptive method of improving successive estimates of $\theta = (\beta_i)$ given the sequence of observations.

We would discuss very briefly the case (c) in its two aspects. For example, revision of prior estimates and the use of robustness as a criterion of estimation. First, if the decision-maker has prior estimates of the parameters $\theta = (\beta_i)$, then he may follow a Bayes approach and search for good decision rules restricted to the class of non-randomized decision rules. For instance, let x be the set of observations for estimating θ and $g(\theta \mid x)$ be the prior distribution subjectively held by the decision-maker conditional on the observation x with a conditional density $dg(\theta \mid x)$; then an optimal Bayes decision rule is to choose a decision rule $d(x)$ for estimating θ which minimizes the posterior conditional expected loss defined by

$$r(D(x), g(\theta)) = \int L(\theta, D(x)) \, dg(\theta \mid x), \tag{53}$$

where

$$L(\theta, D(x)) = \text{the random loss function},$$

under the restriction that the decision rule $D(x)$ belongs to a class D^0 specifying the deterministic restrictions on $D(x)$. Sometimes the minimum in Eq. (53) may not exist, then we seek an optimal decision rule $D(x)$ which comes within ϵ of achieving the minimum of $r(\cdot)$ in Eq. (53). Such decision rules are called optimal Bayes rules to order ϵ. For an example, assume the distribution of x given θ to be normal $N(\theta, 1)$ with mean θ and variance unity. Let the loss function $L(\theta, D(x))$ be equal to $(\theta - D(x))^2$ and the prior distribution of θ be normal $N(0, \sigma^2)$ with zero mean and variance σ^2 and assume for simplicity that θ is a scalar (i.e., one parameter case), then the joint density $h(\theta, x)$ of θ and x, the marginal density $f(x)$ of x and the posterior density of θ given x can be derived as (Wetherill, 1966; Sworder, 1966; Weiss, 1961):

$$h(\theta, x) = \frac{1}{2\pi\sigma} \exp\left[-\frac{(x-\theta)^2}{2} - \frac{\theta^2}{2\sigma^2}\right],$$

$$f(x) = [2\pi(1+\sigma^2)]^{-1/2} \cdot \exp\left[-\frac{x^2}{2(1+\sigma^2)}\right],$$

$$dg(\theta \mid x) = \left(\frac{1+\sigma^2}{2\pi\sigma^2}\right)^{1/2} \cdot \exp\left[-\frac{1+\sigma^2}{2\sigma^2}\left(\theta - \frac{x\sigma^2}{1+\sigma^2}\right)^2\right], \tag{54}$$

that is,

3.3 Economic Applications of Stochastic Control

$$N\left(\frac{x\sigma^2}{1+\sigma^2}, \frac{\sigma^2}{1+\sigma^2}\right).$$

Using these in Eq. (53) and minimizing r in Eq. (53) with respect to $D(x)$ we obtain the optimal Bayes rule

$$D(x) = \frac{x\sigma^2}{1+\sigma^2}, \text{ where } r \text{ becomes } \frac{\sigma^2}{1+\sigma^2}.$$

The above case can be generalized to cases when θ is a vector and the loss function $L(\cdot)$ is positive semi-definite quadratic. In this case the Bayes estimate of θ which was $x\sigma^2/(1+\sigma^2)$ in Eq. (54) becomes θ^*, that is, the mean of θ conditioned on the set X of observations

$$\theta^* = \frac{1}{f(X\mid\theta)} \cdot \int \theta f(X\mid\theta) f(\theta)\, d\theta, \tag{55}$$

where by Bayes' theorem

$$f(\theta\mid X) = \text{conditional density of } \theta = \frac{f(X\mid\theta)f(\theta)}{f(X)}, \tag{56}$$

where

$f(X) = \int f(X\mid\theta) f(\theta)\, d\theta$ is the normalizing factor,

$f(\theta) = $ prior distribution multivariate normal.

However, in the general case, it may be very difficult to evaluate the conditional densities $f(X\mid\theta)$ as they involve the mean value for X which would depend on time also. However, when the posterior distribution $f(\theta\mid X)$ of θ given X is unimodal and symmetric, then the Bayes estimate of θ as the conditional mean becomes equivalent to the mode. That is, that $\hat{\theta}$ which will maximze

$$\max_{\theta} f(X\mid\theta) f(\theta) = f(X\mid\hat{\theta}) f(\hat{\theta}). \tag{57}$$

However, for most economic problems in growth planning, information in the form of prior distribution of θ is frequently not available. Also the fact that the Bayes estimate depends on the form of the loss function $L(\cdot)$ and prior distribution of θ points to the need for analyzing the effects of departure from the assumptions of normality of prior distributions. Some work on the robustness or sensitivity of alternative estimates of parameters in simple one variable linear models are available in the economic literature (Zellner and Geisel, 1968;

Orcutt, 1968). In the general multivariable case, more work is needed to establish the computability of Bayes estimates (which are known to be computationally easier than minimax type decision rules) under varying information on the set X. Simulation methods (Wetherill, 1966; Hoggatt and Balderston, 1963; Orcutt, 1968) readily come to mind here.

Now consider some other particular cases of the model of Eq. (52) mentioned before, where β_i's are assumed to be random. Assume the initial value y_0 of income to be known and fixed and then define

$$g_t = \frac{1}{t} \log(y_t/y_0) = \log(1 + sr - n), \qquad r = \sum_i \lambda_i \beta_i \qquad (58)$$

$$\bar{g}_t = \frac{1}{t} E \log(y_t/y_0), \qquad E = \text{expectation}. \qquad (59)$$

Here $\bar{g}_t$ is the expected future rate of income growth for the next t years and it is explicitly computable if the probability distribution of r is given. The decision rule which maximizes $\bar{g}_t$ in each period t can be easily formulated as a nonlinear programming problem, that is,

$$\underset{\{\lambda_i\}}{\text{maximize}} \; \bar{g}_t \qquad (60)$$

in Eq. (59) subject to

$$\lambda_i \geq 0, \qquad \sum_i \lambda_i = 1.$$

This type of formulation has been investigated by Murphy (1965) for protfolio allocation problems of a particular investor, when the expectation E in Eq. (59) is interpreted as subjective (i.e., in terms of the subjective probability of $r = \sum \lambda_i \beta_i$).

To indicate other formulations alternative to that suggested in Eq. (59), assume that $(sr - n)$ is small in magnitude so that the approximation $\log(1 + sr - n) \doteq sr - n$ holds and let r be assumed normal with mean $\sum \lambda_i \bar{\beta}_i$ and variance $\lambda' \beta \lambda$, where β is the variance-covariance matrix. Then g_t defined in Eq. (58) is approximately normal with mean $(s \sum_i \lambda_i \bar{\beta}_i - n)$ and variance $(s^2 \cdot \lambda' \beta \lambda)$. Using these we could set up the traditional mean-variance type analysis (Markowitz, 1959; Tobin, 1965; Feldstein, 1969, Farrar, 1962):

$$\underset{\{\lambda_i\}}{\text{maximize}} \; s \sum \lambda_i \bar{\beta}_1 - n$$

subject to

$$\lambda_i \geq 0, \qquad \sum_i \lambda_i \leq 1, \qquad (61)$$

$$s^2 \lambda' \beta \lambda \leq v_0 \qquad \text{(pre-assigned)}.$$

3.3 Economic Applications of Stochastic Control

This formulation is possible in an asymptotic sense, even if the approximation $\log(1+a) \doteq a$ does not hold, since the quantity $\log(1+sr-n)$ may tend to be normally distributed (i.e., $1-n+sr$ is log-normally distributed) under the conditions of the central limit theorem.

Another alternative formulation is the safety-first approach (Roy, 1952; Sengupta, 1968, 1969a) which has been applied to various economic problems (Sengupta and Fox, 1969). In this case we define a pessimistic level g^0 say of relative income growth which the decision-maker intends to avoid. However, the form of the probability distribution of g_t (or r) is not known. In this case we use the Tchebycheff inequality

$$\text{prob}\,[g_t \leq g^0] \leq \frac{v}{(m-g^0)^2}, \qquad \begin{array}{l} m = \text{mean of } g_t \\ \\ v = \text{variance of } g_t \end{array} \qquad (62)$$

and in default of knowing the probability on the left-hand side of Eq. (62) we minimize the right-hand side according to the safety-first approach, which only requires the mean m and variance v of g_t to exist. This is equivalent to

$$\underset{\{\lambda_i\}}{\text{maximize}}\,(m-g^0)^2/v, \qquad \text{subject to } \lambda_i \geq 0, \quad \sum_i \lambda_i = 1. \qquad (63)$$

This leads to the area of stochastic and chance-constrained programming which is discussed in great detail in Chapter 4.

Note, however, that we have considered in Eq. (52) only the scalar case. However, the general case is not difficult to formulate in principle. For example, a linear logarithmic model due to Radner (1963, 1966) is discussed in Section 4.1 where the output $z_j(t)$ of commodity j at end of period t is assumed to be given by a Cobb–Douglas type production function with inputs $f_{ij}(t)z_i(t-1)$ and parameters b_j and a_{ij}, that is,

$$\log z_j(t) = b_j + \sum_{i=1}^{n} a_{ij} \log(f_{ij}(t)z_i(t-1)); \qquad i,j = 1,\cdots,n, \qquad (64)$$

where $f_{ij}(t)$ denotes the nonnegative allocation ratios satisfying $f_{ij}(t) \geq 0$, $\sum_i f_{ij}(t) \leq 1$ for all t. The instantaneous one-period welfare function used in one version of the model is

$$\lim_{t \to \infty}(v_t/t),$$

where

$$v_t = \sum_{j=1}^{n} w_j \log z_j(t), \qquad w_j = \text{fixed weights}.$$

In the stochastic case, however, a_{ij} would be random and if the initial condition $\log z_j(0)$ is given in this case and the welfare function Ev_t/t is maximized with respect to $f_{ij}(t)$ under its allocation restrictions, then this would be very similar to the approach followed in Eq. (60), but it would be much more difficult in a computational sense. Similarly, other approaches like Eqs. (61) and (63) can, in principle, be investigated for the general system Eq. (64).

(C) Adaptive Two-level Optimization

For large-scale programming problems, particularly linear and quadratic programming at the static level, methods of decomposition or decentralization have been recently emphasized (Dantzig and Wolfe, 1960; Kornai and Liptak, 1965; Sengupta and Fox, 1969) as techniques for solving the optimization problem in two stages interlinked in some fashion. At the first stage, a set of smaller sub-problems are solved conditional on a set of shadow prices (or provisional allocations) supplied by the central coordinating agency. At the second stage, the central coordinating agency solves an aggregated optimization model and recomputes the shadow prices (or the revised allocations) to be passed on to the sectors which are responsible for solving the sub-problems. The sequential revision of the shadow prices (or the allocated quotas) is continued till we reach the optimum optimorum; that is, when the global maximum of the master problem is reached.

For large-scale programming models applied to national planning or to the determination of optimal economic policy, the decomposition techniques certainly have a great interest for the economists, because the method of successive revision of shadow prices (or quota allocations) appears to be strikingly similar to the tatonnement processes of a competitive market. However, it is important to recognize that there are very many basic questions which are yet unsolved in these decentralization techniques. (a) The convergence of the successive process of revision may be neither assured nor monotonic. (b) The entire process may be very inefficient, since the convergence may be very slow at the start and there are more than one type of rule of adjustment like gradient methods, the generalized simplex, and so on. (c) The economic meaning of imputing a shadow price to a non-market resource is not yet very clear. (d) The method assumes a perfect information channel, where price signals given by the center are correctly interpreted by the sectors.

At the deterministic level, optimal control models are also amenable to decomposition algorithms based on gradient techniques and a number of attempts (Pearson, 1966; Mesarovic et al., 1965) and economic applications (McFadden, 1969; Kulikowski, 1969) are reported in the literature. In the approach by Pearson, the system model is linear, for example,

$$\dot{y}_i = dy_i/dt = A_i y_i + B_i x_i + c_i u_i, \qquad i = 1, \ldots, N, \tag{65}$$

3.3 Economic Applications of Stochastic Control

$$x_i = N_i d_i + \sum_j N_{ij} y_i, \quad N_{ii} = 0, \quad j, i = 1, \cdots, N, \quad (66)$$

with N sub-systems, where y_i (state vector), u_i (control vector) and x_i (interaction vector) are vectors for each i; A_i, B_i, C_i are appropriate constant matrices, N_i, d_i, N_{ij} are appropriate constants. The optimal control problem is to choose a set of controls u_i, $i = 1, \ldots, N$ over the planning horizon $[0, T]$ which minimize a performance criterion of the form

$$J = \sum_{i=1}^{N} \left[g_i(y_i(T)) + \int_0^T f_i(y_i, x_i, u_i) \, dt \right] \quad (67)$$

subject to inequality constraints: $R_i(y_i, x_i, u_i) \leq 0$.
Assuming the functions g_i, f_i, R_i convex with continuous second derivatives for its arguments and the absence of singular cases, the optimal control $u_i(t)$ with piecewise continuity may be shown to exist under certain conditions (Lee and Markus, 1961). Assuming that optimal controls exist, the dynamic optimization problem can be decomposed into two stages.

Stage one: Here N parametric sub-problems are solved using provisional shadow price vectors (i.e., vector of Lagrange multipliers) associated with Eq. (66).

State two: Here the N independent sub-problems are coordinated by an aggregate model which recomputes the shadow price vectors in order to search for better solutions in terms of lower values of the performance criterion.

Here there is one sub-problem for each subsystem and one coordinating problem for each group of interacting sub-systems. However, the computational experiences with such two-level dynamic optimizing techniques are not yet very promising and the effects of truncation in the iterative process are yet unknown. Note that if the model in Eq. (65) and (66) contains any noise or random elements, then the objective function in Eq. (67) may be interpreted as expected performance criterion; the two-level computation procedure could be characterized in such cases also (Morris, 1968). But the dimensionality of the sub-problems and the central problem would be considerably increased.

The latter problem can be better understood in terms of a discrete time control problem. For example,

$$\text{minimize } J = EL(x_N, u_{N-1}) = E\left[g(x_N) + \sum_{i=0}^{N-1} f(x_i, u_i, \xi_i) \right]$$

subject to

$$x_{i+1} = F(x_i, u_i, \xi_i), \quad i = 0, 1, \ldots, N-1$$
$$x_0 = \text{constant}, \quad N \text{ fixed}, \quad (68)$$
$$R(x_i, u_i, i) \leq 0,$$

(E: expectation; all x_i, ξ_i: $n \cdot 1$; all u_i: $m \cdot 1$).

178 3. Stochastic Control Theory

Here $i = 1, \ldots, N$ is the index of time, x_i is the state vector (which comprises both the central and the sectoral states) and u_i is the control vector (which comprises both the central and sectoral controls); it is assumed that the functions g, f, F, and R are convex (f strictly convex) with continuous second derivatives for its arguments and the random vector ξ_i must satisfy the regularity conditions mentioned before in Sections 3.2 and 3.3. Note that the inequality constraint $R \leq 0$ does not have random components. Defining the vectors

$$y = (x_1, \ldots, x_N, u_0, \ldots, u_{N-1})'$$
$$F_1(y, \xi) = [f(x_0, u_0, \xi_0) - x_1, \ldots, f(x_{N-1}, u_{N-1}, \xi_{N-1}) - x_N]',$$
$$F_2(y) = [R(x_0, u_0), \ldots, R(x_{N-1}, u_{N-1})]',$$

the above problem can be written as the following nonlinear program, where ϕ is a suitable function derived from Eq. (68)

$$\text{minimize } J = E\phi(y, \xi) \tag{69}$$

subject to

$$F_1(y, \xi) = 0, \quad F_2(y) \leq 0.$$

The gradient method as developed by Arrow and Hurwicz (1959) when applied to this nonlinear program has a decomposition interpretation (Sengupta and Fox, 1969) well known in the theory of decentralization of economic systems. Note that this formulation Eq. (69) of decomposition framework is different from that of Pearson's approach in two respects. First, this formulation seeks to define a decomposition procedure (e.g., Kornai and Liptak, 1965) even when the state and control vectors are not distinguished at the sectoral and the central levels to start with [e.g., Pearson's approach is basically the Dantzig–Wolfe (1960) type decomposition algorithm applied to optimal controls]. Second, the fact that non-randomized optimal controls are sought as solutions is clearly indicated here. Introducing appropriate slacks if necessary, the above problem may be rewritten in a slightly different form as

$$\text{minimize } J = E\phi(X, U, \xi), \tag{70}$$

subject to

$$F(X, U, \xi) = 0,$$

where X, U contains respectively the state and control vectors and the function F subsumes the two sets of restrictions of Eq. (69). Now suppose the random vector ξ is discrete and it takes the value $\xi^{(k)}$ with probability p_k, $k = 1, \ldots,$ M. Then for any given k, we solve the deterministic problem

$$\text{minimize } J^{(k)} = \phi(X, U) \quad \text{subject to} \quad F(X, U, \xi^{(k)}) = 0.$$

Let $H^{(k)} = \phi(X, U) + \lambda^{(k)'} F$, where $\lambda^{(k)}$ = Lagrange multiplier vector, then the necessary conditions of optimality are

$$\partial H^{(k)}/\partial X = 0, \quad \partial H^{(k)}/\partial U = 0, \quad F(X, U, \xi^{(k)}) = 0. \quad (71)$$

Now since each $\xi^{(k)}$ appears with probability p_k; therefore, the necessary condition for the entire problem must satisfy

$$\sum_{k=1}^{M} p_k [\partial H^{(k)}/\partial X] = 0, \quad \sum_{k=1}^{M} p_k [\partial H^{(k)}/\partial U] = 0, \quad F(X, U, \xi) = 0. \quad (72)$$

The great computational problem is how to start from arbitrary but feasible control vectors $U(t)$ at computation time t and converge to the optimal one U^* in finite horizon. The gradient-type algorithms based on the gradient vector $\partial H(t)/\partial U(t)$ search for a process defined in terms of a system of difference equations

$$U(t + 1) = U(t) - h(t) \cdot [\partial H(t)/\partial U(t)], \quad (73)$$

where the gradient $\partial H/\partial U$ is to be evaluated from an initially chosen sample ξ and the necessary conditions of Eq. (71) and $h(t)$ is a scalar representing the size of the correction at each step which must be such as to make the new control $U(t + 1)$ feasible. For convergence of this process in Eq. (72) to optimal U^* several conditions are required. For instance, $h(t) \to 0$ as $t \to \infty$, $\sum_{t=1}^{\infty} h(t) = \infty$, $\sum_{t=1}^{\infty} h^2(t) < \infty$, finite variance for each component of ξ. This leads us to the theory of stochastic approximation characterized by the Robbins–Monroe process (Wetherill, 1966; Leondes, 1965), which specifies the detailed mathematical conditions needed for the stochastic convergence of $U(t, s)$ to U^* in the sample space indexed by s.

Two general remarks may be added. First, if decomposition and aggregation are viewed as two facets of an optimization process, then it has been argued in discussions of national planning (Ellman, 1969; Wolfe, 1967) that aggregation in real life is not always perfect and that it is a compromise to avoid inconsistency. In programming terms, this means that if the submodels are given (i.e., their restrictions and performance criteria are given), then the aggregate model should attempt to obtain aggregate solutions like least squares solutions, for instance, by minimizing the squared deviation from some desired goal in aggregate terms (Sengupta, 1970; Kuhn, 1963). Sometimes, of course, no feasible solution may exist because the restrictions in different submodels may be mutually conflicting. In this case, the aggregate solution may be sought as a least squares-like or least absolute deviations-like solution from a desired value (Sengupta, 1970c). Also, if the restrictions of the submodels or the center are probabilistic in the sense of chance constraints, such approaches may be profitable to consider. This sort of problem may arise typically in programming models for non-market institutions like educational system planning (Sengupta and Fox, 1970; Alper and Smith, 1967; Alper, 1967). Second, the decomposition algorithms usually assume two sets of decision-makers, one at the level of sectors or sub-problems and the other at the center as coordinator. However, if there are some parameters like θ in the models which is not completely fixed, then it is very important to analyze the characteristics of the information channel, through which the estimates of θ held at one level are transmitted to other levels. This leads in a general framework to problems known as team decisions (Radner, 1962; Marschak, 1966), where again Bayes-type estimates may be characterized (given the prior estimates) and compared with minimax type estimates.

3.4 Problems in Stochastic Control

This section discusses very briefly some basic operational problems of applying stochastic control problems, particularly in economic systems and organizations for problems like dynamic resource allocation, growth planning and economic coordination. The problems are classified into three interrelated groups: specification and modeling, estimation and filtering, and computation and attainability problems.

3.4.1 SPECIFICATION PROBLEMS

(A) Distributed Parameter Systems

All the control models discussed so far, whether deterministic or stochastic, have assumed the form of specification as a set of differential equations

$$\dot{x} = dx/dt = f(x, u, \theta, \xi), \qquad (1)$$

where $x =$ state, $u =$ control, $\theta =$ parameter, $\xi =$ random noise, and where the derivative of the state variable which appears on the left-hand side of Eq. (1) indicates the time rate of change of the state vector x. However, an ordinary vector differential equation may not represent an adequate mathematical model for dynamic systems which are characterized by an implicit system of equations as

3.4 Problems in Stochastic Control

$$F(x, \dot{x}, u, \dot{u}, \theta, \xi) = 0, \tag{2}$$

for which a set of partial differential equations can, of course, be defined by expanding F in its arguments x, $\dot{x}$, u, $\dot{u}$. These systems of partial differential equations offer much more general formulations than those in Eq. (1). The former system is called distributed parameter systems in control theory (Wang, 1964; Sakawa, 1966) as distinct from the latter, which is called lumped parameter system.

The reason why distributed parameter systems offer an important field of research and application can be better understood in terms of the so-called problem of indeterminacy arising in models of optimal growth with several heterogenous capital goods (Hahn, 1966; Samuelson, 1967). Suppose we have n capital goods K_i ($i = 1, \ldots, n$) and m consumption goods C_j ($j = n+1, \ldots, m$) and we set up a Ramsay-type optimum growth model which seeks to

$$\text{maximize } J = \int_0^\infty U(C_{n+1}, \ldots, C_{n+m}, t)\, dt$$

subject to

$$\begin{aligned}
&G[\dot{K}_1, F^{(1)}(K_{11}, \ldots, K_{n1}; L_1)] = 0,\\
&\quad\vdots\\
&G^{(n)}[\dot{K}_n, F^{(n)}(K_{1n}, \ldots, K_{nn}; L_n)] = 0,\\
&G^{(n+1)}[C_{n+1}, F^{(n+1)}(K_{1,n+1}, \ldots, K_{n,n+1}; L_{n+1})] = 0,\\
&\quad\vdots\\
&G^{(m)}[C_m, F^{(m)}(K_{1m}, \ldots, K_{nm}; L_m)] = 0,
\end{aligned} \tag{3}$$

where $U(\cdot)$ is a suitable concave instantaneous social welfare function, $G^{(s)}$ ($s = 1, \ldots, n, n+1, \ldots, m$) are "implicit functions" with arguments like $\dot{K}_i$ = rate of change in capital of type i, C_j = consumption good of type j and the capital allocations K_{ij} in the production functions $F^{(s)}(\cdot)$ for the two sets of goods. The implicit functions $G^{(s)}$ are by assumption reduced to explicit forms as

$$\dot{K}_i = F^{(i)}; \quad C_j = F^{(j)}; \quad i = 1, \ldots, n; \quad j = n+1, \ldots, m \tag{4}$$

in the economic formulation (Kurz, 1968) and then solved. However, this assumption implies that explicit production functions of the form in Eq. (4) are derivable from the implicit relations of Eq. (3). This derivation may be very hard, if not impossible, to obtain through statistical estimation. Also, as Frisch

(1965) has indicated in his theory of the multiproduct firm, the parameters like θ belonging to the implicit system of Eq. (3) would not preserve their unique and invariant meanings unless a suitable technique called "reparametrization" (i.e., lumping of distributed parameters in control theory language) is imposed on the system. However, the distributed parameter systems raise very fundamental problems of computation.

(B) Approximation of Deterministic and Stochastic Parts:

In general, a nonlinear performance criterion (beyond the quadratic) and a nonlinear dynamic model with nonlinear ordinary differential equations are difficult to operate with as regards the computations and the satisfaction of both the necessary and sufficient conditions of optimality. Here there is a great need for attempting several alternative approximations to both the deterministic and stochastic parts, before particular optimal controls are recommended as robust. While linear approximations to the deterministic part would be subject to considerations of deterministic stability, those for the stochastic part would have to be guided by stochastic stability considerations. In economic models, the latter aspects have not much been emphasized so far.

(C) Control Applications in Non-market Framework:

Increasingly control theory methods are applied to non-market institutions and their dynamic problems of resource allocation for producing services like education, health, reduction of water and air pollution, and so on, where there exist no definite market categories in the same sense as the market for tangible agricultural and industrial goods. Recently, methods of stochastic control (e.g., dual control) have been considered for application to educational system planning (Alper and Smith, 1967: Sengupta and Fox, 1969; Alper, 1967; Armitage and Smith, 1967; Stone, 1965). In this field there is some need in our view in developing alternative formulations and simplifications for the complex system by identifying the important levels of decision-making and budgeting and then defining controls which are better than a standard. An optimizing direction rather than the optimum itself should be of more practical interest here. Also, techniques are needed which through "imaginary interviewing" (van Eijk and Sandee, 1959) or otherwise (e.g., imputing a class of loss functions; Theil, 1964) help in the process of deriving the barter terms of trade between the different target variables considered by the policy-makers (Fox et al., 1966).

3.4.2 PROBLEMS OF ESTIMATION

In multivariable control methods to economic models like national planning and growth, problems of statistical estimation are more complicated due to the fact that, unlike in physical models, economic observations cannot be easily generated under experimental and controlled conditions. Also it is very difficult

3.4 Problems in Stochastic Control

to know whether a particular set of statistical assumptions about the random elements (e.g., whether the noise in the system is subject to a Gaussian stochastic process or not) are satisfied by a given set of observed time series data that are non-experimental. Several partial attempts are available for tackling such difficulties. For example, (a) the use of central limit type theory which asymptotically generates under certain conditions a tendency to normality and a Gaussian process (b) the methods of simulation which vary the underlying distribution assumptions to see the effects on the solution, and (c) the reliance on feedback rather than on optimal controls. However, under simplified framework estimation procedures exist which in effect are based on sequential and iterative calculations (Bharaucha–Reid, 1960; Leondes, 1965).

3.4.3 Problems of Computation

The most important problem from a practical viewpoint is the computational realizability of stochastic optimal controls. Our brief remarks will be restricted to dynamic systems with a small number of state and control variables, since most applied economic problems are usually of this type.

(A) Characterization Problems

In our discussion of deterministic models of optimal growth in Section 3.1 of Chapter 3, we emphasized the fact that quasi-stationary solutions on the optimal trajectory are much more easily computable than the entire time profile of optimal trajectory. In problems of stochastic optimal control, such techniques are attempted in different forms. For example, forms like substiution of limited feedback in an otherwise optimal control, defining a modified performance criterion which emphasizes the steady state attained at the infinite planning horizon, suitable approximations and comparisons with known solutions of standard models (e.g., standard model may be a feedback or deterministic model) have great potentiality of application in economic models.

To illustrate, consider the reference model as a linear-quadratic regulator problem which is well known in control literature (Athans and Falb, 1966)

$$\text{minimize } J = E \int_0^T Q(x, u)\, dt$$

subject to

$$\dot{x} = [A(t) + \hat{A}(t)]x + [B(t) + \hat{B}(t)]u + C(t)\xi(t), \tag{5}$$

$$x(0): \text{given}, \quad T: \text{fixed}, \quad E: \text{expectation},$$

where x, u, ξ are the state, the control, and the noise vectors, $Q(\cdot)$ is a suitable quadratic function of x, and u and the matrices $\hat{A}(t)$, $\hat{B}(t)$ are white noise (e.g., Gaussian or otherwise) like the vector $\xi(t)$ (i.e., its elements have zero means and stationary independent increments in any time interval). Note that although $\hat{A}(t)$ and $\hat{B}(t)$ [which represent random disturbances and fluctuations around the deterministic coefficients $A(t)$ and $B(t)$] are perturbations which may be small in a small time-interval, these when associated with large deviation in the state vector or large control amplitudes may generate worse fluctuations on the optimal trajectory. For this reason, optimal controls are restricted to feedback controls of the form $u = \phi(t, x)$ as a compromise (such a compromise is basically intended to combine stability and optimality aspects). With this restriction $u = \phi(t, x)$ the problem is known to have a solution in the form

$$u^* = \text{optimal } u = K(t) \cdot x \quad \text{(i.e., linear control)}, \tag{6}$$

where the elements of the matrix $K(t)$ depend on the statistical parameters of $\hat{A}$ and $\hat{B}$ and the deterministic parameters A, B, but not on the random elements of $C \xi(t)$. However, if the perturbations in $\hat{A}$ are large, the controlling matrix $K(t)$ in Eq. (6) must also be large for imparting stability, but a large $\hat{B}$ indicates that the control vector u may have high noise elements which may be undesirable in the control rule of Eq, (6). These trade-offs between stability and optimality offer interesting economic applications in the designing of optimal stabilization and growth policies using a macrodynamic model (Sengupta, 1970; Fox et al., 1966). Similary other methods of conditioning on the noise elements in $\hat{A}(t)$, $\hat{B}(t)$ so that they do not get too destabilizing may be tried as methods of compromise. Note in particular that if $\hat{A}(t)$ and $\hat{B}(t)$ are too large and destabilizing, no solution may exist for the optimal stochastic control problem in Eq. (5).

(B) Computation Algorithms

Although there are several computational techniques available now for control problems (e.g., conjugate gradient and other reduced gradient techniques using second variation of the Hamiltonian, sequential unconstrained minimization techniques, successive approximation and quasi-linearization through dynamic programming, etc.), it is very difficult to compare their efficiencies. Most continuous time problems need discretization for numerical computation. Hence, the control problem is finally reduced to a general problem of nonlinear programming with, of course, increased dimensions. As yet there is no particular blanket technique which could be applied for all practical cases of nonlinear programming. Also, there are great problems in evaluating the effects of truncation in the sequence of iterative calculations (Bellman and Kalaba, 1965; Fiacco and McCormick, 1968; Balakrishnam and Neustadt, 1964; Plant, 1968).

(C) Simulation in Economic Applications

The recent trend in computation of stochastic optimal control problems has taken two different directions. The first is the development and comparison of

3.4 Problems in Stochastic Control

multivariable optimization methods (Tapley and Lewallen, 1967; Kopp and McGill, 1964; Balakrishnan and Neustadt, 1964; Plant, 1968). The second is the development of simulation techniques based on (a) random samples drawn from different distributions, (b) the alternative specification of stochastic convergence of the iterative calculations, and (c) the search for control policies which are in some sense robust or insensitive to some specified perturbations in the original model. Some of these simulation techniques have attracted considerable attention in current research in control engineering (Balakrishnan and Neustadt,1964; Tapley and Newallen, 1967; Plant, 1968) and current applications to economic models of growth and stabilization (Orcut, 1968; L. Buchanan and Stubberud, 1969; Kuhn, and Szego, 1969; Naylor *et al.*, 1968, 1969). However, there are very deep and fundamental questions in the approach of simulation, its criteria of comparison and usefulness for future situations. Economists are increasingly being interested in these techniques so that the design of an optimal macrodynamic policy can be better understood, analyzed, and appreciated. The days may not be very far off when a responsible dialogue would ensue between the physicists, the control engineers, and the economists for solving the problems of an optimum economic and social system.

4 STOCHASTIC PROGRAMMING METHODS WITH ECONOMIC APPLICATIONS

The theory of stochastic programming may be conceived in several different ways. As a method of programming it analyses the implications of probabilistic variations in the parameter space of linear or nonlinear programming model. The generating mechanism of such probabilistic variations in the economic models may be due to incomplete information about changes in demand, production and technology, specification errors about the econometric relations presumed for different economic agents, uncertainty of various sorts and the consequences of imperfect aggregation or disaggregation of economic variables, and so on. A second way to approach stochastic programming is to consider the theory of games, particularly matrix-games with mixed strategies when the value of the game is stochastically varying. Game-theoretic models have been discussed in economic framework in several important situations, for instance, specification of alternative market structures (Shubik, 1959), allocation of resources under alternative forms of coalition and/or competition (Luce and Raiffa, 1957), and models of decision-making under uncertainty (Borch, 1968; Ozga, 1965; Sengupta, 1969; Fellner, 1965). These methods have also been recently applied to characterize or solve the problem of optimal resource planning through decentralized decisions following some iteration sequences (Kornai and Liptak, 1965; Sengupta and Fox, 1969). Also models of optimal control have been analyzed through differential games (Isaacs, 1965; Ho, 1965). A third approach to stochastic programming is through the simulation techniques, which are essentially methods for analyzing the sensitivity and robustness of particular assumptions, solutions, or policies defined in a particular economic framework (Naylor et al., 1968; Sengupta, 1970b). In economic fields the scope of application of such techniques is most important in developing optimal economic policies and decision rules on the basis of quantitative models.

In recent applications of stochastic programming considerations to economic models, three aspects have appeared very prominent in the current literature, for instance, measures of risk aversion in utility functions and their relationships to consumer equilibrium (Borch, 1969; Stone, 1970), dynamic portfolio models of asset selection and capital-market equilibrium (Mossin, 1966; Lintner 1965; Fama, 1971), and the implications of a competitive general equilibrium model under uncertainty, when some of the conditions of Pareto optimality fail to hold (Stigum 1969, Radner 1968, Stiglitz 1970). While we have analyzed in appropriate places each of the above specific aspects, we preferred a more general framework of stochastic programming emphasizing the following in particular: (1) the set of formulations available in other fields like operations research and the general theory of decision-making under uncertainty, (2) the implications of alternative decision rules in the static and dynamic framework of linear and nonlinear models, arising out of the theory of quantitative economic policy (Sengupta and Fox, 1969), and (3) the problems of incomplete specification of the probability distribution of the outcomes underlying a linear model structure and, in particular, the need for caution exercised through adoption of distribution-free and robust solutions.

Our discussion proceeds as follows: the general framework of economic model specifications using programming methods is outlined in Section 4.1, followed by a general survey and appraisal in Section 4.2 of the current state of the theory of applied stochastic programming. Economic applications from an empirical standpoint are outlined in Section 4.3, which illustrates the following: (1) the implications of the statistical distribution of the maximand in terms of the mean variance characteristics, (2) the decision rule theory of operations research under the chance-constrained model and (3) a new method of incorporating reliability measures into a systems reliability model. Noting that the Kuhn–Tucker necessary conditions for an optimum can be interpreted in appropriate cases as conditions of equilibrium as understood in economic theory (with appropriate Lagrange multipliers interpreted as prices), the theory of applied linear stochastic programming treated in Sections 4.2 and 4.3 may be considered as models for market equilibrium under alternative types of "markets" characterized by the activity vectors constrained by various decision rules subject to risk-aversion, diversification, and decentralization. Section 4.4 analyzes the implications of dynamic portfolio models and the truncated Pareto-optimal framework of the second best. The formulations in this section appear to us to be somewhat incomplete in the sense that empirical and operational verifications are not yet available.

4.1 Programming Framework in Economic Analysis

From an operational viewpoint the methods of deterministic programming have been applied to economic models in three most important areas e.g. the theory of economic policy, problems of optimum growth and decentralized systems of resource planning and the theory of the multiproduct firm. Since stochastic programming attempts at generalizing the framework of deterministic programming, it is natural to expect that stochastic programming techniques would be most important and productive in those areas of economic applications where deterministic programming are already somewhat successful. We intend therefore to analyze in this section a few selected economic models and their programming framework.

The macro-economic models of economic growth and planning offered in the last two decades perhaps the most important field of application of various techniques of linear and nonlinear programming. At the macro-economic level the essence of a planning model consists of the integration of a model of economic growth (or development) and a policy model. In the very simplest terms, this involves adding to the growth model itself a criterion or objective function (also called preference or utility function) by means of which alternative policies can be compared and evaluated by the policy-maker or policy-makers or the planning board as the case may be. The theory of economic policy developed by Tinbergen (1952, 1955) and his followers emphasizes at this point an important distinction between the different variables of the economic model of growth, that is, between the instrument variables (also called decision variables) which are under the control of the policy-maker or planner and the target variables, which, though not directly controlled by the planner, specify the desires and overall aims of the national economy. Furthermore, Tinbergen's approach supports a strictly operationalist attitude in specifying the objective function of the policy-maker or the planners, without entering into the whole question of deriving the social welfare function from individual utility functions as in the theory of welfare economics (Arrow, 1951, 1952; Debreu, 1951, 1954; Koopmans and Bausch, 1959). In Tinbergen's policy framework a feasible policy is usually defined by a set of values of the instrument variables, which satisfies all the restrictions of the economic model (which themselves are partly formulated from relationships derived from economic theory) and/or the constraints imposed by policy-making itself e.g. using an instrument variable (i.e., government expenditure) beyond a maximum permissible level may not be politically feasible. An optimal policy then is an optimal choice (i.e., maximizing as the case may be) between alternative feasible policies by means of the criterion function of the policy-maker. If the criterion function and all the restrictions are linear, the problem of determining an optimal policy is the problem of linear programming; otherwise we may have nonlinear programming problems. The method of linear programming and related methods [e.g., input–output analysis, nonlinear problems which could be solved by simplex-like algorithms, or linear programming (LP)], have proved themselves very well in applications to concrete economic situations. Koopmans (1951), Dantzig

(1957) and others have shown how they can be used in various problems of economic planning. The interesting work on *LP* methods by the Russian mathematician Kantorovich (1959) shows the versatile applications of this technique to various problems of production, allocation, transportation, and even optimal routing of traffic. Applications to development planning models have been conceived in the economic policy models discussed by Frisch (1955), Mahalanobis (1955), Tinbergen (1956), Theil (1958), Chenery (1963), Sengupta and Fox (1969) and many others.

Recent developments in computational algorithms for nonlinear programming problems, especially the methods of geometric and generalized polynomial programming (Duffin *et al.*, 1967; Wilde and Beightler, 1967), sequential unconstrained minimization techniques (SUMT) (Fiacco and McCormick, 1968) and special methods like conjugate gradient techniques and other procedures (Balakrishnan and Neustadt, 1964; Lasdon *et al.*, 1967) used in control theory suggest that in the coming decades programming techniques would be increasingly applied to various problems of economic policy, growth, and development planning.

The current models of economic growth, both aggregative and disaggregative (Morishima, 1969; Hicks, 1965; Kaldor, 1961; Hahn and Mathews, 1964) have succeeded at least in raising some of the fundamental questions as to the specification of optimal economic policies and planning procedures. From the viewpoint of applications to various models of investment planning in different countries, one could identify some important influences of currently developed growth models. First, considerable emphasis has been given to problems of specification of intertemporal paths of capital accumulation under single and multi-sector breakdown of the overall economy. Since the pioneering work of Ramsay (1928) on the problem of optimal savings, several types of attempts have been made in recent times to generalize Ramsay-type results (Koopmans, 1960, 1967; Tinbergen, 1960; Samuelson, 1967; Hahn, 1966; Uzawa, 1966) as regards the multi-sector economy, stability of the golden-rule path, and policy interpretations. The implications of utility maximization over infinite horizons (Radner, 1967), the choice of appropriate criteria for ordering intertemporal consumption programs (Koopmans, 1967; Weizsacker, 1965), and the formulation of necessary and sufficient conditions of optimal growth under simple types of utility functionals (Shell, 1967; Gale, 1967; Mirrlees, 1967) subject to various constraints of production and the assumptions of appropriate convexities and non-saturation, are perhaps some of the most important aspects of this generalization. In the context of development planning, Goodwin (1961) has applied this framework to define an optimal growth path for an underdeveloped economy, optimal growth defined as the maximum of real per capita consumption discounted and summed over a finite planning horizon. The sensitivity, stability, and the phase-changes associated with an intertemporal optimal growth-path have been especially emphasized in Goodwin's model; these features have also been noted in multi-sector frameworks (Bruno, 1967; Sengupta, 1968). The mathematical techniques commonly used here have been calculus of variations and its extensions using Pontryagin's maximum principle (Pontryagin, 1961; Sengupta and Fox, 1969), the principle of optimality of dynamic programming (Bellman, 1957), and various methods of nonlinear control (Lee and Markus, 1961; Aris, 1963; Sengupta and Fox, 1969). Several stochastic aspects of the general problem of optimal capital accumulation have been reported in the literature [e.g., Yaari (1965) has treated the problem of uncertainty introduced by a random planning horizon, Phelps (1962), Mirrlees (1965), Levhari, and Srinivasan (1969) have analyzed the case where returns to investment and hence the sequence of capital accumulation are stochastic with a given probability generating process; Naslund (1967) has applied the approach of chance-constrained linear programming to discuss the optimal portfolio problem and the problem of optimal investment allocation over time]. In a somewhat general context, Morishima (1965) has investigated the conditions of ergodicity (i.e., a type of stationarity) of the optimal growth process in a von Neumann type model with stochastic considerations; problems of optimal stochastic control and their stability aspects have also been investigated (Buchanan, 1968, Kurz, 1968) although in a very limited context, so far as economic theory and policy are concerned.

A second important influence of the theory of economic policy applied to models of economic growth, when it is viewed in an operational context, has been to reconsider and perhaps generalize those parts of a growth model which would allow more degrees of freedom to the policy-makers in the sense that the number of instrument (or control) variables should preferably exceed the number of target variables (Tinbergen and Bos, 1962; Fox *et al.*, 1966). Problems of inter-sectoral allocation of investment (Mahalanobis, 1955; Frisch, 1955; Lange, 1957; Sandee, 1960; Kornai and Liptak, 1965) by using the input-output system directly or indirectly, the specification of detailed behavior equations rather than inequalities in terms of demand, supply and equilibrium relations and the estimation of such econometric models which preserve the property that the total number of possible instrument variables exceeds the total number of target variables (Chenery and Bruno, 1962; Klein and Goldberger, 1955; Klein and others, 1961; Netherlands Central Planning Bureau, 1956; the Brookings model 1969 of the U. S., Chenery 1961) and finally the problems of combining input-output systems with econometrically estimated behavior equations and of integrating short and medium term models with long run models (Johansen, 1960, 1967; Stone and Brown, 1962; Hart *et al.*, 1964 and others) and also of regional and partial equilibrium models with overall models of the whole economy (Fox, 1963; Leontief, 1963; Isard, 1951; Lefeber, 1958; Verdoorn, 1956; Netherlands Central Planning Bureau, 1956,

4.1 Programming Framework in Economic Analysis

1961; Chenery, 1961; Sandee, 1960; and many others). The mathematical techniques commonly used in the above problems relate to econometric and other estimation methods coupled with methods for solving simultaneous difference or differential equations (Ichimura, 1960; Radner, 1963; Fox *et al.*, 1966). Stochastic considerations are introduced in such approaches through estimation procedures and consistency checks and sometimes decision rules are followed which are either approximate or related to extraneous assumptions such as certainty equivalence (Fox *et al.*, 1966; Theil, 1964, 1967)

A most important contribution of programming techniques to the theory of growth and planning has been to develop the concept of shadow prices to sustain the sequence of optimal dynamic decisions related to optimal growth and to apply it to decentralized decision-making. The meaning of shadow price sequences in the absence of a market framework (Portes, 1969; Sengupta, 1970c), their uses in transferring resources between different divisions (or sectors) of a large corporate enterprise (or the public sector as a whole) discussed, for instance, by Malinvaud (1967), Heal (1969) and others, and their implications for continual planning revision and the stability of optimal policy (Goldman, 1968; Cass, 1965; Kurz, 1968; Sengupta, 1970d) are some of the important aspects worth mentioning. The use of provisional or interim shadow prices to allocate central resources to the sectoral needs has been compared and contrasted with several types of signalling devices, such as the decomposition algorithms of linear and nonlinear programming (Arrow and Hurwicz, 1959; Kornai and Liptak, 1965; Marschak, 1966; Hass, 1968; Sengupta and Fox, 1969), tatonnement processes of a market framework (Hahn and Negishi, 1962; Nikaido, 1964; Arrow, 1964), approximate consistency in central planning based on aggregative quantities or norms (Kornai and Liptak, 1962; Martos and Kornai, 1966; Weizsacker, 1967; Ellman, 1969), and even planning without prices but based on exchange of information about input allocations and output targets (Portes, 1969; Heal, 1969). Stochastic uses and applications of sequential pricing through shadow prices are yet very limited in economic fields, although various aspects of stochastic control and its decomposition over time and the characterization of adaptive control policies under uncertainty (Murphy, 1965; McFadden, 1969; Sworder, 1966) are reported in the literature. The decomposition procedures utilized in the active approach of stochastic linear programming (Sengupta and Fox, 1969; Sengupta and Tintner, 1970), which is analyzed in Section 4.2 are also relevant in this connection.

We consider now a few selected models for illustrating the programming framework in economic analysis. Three types of models are discussed, a model of optimal growth due to Radner (1963, 1966), a model of two-level planning due to Kornai and Liptak (1965), and a model of multi-product firm due to Dhrymes (1964). Although these models are used for illustrative purposes, they are nevertheless capable of being applied to realistic situations with minor modifications; also these models are capable of being extended without much difficulty to a stochastic framework.

First, we consider a model due to Radner (1963, 1966), which assumes an economy where currently produced goods are linear logarithmic functions (i.e., Cobb–Douglas type) of inputs and the maximizing criterion defining an optimal growth is one of several plausible functions, each of which defines a sequence of measures of "one-period welfare", welfare in a single period being a linear function of the logarithms of various goods produced or consumed. Denote the output of commodity j at period t by $z_j(t)$ and let $x_{ij}(t)$ denote the amount of commodity i that is to be allocated for production of commodity j, that is

$$x_{ij}(t) = f_{ij}(t) z_i(t-1) \tag{1}$$

where $0 \leq f_{ij}(t) \leq 1$ is the allocation ratio and the relation of Eq. (1) assumes one period time-lag. The output $z_j(t)$ of commodity j at the end of period t is assumed to be given by a Cobb–Douglas function of inputs $x_{ij}(t)$ with parameters b_j and a_{ij} that is,

$$\log z_j(t) = b_j + \sum_{i=1}^{M} a_{ij} \log(f_{ij}(t) z_i(t-1)), \tag{2}$$

$$i, j = 1, 2, \ldots, M \quad \text{and} \quad t = 1, 2, \ldots, T$$

By definition, the allocation ratios must satisfy $\sum_{j=1}^{M} f_{ij}(t) \leq 1$ apart from nonnegativity and the assumption is made that the production of each commodity exhibits nonnegative, nonincreasing marginal productivity with nonincreasing returns to scale, that is, $a_{ij} \geq 0$ and $\sum_{i=1}^{M} a_{ij} \leq 1$. Let $z(t)$ be the (M·1) vector of outputs $z_j(t)$ and let $u(z(t))$ be the real-valued nonnegative utility function which is assumed to be a linear function of the logarithms of each coordinate $z_j(t)$ of the vector $z(t)$, that is,

$$u(z(t)) = \sum_{j=1}^{M} w_j \log z_j(t), \tag{3}$$

where w_j ($j = 1, 2, \ldots, M$) are nonnegative weights (i.e., $\sum_{j=1}^{M} w_j = 1$, $w_j \geq 0$). This function $u(z(t))$ given in Eq. (3), also denoted by v_t defines one-period welfare for each single time-period t out of a planning horizon $t = 1, \ldots, T$ and the social welfare function (v) (i.e., the objective function of the planner as the case may be) over the whole planning horizon is defined over the set $v_t = u(z(t))$, $t = 1, 2, \ldots, T$. Given an initial output vector $z(0)$ at at time $t = 0$, and assuming the absence of primary resources (or exogenously available commodities), a time profile of output $(z(1), z(2), \ldots, z(T))$ is defined to be a feasible program for the linear logarithmic economy, if it satisfies in every period, the production restrictions of Eq. (2) with nonnegative allocation

4.1 Programming Framework in Economic Analysis

ratios $f_{ij}(t)$ satisfying $\sum_{j=1}^{M} f_{ij}(t) \leq 1$. Any feasible program which maximizes the social welfare function v defines an optimal program of growth. Four types of social welfare functions are distinguished.

maximize $v = \sum_{t=1}^{T} r^{t-1} v_t$; $r > 0$ is the constant rate of discount. (4)

maximize v_T = maximize v_T of the final output vector $z(T)$, given $z(0)$ and T (finite). (5)

Let the limiting value of the ratio $\lim_{t \to \infty} (v_t/t)$ exist for a class of feasible programs, then we maximize $v[\lim_{t \to \infty} (v_t/t)]$ which is nothing but maximizing the long run rate of average welfare (v_t/t). (6)

Let there be a class of feasible programs having the same value of $s = \lim_{t \to \infty}(v_t/t)$ and let the quantity $v = [\lim_{t \to \infty} (v_t - ts)]$ exist for this class of feasible programs. Then another type of optimal growth is defined by maximizing the social welfare function $v = \lim_{t \to \infty}(v_t - ts)$. (7)

Note however that the stock $z_i(t-1)$ of commodity i could be considered at the end of the period to be divided into two parts, $c_i(t)$ = consumption and $x_i(t)$ = inputs for production, that is, investment

$$c_i(t) + x_i(t) = z_i(t-1), \quad c_i(t) \geq 0, \quad x_i(t) \geq 0. \quad (8)$$

In other words, consumption and investment at time t are limited by output variable at the end of the previous period, assuming of course that any or all outputs can serve for both purposes, consumption or investment. Whenever such division of output into consumption goods and investment goods is economically meaningful, a more appropriate way to define the social welfare functions (4) through (7) to be maximized is in terms of consumption goods only. For example, taking the social welfare function (4), the decision problem now becomes

$$\text{maximize } v = \sum_{t=1}^{T} r^{t-1} (\sum_{j=1}^{M} w_j \log (c_j(t)))$$

under the conditions of Eq. (8) and

$$\log z_j(t) = b_j + \sum_{i=1}^{M} a_{ij} \log (f_{ij}(t) x_i(t)), \quad (9)$$

$z_j(t)$ at $t = 0$ is given,

where the optimal values of allocation ratios $f_{ij}(t)$ are to be determined. In the general case, this becomes a problem of nonlinear programming. However, it is apparent that if the utility function $v_t = u(z(t))$ of Eq. (3) is defined as a linear function of $z_j(t)$ (instead of a log-linear function) and similarly the production of Eq. (2) is taken in a linear form, we would have a problem of linear programming over time. Again, taking log $f_{ij}(t)$ as the decision variables, the optimizing problem with the social welfare function of Eq. (4) using Eq. (3) and the restrictions of Eqs. (2), (3), and (1) is in the form of a linear programming problem over time which could be solved by the well known techniques. Radner has derived the characteristics of the optimal programs under alternative social welfare functions of Eqs. (4) through (7) and in particular the implications of shadow prices for infinite horizon programs (Radner, 1967; Gale, 1967; McFadden, 1967). By the overtaking principle (Weizsacker, 1965) commonly used for ordering infinite programs, a feasible production program $(z(t), t = 1, 2, \ldots)$ is defined to overtake (or prove better than) another feasible production program $(\bar{z}(t), t = 1, 2, \ldots)$, if starting from the same initial vector $(z(0) = \bar{z}(0))$ there exists a time point T_0 such that for all time points $T \geq T_0$ the following holds

$$\sum_{t=1}^{T} \sum_{j=1}^{M} w_j r^{t-1} (\log z(t) - \log \bar{z}(t)) > 0. \tag{10}$$

If the feasible production program $(z(t), t = 1, 2, \ldots)$ overtakes all other feasible programs in a certain class of programs, then in that class the infinite program $(z(t), t=1, 2, \ldots, \infty)$ is defined to be optimal. The implications of other criteria of ordering infinite programs (Koopmans, 1960, 1967; Weizsacker, 1965, 1967), the stability of the optimal path when it exists (Kurz, 1968; Hahn, 1966) and its decentralization aspects (Malinvaud, 1967) have been investigated in the literature.

A few words about the generality of the above allocation model may be mentioned. First, by considering continuous instead of discrete time points and applying optimality conditions of control theory (e.g., Pontryagin's maximum principle, Bellman's optimality principle) most of the results derived in models of optimal inter-sectoral investment allocation (Bruno, 1967; Uzawa, 1966; Fox et al., 1966) can be obtained. Second, stochastic consideration may be introduced at two important stages, apart from the fact that the parameters b_j, a_{ij} of the production function have to be statistically estimated, for instance, the initial conditions $z_j(0)$ may be random (Kushner, 1967), or the production relation may be chance-constrained (Naslund, 1967) as follows

$$\text{prob} \, (b_j + \sum_{i=1}^{M} a_{ij} \log \, (f_{ij}(t) x_i(t)) \leq \log z_j(t)) \geq \alpha_j(t), \tag{11}$$

where $\alpha_j(t)$ $(0 \leq \alpha_j \leq 1)$ is the tolerance measure up to which constraint violations due to the randomness in production outcomes $z(t)$ are allowed for. Third, this allocation model is computationally very convenient for applying

4.1 Programming Framework in Economic Analysis

the techniques of geometric and generalized polynomial programming (Wilde and Beightler, 1967).

For illustration we consider a two-sector model with a private and a public sector due to Uzawa (1966) and show how geometric programming techniques could be applied (Sengupta and Portillo–Campbell, 1970a) in this framework. The outputs $Y_c(t)$, $Y_v(t)$ of the private and the public sectors depend on the respective amounts of labor $L_c(t)$, $L_v(t)$ and capital $K_c(t)$, $K_v(t)$ by the assumption of production functions subject to constant returns to scale (i.e., homogeneous of order one) and to diminishing marginal rate of substitution between capital and labor, that is,

$$Y_c(t) = F_c(K_c(t), L_c(t)),$$
$$Y_v(t) = F_v(K_v(t), L_v(t)).$$

The production functions F_c, F_v are assumed strictly quasi-concave with positive marginal products everywhere. The demand for output of private sector (i.e., private goods) is composed of consumption $C(t)$ and investment $Z(t)$ so that

$$C(t) + Z(t) \leq F_c(K_c(t), L_c(t)), \quad \text{(private sector)}. \tag{12}$$

The output $X(t)$ of the public sector (i.e., public goods) is assumed to be entirely consumed so that

$$X(t) \leq F_v(K_v(t), L_v(t)) \quad \text{(public sector)}. \tag{13}$$

The accumulation of total capital $K(t)$

$$K_c(t) + K_v(t) \leq K(t) \tag{14}$$

is related to gross investment $Z(t)$ through depreciation

$$\dot{K}(t) = Z(t) - \mu \cdot K(t), \text{ where } \dot{K}(t) = dK(t)/dt \tag{15}$$

and μ is the rate of depreciation assumed constant. The growth of labor $L(t)$

$$L_c(t) + L_v(t) \leq L(t) \tag{16}$$

is assumed exogenous with a constant proportional rate of growth n,

$$\dot{L}(t) = nL(t), \quad \text{where } \dot{L}(t) = dL(t)/dt. \tag{17}$$

It is assumed that the social welfare function is representable as the discounted sum of instantaneous utilities over time

$$W = \int_0^\infty u\left(\frac{C(t)}{L(t)}, \frac{X(t)}{L(t)}\right) e^{-\delta t} dt, \tag{18}$$

where δ is the exponent of the discounting function assumed constant and the utility function u depends on per capita consumption of private and public goods such that u is assumed strictly concave, twice differentiable with positive marginal utilities. Given initial capital $K(0) \geq 0$ and labor $L(0) \geq 0$ and the nonnegativity of all the variables, the optimum path of economic growth is defined by that feasible time path of $(K_c(t), K_v(t), L_c(t), L_v(t), C(t), Z(t), X(t))$ for which the utility functional W in Eq. (18) is maximized subject to the constraints given in Eqs. (12) through (17). If we use small letters to indicate the quantities per capita and omit the time suffix, the problem of optimum growth becomes

$$\text{maximize } W = \int_0^\infty u(c, x) e^{-\delta t} dt, \tag{19}$$

subject to the constraints

$$c + z \leq f_c(k_c) \cdot l_c \tag{20}$$

$$x \leq f_v(k_v) \cdot l_v \tag{21}$$

$$k_c l_c + k_v l_v \leq k \tag{22}$$

$$l_c + l_v \leq 1 \tag{23}$$

$$\dot{k} = z - (n + \mu)k \tag{24}$$

$k(0) \geq 0$ given, all variables nonnegative.

For numerical computation we introduce the following simplifying assumptions:

(a) $z \geq rk + (n + \mu)$,
where it is assumed $\dot{k}/k = r = 0.05$, $n = 0.04$, $\mu = 0.06$, and $k = 10$ in some units.

(b) The objective function W in Eq. (19) is replaced by a simpler comparative-static form, that is,

$$W = c^\alpha \cdot x^\beta, \quad \text{with} \quad \alpha = 0.6, \quad \beta = 0.4.$$

(c) The production functions are

$$f_c(k_c) = k_c^{0.8}, \quad f_v(k_v) = k_v^{0.9}.$$

Denoting the variables as $x_1 = c$, $x_2 = x$, $x_3 = z$, $x_4 = k_c$, $x_5 = k_v$, $x_6 = l_v$ and $x_7 = l_c$ the programming problem is then reduced to the following geometric program

4.1 Programming Framework in Economic Analysis

$$\text{minimize } V = x_1^{-0.6} \cdot x_2^{-0.4} \tag{25}$$

subject to the constraints

$$x_1 x_4^{-0.8} x_7^{-1} + x_3 x_4^{-0.8} x_7^{-1} \leq 1, \tag{26}$$
$$x_2 x_5^{-0.9} x_6^{-1} \leq 1, \tag{27}$$
$$0.1 x_4 x_7 + 0.1 x_5 x_6 \leq 1, \tag{28}$$
$$x_6 + x_7 \leq 1, \tag{29}$$
$$1.5 x_3^{-1} \leq 1, \tag{30}$$
$$0.1 x_1^{6.0} x_2^{0.4} \leq 1, \tag{31}$$
$$0.1 x_1^{-0.6} x_2^{-0.4} \leq 1, \tag{32}$$

where the last two restrictions are used only to restrict the domain of search within a reasonable range of the values of the objective function. This is now a standard geometric programming problem (Duffin et al., 1967) with three degrees of difficulty, eleven terms minus seven variables minus one. Note that all the restrictions are posynomials (i.e., polynomials with positive coefficients). The dual problem associated with this primal geometric program is one of maximizing the dual function

$$v(\delta) = \left[\left(\frac{1}{\delta_1}\right)^{\delta_1} \left(\frac{1}{\delta_2}\right)^{\delta_2} \left(\frac{1}{\delta_3}\right)^{\delta_3} \left(\frac{1}{\delta_4}\right)^{\delta_4} \left(\frac{0.1}{\delta_5}\right)^{\delta_5} \left(\frac{0.1}{\delta_6}\right)^{\delta_6} \left(\frac{1}{\delta_7}\right)^{\delta_7} \right.$$
$$\cdot \left(\frac{1}{\delta_8}\right)^{\delta_8} \left(\frac{1.5}{\delta_9}\right)^{\delta_9} \left(\frac{0.1}{\delta_{10}}\right)^{\delta_{10}} \left(\frac{0.1}{\delta_{11}}\right)^{\delta_{11}} (\delta_2 + \delta_3)^{\delta_2 + \delta_3} (\delta_4)^{\delta_4}$$
$$\left. \cdot (\delta_5 + \delta_6)^{\delta_5 + \delta_6} (\delta_7 + \delta_8)^{\delta_7 + \delta_8} \cdot (\delta_9)^{\delta_9} (\delta_{10})^{\delta_{10}} (\delta_{11})^{\delta_{11}} \right]$$

subject to the nonnegativity conditions $\delta_j \geq 0$ ($j = 1, 2, \ldots, 11$) and the following normality and orthogonality conditions

$$\delta_1 = 1,$$
$$-0.6\delta_1 + \delta_2 \qquad\qquad + 0.6\delta_{10} - 0.6\delta_{11} = 0,$$
$$-0.4\delta_1 \qquad + \delta_4 \qquad\qquad + 0.4\delta_{10} - 0.4\delta_{11} = 0,$$
$$\delta_3 \qquad\qquad - \delta_9 = 0,$$
$$-0.8\delta_2 - 0.8\delta_3 \quad + \delta_5 = 0,$$
$$-0.9\delta_4 + \delta_6 = 0,$$
$$-\delta_4 + \delta_6 + \delta_7 = 0,$$
$$-\delta_2 - \delta_3 + \delta_5 \qquad\qquad + \delta_8 = 0.$$

The numerical solutions of the optimal primal and the dual variables are approximately as follows, up to 18th iteration

Primal variable	Value	Dual variable	Value
x_1	2.9519	δ_1	1.0000
x_2	62.3525	δ_2	7.4081×10^{-8}
x_3	1.5000	δ_3	3.7645×10^{-8}
x_4	6.8360	δ_4	3.1831×10^{-8}
x_5	355.8593	δ_5	7.5179×10^{-8}
x_6	0.0065	δ_6	2.7097×10^{-8}
x_7	0.9436	δ_7	5.8393×10^{-10}
$V = v(\delta)$	0.1000	δ_8	8.4336×10^{-8}
$W = c^{0.6} x^{0.4}$	10.0000	δ_9	9.0864×10^{-8}
		δ_{10}	1.0000
		δ_{11}	8.0045×10^{-10}

Note that the optimal values of the dual variables are approximate here, since the iteration is terminated at step 18, when the difference between primal and dual objective functions is found to be less than one percent.

This problem has three degrees of difficulty but in the general dynamic case when the simplifying assumptions about the time-horizon are not made, computation problems would be much more complicated.

As a second illustration we consider a model of two-level planning due to Martos and Kornai (1966) who emphasized that aggregation aspects are as important as decentralization methods in models of optimal allocation of central resources. The decomposition and decentralization aspects of large scale programming models are discussed by Kornai and Liptak (1962) and several others (Sengupta and Fox, 1969; Kulikowski, 1969). The Martos–Kornai model defines two levels of planning linked through the intersectoral allocation vector u_i for the central resources. The first level sets up the sectoral linear porgramming (LP) model for each sector $i = 1, 2, \ldots, n$

$$\text{maximize } c_i' x_i, \quad c_i, x_i = n_i \cdot 1$$

subject to

$$A_i x_i \leq u_i \text{ (central constraint)},$$
$$B_i x_i \leq b_i \text{ (sector } i - \text{constraints)}, \tag{33}$$
$$x_i \geq 0, \quad A_i = m \cdot n_i, \quad B_i = m_i \cdot n_i, u_i = m \cdot 1, b_i = m_i \cdot 1.$$

The vector u_i of allocation of the central resources to sector i must satisfy two conditions. First it should satisfy any official activity vector x_i^0 that may be preassigned by the central authority

4.1 Programming Framework in Economic Analysis

$$A_i x_i^0 \leqq u_i.$$

Second, the allocation vectors u_i are to be sequentially revised to effect transfer of central resources between sectors till the second level LP model is solved. For a given set of feasible allocation vectors $(u_i, i = 1, \ldots, n)$ let the optimal feasible solutions x_i^* of the LP problem in Eq. (33) be combined to define the vector

$$x^* = (x_1^*, \ldots, x_n^*)'$$

The second level combines the sectoral LP models into a single large scale LP model at the national level, for example,

$$\text{maximize} \sum_{i=1}^{n} c_i' x_i$$

subject to

$$\sum_{i=1}^{n} A_i x_i = b_0, \text{ where } b_0 = \sum_i u_i = \sum_i A_i x_i^0, \tag{34}$$

$$B_i x_i \leqq b_i, i = 1, \ldots, n,$$

$$x_i \geqq 0.$$

Note that although for a given set of feasible allocation vectors $\{u_i, i = 1, \ldots, n\}$ both the official pre-assigned activity-mix $x^0 = (x_1^0, \ldots, x_n^0)'$ and the combined one $x^* = (x_1^*, \ldots, x_n^*)'$ are feasible to the master LP model of Eq. (34), yet in terms of the objective function of Eq. (34), x^* defines a better solution than x^0. Second, the optimal choice of allocation vectors $\{u_i\}$ is reached only if in the set of all feasible allocation vectors it attains the highest objective function value in Eq. (34). These decomposition aspects have also been discussed in the framework of nonlinear programming and also integer programming (Hass, 1968; Pearson, 1966). The methods of two-level planning have been discussed in relation to profit sharing plans of public enterprises by Kornai and Liptak (1962).

As a third illustration we consider a nonlinear programming model of a multiproduct firm under uncertainty (Dhrymes, 1964). A firm is assumed to produce n products $(Q_1, \ldots, Q_n)$ with m inputs $(x_1, \ldots, x_k; x_{k+1}, \ldots, x_m)$, of which the first k inputs are available with competitive factor prices $w_j (j = 1, \ldots, k)$ and the remaining $(m - k)$ inputs are fixed and limited in quantites. Let $x_{ij}, j = k + 1, \ldots, m$ denote the amount of jth fixed factor attached to the plant producing ith good, then the following constraints are assumed for the fixed inputs

$$x_j - \sum_{i=1}^{n} x_{ij} \geqq 0, \quad j = k + 1, \ldots, m, \tag{35}$$

which allow disposability and divisibility in a limited sense. The fixed factors have additional costs due to shifting from one product to another and these costs are termed 'relocation costs' S,

$$S = h(x_{1(k+1)}, \ldots, x_{1m}; x_{2(k+1)}, \ldots, x_{2m}; x_{n(k+1)}, \ldots, x_{nm}), \tag{36}$$

which satisfy the conditions

$$\partial S/\partial x_{ij} \equiv 0, \quad \text{all} \quad i = 1, \ldots, n \quad \text{and} \quad j = 1, 2, \ldots, k. \tag{37}$$

The total cost of production (C) to the firm is given by

$$C = V + S + F, \tag{38}$$

where

$V = $ total variable costs $= \sum_{i=1}^{n} \sum_{j=1}^{k} x_{ij} w_j$,

$S = $ relocation costs defined in Eq. (36),

$F = $ total fixed costs which depend on x_j, $j = k+1, \ldots, m$.

The production of each commodity is assumed to follow a production function

$$Q_i = f_i(x_{i1}, x_{i2}, \ldots, x_{ik}; x_{i(k+1)}, \ldots, x_{im}) \tag{39}$$

(strictly) concave in its arguments (this implies that production of every commodity takes place under (diminishing) non-increasing marginal productivity for each factor. The demand for the ith output of the firm is subject to the following equation of prices p_i, $i = 1, \ldots, n$

$$p_i = g_i(Q_1, Q_2, \ldots, Q_n) + u_i; \quad i = 1, 2, \ldots, n, \tag{40}$$

where $u = (u_1, u_2, \ldots, u_n)'$ is the random vector of errors having a multivariate distribution with mean vector zero and covariance matrix $K = [\sigma_{ij}]$. The profit $\prod(x, u)$ of the firm then becomes

$$\prod(x, u) = R(Q_1, \ldots, Q_n) - C(x_1, \ldots, x_n) + \sum_{i=1}^{n} u_i Q_i, \tag{41}$$

with expected value and variance as

$$E\prod(x, u) = R(Q_1, \ldots, Q_n) - C(x_1, \ldots, x_n),$$
$$\text{var} \prod(x, u) = Q'KQ \quad \text{where} \quad Q = (Q_1, \ldots, Q_n)', x = (x_1, \ldots, x_n)'$$

4.1 Programming Framework in Economic Analysis

assuming $E \sum_{i=1}^{n} u_i Q_i = 0$. Here $R(Q)$ is the revenue function

$$R(Q(x)) = \sum_{i=1}^{n} g_i(Q_1, \ldots, Q_n) \cdot Q_i(x), \quad (42)$$

assumed concave in $Q = (Q_1, \ldots, Q_n)'$. Because the profit function is stochastic, some criteria of ordering uncertain prospects must be introduced in the firm's decision space. A number of approaches to such a problem is discussed in Section 4.2 in our review of stochastic programming.

Dhrymes (1964) solves this problem in a particular way (Markowitz, 1959) by introducing a quadratic utility function U for the firm

$$U = E\prod(x, u) + \tfrac{1}{2}\alpha\{\prod(x, u) - E\prod(x,u)\}^2, \quad (43)$$

where the scalar parameter α implies risk aversion (attraction) if it is negative (positive). The optimal decision problem for the firm is then set up as a nonlinear program for maximizing the utility function in Eq. (43) subject to a variance constraint

$$v \geq Q'KQ \quad (44)$$

and the constraints in Eqs. (38), (39), and (40). The solution of this problem could be decomposed into two stages. At the first stage it is assumed that the output vector Q^0 is preassigned, so that the maximization of the utility function in Eq. (43) becomes equivalent to minimization of total costs $C(x)$ in Eq. (38) under the constraints

$$Q_i^0 = f_i(x), \quad i = 1, 2, \ldots, n \quad (45)$$

$$x_j \geq \sum_{i=1}^{n} x_{ij}, \quad j = k+1, \ldots, m, \quad (46)$$

$$x_{ij} \geq 0. \quad (47)$$

This is a typical convex programming problem (Pfouts, 1961), where the necessary and sufficient conditions for the minimum are easily derived from the Kuhn–Tucker (1951) theorem

(a) $(\partial C/\partial x_{ij}) - \lambda_i(\partial f_i/\partial x_{ij}) \geq 0, \quad i, j = 1, 2, \ldots, k,$
(b) $\partial C/\partial x_{ij} = \mu_j - \lambda_i(\partial f_i/\partial x_{ij}) \geq 0, \quad i, = k+1, \ldots, m,$
(c) $\sum_i \sum_j x_{ij}^0 (\partial L(x^0, \lambda^0, \mu^0)/\partial x_{ij}) = 0,$
(d) $\partial L/\partial \lambda_i \leq 0, i = 1, \ldots, n; \lambda_i \geq 0,$ \quad (48)
(e) $\partial L/\partial \mu_j \leq 0, j = k+1, \ldots, m; \mu_j \geq 0,$

(f) $\sum_{i=1}^{n} \lambda_i^0 (\partial L(x^0, \lambda^0, \mu^0)/\partial \lambda_i) = 0$,

(g) $\sum_{j=k+1}^{m} \mu_j^0 (\partial L(x^0, \lambda^0, \mu^0)/\partial \mu_j) = 0$,

where $L = L(x, \lambda, \mu)$ is the Lagrangean function associated with the convex program

$$L(x, \lambda, \mu) = C(x) + \sum_{i=1}^{n} \lambda_i(Q_i^0 - f_i(x)) + \sum_{j=k+1}^{m} \mu_j(\sum_{i=1}^{n} x_{ij} - x_j)$$

λ_i, μ_j being Lagrange multipliers and the variables with zero superscripts indicate equilibrium (or optimal) magnitudes. Note that at the optimal solution of this problem the optimal values of the multipliers λ_i and μ_j can be interpreted as implicit (or shadow) prices for the pre-assigned outputs Q_i^0 and the fixed factors $x_j, j = k + 1, \ldots, m$, respectively. Also we get the familiar results:

Equation (48a) implies the equivalence of marginal rate of substitution to factor price ratio appropriately defined.

Equation (48c) implies that a variable factor will not be used if its price exceeds its marginal value productivity for a given output.

Equation (48b) implies that the marginal cost of relocation equals the marginal value productivity of the relocated input.

Once the first stage problem is solved for a pre-assigned but feasible output vector Q^0, the second stage problem is set up as one of maximizing the function

$$R(Q^0) - C(Q^0) + \tfrac{1}{2}\alpha Q^{0\prime} K Q^0 \tag{49}$$

subject to the constraints $Q^0 \geq 0$ and the entire domain of variation of Q^0. In other words, by parametric variation of pre-assigned Q^0 we induce a parametric variation of minimal costs $C(Q^0)$ and the associated revenue function $R(Q^0)$ and finally select that vector Q^{00} to be the optimal which maximizes Eq. (49), that is, attains the maximum maximorum of profits.

Some comments about this type of nonlinear programming model of the multi-product firm may be added at this stage. First, some of the assumptions of this model could be generalized, for example, the production function of Eq. (39) and the revenue function of Eq. (42) need be only quasi-concave (Arrow and Enthoven, 1961; Sengupta and Fox, 1969; Mangasarian, 1969). Also the criterion of maximizing expected utility of profits assuming a quadratic utility function has some limitations, for example, it ignores the characteristics of the probability distribution of the random vector u and the fact emphasized by Arrow (1965), Pratt (1964), and others that a (concave) quadratic utility function has an increasing degree of risk aversion in the sense that the ratio $(-\partial^2 U/\partial \Pi^2)/(\partial U/\partial \Pi)$ is an increasing function of profits Π. Second, the specific role of fixed factors in the medium and long-run expansion of capacity of the firm and the related problems of investment are neglected in this formultaion. The first aspect has been discussed in a two-level profit allocation model by Kornai and Liptak (1962) and the second aspect has been discussed in the literature under specific capacity expansion models (Sengupta and Fox, 1969; Manne and Veinott, 1967; V. Smith, 1961). Third, a most fundamental question has been raised by Frisch (1965) that the assumptions of univalued functions in the production relations of Eq. (39) and cost relations of Eq. (36) are too simplifying; these in effect allow the multiproduct model to be viewed exactly in the same manner as a single product model. A multi-valued production function for each output, although very awkward from a mathematical viewpoint explicitly allows additional degrees of freedom and diversity in firm's resource allocation models especially when the fixed factors may be transferred between different outputs. Fourth, the above model implies that any fixed factor which is in excess over requirement at the optimal solution must have a zero implicit (or shadow) price. Applying the approach of chance-constrained linear programming and following some results in organization theory, Naslund (1967) has argued against this type of imputation in a framework of uncertainty. This leads us to the decision models under stochastic programming and the related approaches to risk and uncertainty.

4.2. The Approach of Stochastic Linear Programming: A Brief Summary

This section outlines a brief survey and appraisal of some of the important operational methods of characterizing and solving stochastic linear programming problems in the context of the general theory of probabilistic programming.

A linear programming (*LP*) problem

$$\max z = c'x, \quad Ax \leq b, \quad x \geq 0 \quad \begin{array}{l} c: n.\ 1 \\ A: m.\ n \\ b: m.\ 1 \end{array} \quad (1)$$

is defined to be probabilistic if one or more of the coefficients in the set (A, b, c) are random variables with some specific probability distribution. If the probability distribution of the parameters (A, b, c) is known or assumed to be specified, then an important class of decision problems can be characterized by the following questions,

(a) How to decide on a decision vector x which is in some sense optimal?

(b) How to characterize the sensitivity of an optimal decision vector when it exists in terms of parameter variations, characteristics of profit distribution and the tolerance measures up to which constraint violations are admitted into the model?

Partial answers to these questions are available in the current literature on probabilistic linear programming. The three most important approaches in this field are the stochastic linear programming (SLP) (Tintner, 1955; Tintner and Sengupta 1964), the method of chance-constrained programming (CCP) (Charnes and Cooper, 1963; Thompson *et al.*, 1963) and the technique of two-stage programming under uncertainty (Dantzig and Madansky, 1961; Madansky, 1962). Broadly speaking these three approaches have the following common characteristics. They incorporate the initial probability distribution of the parameters in order to convert a probabilistic linear program into a deterministic form and then define a set of decision rules having some optimality properties. Methods of incorporating probability distributions and specifying decision rules are of course different in different approaches.

If the initial distribution of the parameters is either unknown or incompletely specified, the problem of characterizing the optimal decision vector becomes much more complicated. The decision rules under uncertainty (Milnor 1954; Pratt *et al.*, 1964) and simulation techniques (Wolfe and Cutler, 1963; Hufschmidt, 1963; Wetherill, 1966) have scope here. Another important line of investigation is provided by looking at the various characteristics of the matrix game associated with a LP model (Shapley, 1953; Weiss, 1961; Soults, 1968).

From a computational viewpoint there are at least three basic problems of incorporating the random variations in (A, b, c) into the framework of an *LP* model, given than the from of the activity vector x is truncated by the inequalities in the constraint space and the points of truncation are not generally known *a priori*. Also a linear function of non-normal random variates is not generally of the same form of distribution as the individual random components and hence the exact specification of truncation restriction does not appear to be very simple. Second, if we interpret that sample observations ($t = 1, 2, \ldots, T$) are available for $A_t = \bar{A} + \alpha_t$, $b_t = \bar{b} + \beta_t$, $c_t = \bar{c} + \gamma_t$, where ($\bar{A}, \bar{b}, \bar{c}$) represents parameter values unkown to us, then the sampling distribution of the activiy vector x becomes dependent on the restriction on the random elements ($\alpha_t, \beta_t, \gamma_t$) imposed by feasibility, the sample design and the form of the population distribution. Problems related to this basic issue have been raised by a number of authors in the context of SLP (Wagner, 1958; Box and Hunter, 1954) and other related probabilistic models (Elmaghraby, 1960; Wets, 1966). Third, if we consider the decision vector x to be non-stochastic in the sense that given the random variations in (A, b, c) and the associated stochastic LP models we have to determine an optimal solution vector x, then the specification of the decision-maker's attitude to risk becomes very important. There are several approaches in this field, for example, the method of certainty equivalents wherever applicable (Theil, 1961). the CCP approach of pre-assigning tolerance measures for the constraints (Charnes and Cooper, 1959) and the non-parametric method based on safety first principles (Roy, 1952). Uses of specific utility functions with pre-assigned risk attitudes (Tobin, 1968; Arrow, 1951; Fishburn, 1968) and of what may be called pessimistic solutions (Medansky, 1963) are also suggested in the literature in the hope that such solutions are more likely to retain feasibility.

From an operational standpoint it is convenient to consider the stochastic approach to LP models by successive stages. For static programming problems under risk [i. e., when the probability distribution of the parameters (A, b, c) is assumed known or estimable] the three most important areas where stochastic considerations are operationally introduced are in the fields of (a) variations in parameters, (b) sensitivity analysis, and (c) stochastic games and related problems.

4.2.1 PARAMETRIC VARIATION IN PROGRAMMING

(A) Variations in vector c

The simplest case of variations in parameters arises when only the vector c in the LP model in Eq. (1) is assumed random with finite mean and covariances. In this case the expected level of profits and its variance can be defined and a portfolio-type model (Markowitz, 1959; Farrar, 1962) could be specified, where the original (random) objective function is replaced by the expected profit function subject to one additional restriction that the variance of profit must be within a specific tolerance limit. By allowing a parametric variation of the tolerance limits a whole set of (E, V) points (E = expected profits and V = variance of profits) can be generated. The associated solutions in the x-space are then candidates for optimal decisions.

If there is any information on the probability distribution of the vector c, it could be profitably incorporated into the logic of specification of optimal decisions. The case when c has a multivariate normal distribution with mean vector m and dispersion matrix V has been analyzed in several respects (Simon, 1956; Sengupta and Fox, 1969). This leads to an interesting approach of fractile programming (Sengupta and Portillo–Campbell, 1970) by which a specified fractile of the distribution of profits is maximized under the usual constraints. For instance this approach would transform the LP model in Eq. (1) as

$$\text{maximize } f(x) = m'x - k \cdot (x'Vx)^{\frac{1}{2}}$$
$$\text{subject to } Ax \leq b; x \geq 0, \qquad (2)$$

where $k = -F^{-1}(\alpha)$ and $F(w)$ is the cumulative distribution of a unit normal variate and $\alpha = 0.99$ say indicates the pre-assigned fractile measure (viz, with $\alpha = 0.99$ the constant k is 2.33 from standard Normal Tables). Note that because of the square root term in the objective function of Eq. (2), the standard algorithms of quadratic programming would not apply. However, given m, V and a nonnegative value of k, the optimal solution x of this fractile program would provide a decision or policy which attains a 100α percent probability level of profits. By considering parametric variation of the coefficient α and for that matter k, the entire decision space of feasible policies can be characterized. The fractile approach seems to have some practical appeal due to two special reasons. First, it may approximately hold in those cases where the vector c is not normal but tends asymptotically to normality under conditions of the central limit theorem. Second, if the unknown population distribution depends only on location and scale parameters, then the limiting distribution of the sample quantiles can be easily worked out (Sarhan and Greenberg, 1962).

In cases where the distribution form of the random vector is not completely known, there are several lines of which two may be briefly mentioned. First, there are safety-first principles and related methods (Sengupta, 1968, 1969, 1969a) based on Tchebycheff-type probability bounds which could be utilized to transform the stochastic objective function into a non-stochastic one. Second, the stochastic objective function $z = c'x$ could be subsumed in a utility function $U(z)$ satisfying certain assumptions which guarantee the ordering condition of a real-valued preference function (Wessels, 1967; Kingman, 1963). If the utility function is analytic in its domain, so that it can be expanded in a Taylor series around a specific value, say the expectation $\bar{z}$ of profits, then the expected utility function $E\,U(z)$ can be written as

$$E\,U(z) = U(\bar{z}) + \tfrac{1}{2}r\sigma_z^2 + o(\cdot), \tag{3}$$

where r denotes a fixed constant defined by $(\partial^2 U/\partial z^2)|_{z=\bar{z}}$ and $o(\cdot)$ denotes the remaining terms. If the utility function is quadratic, then of course the term $o(\cdot)$ disappears and we have, if $r < 0$ [i. e., concave $E\,U(z)$], a typical quadratic programming problem in the decision vector x. Also, if utility function is linear and the profit z is normally distributed (i. e., vector c being normal), a suitable equivalent problem in quadratic programming can be formulated. Otherwise the incidence of nonnormality and the sensitivity of risk aversion measure (Pratt, 1964) on the solution vector have to be explored.

Note that if in fact profits are normally distributed either exactly or asymptotically whenever the terms $(\partial^i U/\partial z^i)_{z=\bar{z}}$ can be neglected for $i \geq 3$, then the utility function $U(z)$ need not be quadratic in order to lead to a quadratic program. For instance, Freund (1956) has used the following form of the utility function

4.2 The Stochastic Linear Progamming Approach

$$U(z) = 1 - \exp(-\alpha z), \qquad (4)$$

where α is a pre-assigned constant indicating risk aversion (or risk taking attitude) if it is positive (or negative). For normal z, the criterion of maximizing expected utility leads in this case to the following quadratic programming problem in the decision vector x if $\alpha > 0$:

$$\begin{aligned} \text{maximize } & f(x) = m'x - (\alpha/2)x'Vx, \\ \text{subject to } & Ax \leq b, \quad x \geq 0. \end{aligned} \qquad (5)$$

Two remarks may be made about this approach. First, the specific form of Eq. (4) of the utility function is known to arise from the strategy preserving property of matrix games (Kemeny and Thompson, 1957) and from a slight modification of Bernoulli's utility axioms (Borch, 1968). Second, note that the maximization of expected utility $E U(z)$, where $U(z)$ is defined in Eq. (4) leads to minimization of $E \exp(-\alpha z)$, but the latter term leads precisely to the moment generating function of the random variable z. This means that even if z is not normal but its moment generating function exists, then a corresponding programming problem could be defined. For instance, if profits z has a gamma density with parameters β and θ such that $Ez = \mu = \theta/\beta$ and var $z = \sigma_z^2 = \theta/\beta^2$ then the maximization of expected utility leads to the following nonlinear program

$$\begin{aligned} \text{maximize } & f(x) = (\mu/\sigma_z)^2 \log(1 + \alpha(\sigma_z/\mu)), \\ \text{subject to } & Ax \leq b, \quad x \geq 0., \end{aligned} \qquad (6)$$

The objective function in this case is more complicated than a quadratic form.

Another advantage of the utility function approach is that the expected utility criterion leads to convenient rules for ordering uncertain prospects.

For instance, if $F(t)$ and $G(t)$ denote two cumulative distribution functions of the random variable profits z such that $F(t)$ lies on or above $G(t)$ for all z in its domain, then the expected utility under the density function $dG(t)$ equals or exceeds that under the alternative density $dF(t)$, provided marginal utilities are positive and certain regularity conditions (Hader and Russell, 1969) hold. Note that this result is independent of any specific form of the utility function. This result has some similarity with other nonparametric bounds and inequalities which would be referred to later.

(B) *Variations in vector b*

Since the dual of the LP model of Eq. (1) would have resource vector b in the objective function, most problems of variations in vector b may be analyzed by methods indicated in the earlier section. However, there are some additional problems in this case due to the inequalities in the constraint space (i.e., $Ax \leq b$, $x \geq 0$). At least three different types of characterizations are available in the literature. First, there is the penalty function approach (Theil, 1961; Sengupta, 1969) which introduces penalties for violating the stochastic constraints and adjoins the expected penalty costs as an additional facet of the objective function. For instance, if there is a constant penalty cost $h_i > 0$ per unit for each violation of the ith constraint, that is $b_i \geq a_i x$ and the constraints have finite probabilities of violation, then total expected penalty cost is $E\Sigma_i h_i y_i$ where E is expectation and y_i is the safety margin (i.e., $a'_i x + y_i = b_i$; $y_i \geq 0$, $i = 1, 2, \cdots, m$), The reduced problem is then

$$\text{maximize } f(x, y) = c'x - Eh'y, \tag{7}$$

$$\text{subject to } Ax + Iy = b; \quad x \geq 0, y \geq 0,$$

where I is the identity matrix. Note that this method is basically related to the approach of two-stage programming under uncertainty (Dantzig, 1955; Williams, 1965) and generally the objective function $f(x, y)$ of Eq. (7) is nonlinear. Computational methods for reducing such problems to quadratic programs are available (Beale, 1961; Elmaghraby, 1959) but they are dependent on approximations in the probability distribution space or in the specification of expected cost function. Algorithms of decomposition (Sengupta and Fox, 1969) have some scope here.

Second, there is the CCP approach which puts a chance-constrained interpretation for the constraints, that is,

$$\text{prob}(b_i \geq a'_i x) \geq u_i, \quad 0 \leq u_i \leq 1, \quad i = 1, 2, \cdots, m \tag{8}$$

by pre-assigning the tolerance measures u_i up to which constraint violations are tolerated. The penalty costs are not present here but suitable tolerance measures have to be pre-assigned. Also with observational data on variations in b_i it is computationally convenient to restrict the solution to a given class of decision rules (Charnes and Cooper, 1962).

Third, there is the truncated distribution approach which interprets that the inequalities $b_i \geq a'_i x$ ($i = 1, 2, \ldots, m$) serve to truncate the probability distribution of each b_i and this fact should be incorporated in the penalty function and CCP approaches above and also in computing the distribution of profits. In case of truncated normal distributions, the reduced problem is one of nonlinear progamming (Sengupta, 1969). Note also that if instead of pre-assigning the tolerance measures u_i in Eq. (8), we want to optimally solve for them by adjoining a system reliability measure (e.g. $\prod_{i=1}^{m} u_i$ or somewhat more simply $\sum_i \log u_i$) to the original objective function, then the truncation points play an important role in reliability programming (Sengupta, 1969b). Computational algorithms (Duffin *et al.*, 1967) based on generalized polynomial programming

4.2 The Stochastic Linear Programming Approach

(Wilde and Beightler, 1967) become relevant in this connection.

(C) Variations in matrix A

The penalty function, CCP and the truncated distribution approach are all applicable to the case when both the coefficients b_i and c_j are random. However, the variations in a_{ij} coefficients of matrix A involve simultaneity and interdependence of relations between activies in a more basic sense than the c_j or b_i coefficients.

If the changes in the coefficient matrix A are observable, then we consider LP models of Eq. (1) with A replaced by $A^{(k)}$, where the superscript k denotes the index set $k = 1, 2, \ldots, N$. we consider the subset $k = 1, 2, \ldots, K$ (with $K \leq N$) for which the feasibility $A^{(k)} x \leq b$, $x \geq 0$ and hence optimality conditions are satisfied. Denote the optimal solutions and optimal profits for this subset $(k = 1, 2, \ldots, K)$ by $x^{(k)}$ and $z^{(k)}$ respectively. Given the probability measures associated with $\{A^{(k)}, k = 1, 2, \ldots, K\}$, the sequence $\{z^{(k)}, k = 1, 2, \ldots, K\}$ determines the distribution of optimal profits. If the underlying distribution can be estimated exactly or approximately, its characteristics can be utilized in decision making in several ways. First, the observations on $\{A^{(k)}, k = 1, \ldots, K\}$ censored (Gupta, 1952) by conditions of feasibility, leading to an estimate of the distribution of optimal profits afford an insight into its characteristics (e. g., whether it is nearly normal or not). The estimated distribution may then be used to set up a new criterion function in terms of expected value or fractile maximization, say.

Second, additional decision rules may be imposed before the estimated distribution characteristics of profits are utilized. For instance, decomposition of the resource vector b in terms of allocation ratios, i. e.,

$$b_i = \sum_{j=1}^{n} b_i u_{ij}, \qquad u_{ij} \geq 0, \quad \sum_j u_{ij} \leq 1, \tag{9}$$

where the allocation matrix $U = [u_{ij}]$ contains the decision variables, may provide the decision-maker with a set of alternative choices. By selecting and preassigning specific allocation matrices, one could compare alternative conditional distribution of profits. The rules for ordering distributions (Lehmann, 1955) mentioned before before become relevant here.

Third, if $A^{(k)}$ be interpreted to represent cross-sectional sample observations, then the problem of obtaining consistent estimates in a specific sense has been considered in the literature. Problems of perfect aggregation (Weil, 1968; Day, 1963) bias and other aspects of decomposition (Theil, 1954; Sengupta, 1970c) are worth mentioning in this connecton.

(D) Variation in all parameters

The general case of variations in all parameters (A, b, c) have been considered in the SLP, CCP and two-stage programming under uncertainty. In the

SLP framework a number or characterizations (Tintner *et al.*, 1963; Bereanu, 1961) of the technical and the decision problem are suggested. Apart from the characterization of the distribution problems (sampling and population distributions), two other aspects are of relevance here. First, we have to have characterization of systematic and non-systematic changes in the parameter set $S_k = (A^{(k)}, b^{(k)}, c^{(k)})$. In empirical applications the importance of this fact is emphasized by methods of recursive programming (Sengupta, 1966; Day, 1963) applied to dynamic models, adaptive programming (Naslund, 1967) and sequential planning (Arrow and Hurwicz, 1959). One interesting line of approach in this respect is to introduce ordering in the parameter space and apply a phase analysis in the sense of recursive programming. To fix ideas assume the set $S = (a_{ijt}, b_{it}, c_{jt}; t = 1, 2, \ldots, T)$ to be nonngative for all t such that the elements are ordered as follows

$$\begin{aligned} a_{ij1} \geq a_{ij2} \geq \ldots \geq a_{ijT}, & \quad i = 1, 2, \ldots, m, \quad j = 1, \ldots, n, \\ b_{i1} \leq b_{i2} \leq \ldots \leq b_{iT}, & \quad i = 1, 2, \ldots, m, \\ c_{j1} \leq b_{j2} \leq \ldots \leq c_{jT}, & \quad j = 1, 2, \ldots, n, \end{aligned} \qquad (10)$$

where T represents the number of observations. Assume that for each fixed $t = 1, 2, \ldots, T$ the LP model based on the specific set of parameters has finite optimal solutions. Then by construction the optimal profits for $(a_{ij1}, b_{i1}, c_{j1})$ is the least and it is the highest for $(a_{ijT}, b_{iT}, c_{jT})$. How do we characterize the situation that the decision maker intends to choose with high probability a decision vector x which gives an optimal profit not lower than that associated with the parameter set $(a_{ijt}, b_{it}, c_{jt})$ for a fixed $t \geq 2$? The use of order statistics and two-point extreme value problems (Sengupta, 1969c) become relevant here.

Second, the characterization of the decision by considering meaningful and operational restrictions is required when there is simultaneous variation in all parameters. Linear decision rules of CCP and allocation matrices under SLP are only specific examples of such meaningful restrictions in the decision space. The general problem is the following: Denote optimal profits by $z = z(x, S)$ by a function of x and the parameters, where S denotes the set $(A^{(k)}, b^{(k)}, c^{(k)}; k = 1, 2, \ldots, K)$. The decision maker intends to choose a policy $x = \bar{x}$ such that $Ez(\bar{x}, S)$ is maximum where E is expectation. In the resource decomposition case x is a function of the allocation matrix U, so that the policy problem is to choose $U = \bar{U}$ such that $Ez(\bar{U}, S)$ is maximum. The characterization of the neighborhood around which this maximality of expected profits is preserved is of special relevance here (Sengupta *et al.*, 1965).

4.2.2 SENSITIVITY ANALYSIS

The analysis of sensitivity of the optimal solution vector and optimal profits has been attempted both in deterministic and stochastic LP models. In the

4.2 The Stochastic Linear Programming Approach

latter case the following methods are worth mentioning

(a) An interesting operational measure of sensitivity that has been applied to problems of production (Saaty and Webb, 1961) and allocation is to derive the sensitivity coefficient k_{ij} by taking the partial derivative of the optimal objective function $z = c'x$ as

$$k_{ij} = -\partial z / \partial a_{ij} = x_j y_i$$

(i, j in optimal basis) for a fixed set of parameter values. Here x_j's are optimal primal variables and y_i's are optimal dual variables ($x_j \geq 0$, $y_i \geq 0$). These sensitivity indices averaged over sample observations on (A, b, c) provide some idea as to how sensitive the objective function is at the specific optimal point when data variations are likely. The need for analyzing the statistical distribution of the sensitivity coefficients k_{ij} is all the more apparent in this framework.

(b) In some cases solution vectors other than the optimal has been computed in order to characterize second best and third best solutions or policies under stochastic variations of data; also their stability properties have been analyzed (Sengupta, 1966).

(c) In situations where a single factor (e. g,. informational basis or technology) relates two specific sets of parameters $S = (A, b, c)$ and $\tilde{S} = (\tilde{A}, \tilde{b}, \tilde{c})$ at each of which the optimal profits of corresponding LP models are $z(S)$ and $z(\tilde{S})$, the concept of "marginal value" (Williams, 1963) defines a sensitivity coefficient $g(h)$ as a function of the scalar quantity h,

$$\frac{dg(0)}{dh} = \lim_{h \to 0^+} \left(\frac{z(S + h\tilde{S}) - z(S)}{h} \right)$$

provided the limit exists. Note that if the existence conditions are fulfilled, one could write approximately

$$g(h) \simeq g(0) + h(dg(0)/dh), \quad h \in [0, h_0]$$

for small perturbations in the closed interval $[0, h_0]$. Under certain regularity conditions this result can be used to specify approximate size of the neighborhood containing a given optimal solution (e. g., the specific optimal solution may be the expected value solution with expectation of all paremeters).

(d) The idea that sensitivity analysis could profitably be made around a suitably chosen optimal solution (e. g., the optimal solution of a LP model after replacing all random parameters by their expected values) has led to some attempts (Hanson, 1960) at replacing the optimal basis equations

$$\bar{B}x = \bar{b}, \quad \bar{B}: m.m, \quad \bar{b}: m.1 \tag{11}$$

of the expected value model by

$$(\bar{B} + \delta B)(x + \delta x) = \bar{b} + \delta b, \tag{12}$$

where the δ-operators denote small errors such that the original basis $\bar{B}$ is preserved. Then under certain regularity conditions and approximation the expected value $E\delta x$ and variance var x can be computed; this would provide some characterization of the neighborhood of the original optimal basis equations. If the size of the neighborhood is large, the optimal solution is stable or robust. However, unless specific distribution assumptions are introduced for the error terms δB, δb, and therefore δc, the specific distribution of optimal solution and profits are not determined in this approach,

(e) In case specific distribution assumptions are imposed on the error terms δB, δb in Eq. (12) and approximations are made to the effect that cross-product terms beyond the second degree are ignored, more explicit results on the mean and variance of optimal profits and the probability of profits being higher than a specific level, can be derived. The cases of normally distributed errors (Babbar, 1955), of using Laplace transforms in the distribution space (Bereanu, 1966) and mixture of distributions (Sengupta, 1969d, e) are reported in the literature. The conditions of continuity and nonsingularity under which the distribution of profits tends to asymptotic normality (Prekopa, 1966) are also known.

4.2.3 Stochastic Game Models and Programming

Since an LP model of Eq. (1) with a fixed set $S = (A, b, c)$ of parameters can be written equivalently as a two-person zero-sum game, stochastic variations in parameters could be analyzed through the characteristics of stochastic games. The literature on stochastic gemes in general is quite extensive and we would not attempt to review it. Instead in this section we briefly mention some of the important results on two-person zero-sum games which may be relevant to our discussion of SLP and related models.

(a) From an economic viewpoint, the two-person zero-sum analog of an LP model provides a very interesting characterization of an adaptive decision process which is competitive. In this class of stochastic games (Rosenfeld, 1964), some of elements ϕ_{ij} of the pay-off matrix are unknown to the two players, the decision-maker and nature when the game is begun. Players are assumed to know only the *a priori* probability distribution of the pay-offs. The true value of the pay-off ϕ_{ij} is not known, until the first player selects strategy i and the second uses alternative j at some step of the infinite process. If the true values of the pay-off matrix were known to the first player (the decision-maker), he could have computed his maximum expected return on this basis.

At a given kth step of the play with N steps, he could therefore compute his expected loss $L^{(k)}$

$$EL^{(k)} = Ev - Er^{(k)},$$

where v is the maximum return (as a function of true values of pay-offs) and $r^{(k)}$ is the expected return to the first player. If the process were to continue for N steps and terminate, an optimal strategy for the first player could be the one that maximizes the total expected return over N steps. This is identical with the strategy which minimizes the total expected loss EL_N, where

$$EL_N = NEv - \sum_{k=1}^{N} Er^{(k)}.$$

The cases of truncation of the process and implications, equal and unequal information available to the two players (Thomas and David, 1967), asymptotic distribution of matrix-game values (Soults, 1968), uniform probability distributions used by the two players, the extension to non-zero sum cases of market models (Shubik, 1959) and the analysis of their stability properties (Radstrom, 1964) are investigated in the literature.

(b) The distribution of the value of perfect information (two-person zero-sum) games with random pay-offs has been extended (Thomas, 1967) for the case where the pay-offs to the first player result from distributions other than uniform. The limiting or asymptotic distribution of the value of the game $v = v(\{x(i_1, i_2, \ldots, i_n)\})$, where $x(i_1, \ldots, i_n)$ denotes the random pay-off to the first player when his selected strategies are $(i_1, \ldots, i_n)$ has been investigated under conditions when $x(i_1, \ldots, i_n)$ is replaced by random variables $X(i_1, \ldots, i_n)$ each with an arbitrary cumulative distribution. It is of some importance to note that the limiting distributions closely resemble extreme value distributions, (Soults, 1968; Sarhan and Greenberg, 1962).

(c) Again since decomposition principles for LP models can be interpreted as alternative ways of analyzing the effects of resource aggregation (centralization) or disaggregation (decentralization) on the convergence of sub-optimal solutions, the game analog of this framework would be the decomposition of stochastic games and the implications of truncated games. The stability and other aspects of such games (Kornai and Liptak, 1965) have been reported.

4.2.4 STOCHASTIC APPROACH IN CCP MODELS

The CCP approach in its recent developments and applications has shown a very operational way of introducing probabilistic considerations in the constraints of an LP model. It is based on the assumption of normality of distribution of the parameters and the tolerance measures are to be suitably pre-assigned by the decision-maker. For economic models where parameters like prices are required to be usually nonnegative, the normality assumption has to be generalized in order to allow for nonnegative distributions (e. g., chi-square). Also it would be considerably helpful if some method is developed for determining in some optimal fashion the tolerance measures instead of arbitrarily pre-assigning them. This leads us to consider the following aspects in the CCP approach: (a) the types and characteristics of distribution problems, (b) methods of building optimal tolerance measures into the model framework, and (c) the analysis of system reliability for the chance-constrained LP system.

(A) Distribution Aspects of CCP

Three types of distributional analysis in the CCP framework may be briefly mentioned in this section: (a) implications of distributions with nonnegative domains, (b) aspects of sampling distributions, and (c) the implications of nonparametric and other distribution-free methods.

(a) Consider for instance the assumption that the parameters (a_{ij}, b_i, c_i) are each independently distributed as chi-square variates. Define $P_k = a'_k x$ where a'_k denotes the kth row of matrix A in LP model of Eq. (1). Since each a_{ik} is by assumption chi-square distributed, it can be written as a square of a normally distributed variate; hence one may denote

$$\xi_{kj}^2 = (n_{kj} r_j)^2, \quad \text{where} \quad r_j = +\sqrt{x_j} \geq 0; \quad k = 1, 2, \ldots, m,$$
$$j = 1, 2, \ldots, n,$$

where n_{kj} are independent normal variates with finite means and variances. The characteristic function $\phi(t)$ of the random variable P_k may therefore be derived as

$$\phi(t) = \prod_{j=1}^{n} (1 - 2itv_{kj})^{-\frac{1}{2}} \cdot \exp\left\{\frac{\sum_{j=1}^{n} itv_{kj} \bar{m}_{jk}^2}{1 - 2itv_{kj}}\right\},$$

where m_{kj} and v_{kj} are mean and variance of ξ_{kj} and $\bar{m}_{kj} = m_{kj}/v_{kj}$. From this characteristic function all the moments can be calculated. Also the distribution of P_k can be very closely approximated by a single non-central chi-square variate. Now since b_i is assumed to be a chi-square variate, the ratio $K_i = b_i / \sum_{j=1}^{n} a_{ij} x_j$ is the approximate distribution of a non-central F variate for which statistical tables are available. Hence the chance-constrained inequality of Eq. (8) in this case reduces to

$$\text{prob}(K_i \geq 1) \geq u_i, \quad i = 1, 2, \ldots, m,$$

where each K_i has a non-central F-distribution. Other implications of non-central chi-square variates in this context and the uses of other nonnegative

4.2 The Stochastic Linear Programming Approach

distributions like the exponential are also available. The incidence of non-normality in the CCP model has also been examined (Sengupta, 1969d) in terms of the assumption that a linear function of non-normal variates is only approximately normal.

(b) The sampling distribution problems associated with the CCP model require more intensive analysis. To illustrate the problem assume that only the resource vector b is random in Eq. (8) with each element b_i having independent normal distributions $N(\mu_i, \sigma_i)$ with population means μ_i and variances σ_i^2 ($i = 1, 2, \ldots, m$). However, in almost all practical cases these population parameters are unknown and all we would have is a sample of T observations $b_{it}(t = 1, 2, \ldots, T)$ for each b_i. Two approaches are conceivable in this case. We may adopt a confidence interval approach and determine two limits b_i^0, ^0b_i ($b_i^0 < {}^0b_i$) such that for the sample interval $(b_i^0, {}^0b_i)$ it can be asserted with $100u_i$ per cent probability that it satisfies the chance-constraints in Eq. (8). A second approach is to replace the pre-assigned tolerance measures u_i in Eq. (8) by profit estimates and adjoin it to the objective function. For instance, the probability that each sample value b_{it} satisfies $b_{it} \geq a_i'x$ for all $t = 1, 2, \ldots, T$ is given by

$$[1 - F(w_i)]^T,$$

where $w_i = a_i'x$ and

$$F(w_i) = (\sigma_i\sqrt{2\pi})^{-1} \int_{-\infty}^{w_i} \exp\left\{\frac{-(b_{it} - \mu_i)^2}{2\sigma_i^2}\right\} db_{it}.$$

Taking logarithms of this joint cumulative probability and adjoining it to the original objective function, we get the following concave programming problem in the variables x_j, μ_i and σ_i:

$$\text{maximize } z = \sum_{j=1}^{n} c_j x_j + T \sum_{i=1}^{m} \log\{1 - F(w_i)\} \tag{13}$$

under the restrictions

$$x \geq 0 \quad \text{and} \quad a_i'x \geq \bar{b}_i.$$

Note that the last restriction has been added to impose the condition that the likelihood estimates must satisfy the constraints at the sample mean (i. e., $\bar{b}_i$ denotes the sample mean of b_{it}). Note that this problem seeks to solve simultaneously the optimal estimates of population parameters μ_i, σ_i and the decision vector. This framework has close similarity with restricted maximum likelihood estimation problems. Note also that the sample size T has a role in this framework. Also if each b_i is distributed like a one parameter exponential distribution then the concave programming problem of Eq. (13) becomes very simple. This in fact makes it suitabe for reliability analysis.

(c) The use of nonparametric methods is motivated by the fact in most practical cases where the data are nonexperimental, we do not have adequate knowledge about one or the other form of the distribution. Use of Tchebycheff-type and other probabilistic bounds, the Kolmogorov–Smirnov type distance statistics and other bounds based on the "increasing hazard rate" (IHR) distributions (Sengupta, 1969) are discussed in the literature.

(B) Reliability Analysis of LP Systems

There are at least two different ways by which optimization considerations can be built into the tolerance measures u_i in the chance constraints Eq. (8). First, one could pre-assign specific levels of u_i and then when optimal feasible solutions exist, analyze the dual variables (i. e., Lagrange multipliers associated with it) in order to allow parametric variations in u_i. Second, one could build a reliability component into the objective function and solve for an optimal set of u_i's. For instance, assume as an example that each b_i follows a one-parameter (λ_i) exponential distribution with $Eb_i = 1/\lambda_i$. Then chance constraints in Eq. (8) become

$$\exp(-\lambda_i a'_i x) - u_i \geq 0, \qquad x \geq 0.$$

Since u_i is nonnegative this can also be written as

$$a'_i x = \sum_{j=1}^{n} a_{ij} x_j \leq -\log u_i / \lambda_i, \qquad i = 1, 2, \cdots, m.$$

If λ_i is known or estimated by the reciprocal of sample mean, then on defining $x_{n+1} = -\log u_i$ and adjoining the product $\prod_{i=1}^{m} u_i$ of the component reliability levels through the function $\sum \log u_i$, we could write a transformed programming problem as follows.

$$\text{maximize } z = w_1 \sum_{j=1}^{n} c_j x_j - w_2 \sum_{i=1}^{m} x_{n+i}$$

subject to

$$\lambda_i \sum_{j=1}^{n} a_{ij} x_j - x_{n+i} \leq 0, \qquad x_j \geq 0,$$

$w_1, w_2 =$ nonnegative constant weights.

Note that this is only a larger LP model. Cases of other nonnegative distributions and a more generalized framework are discussed in the literature. In the

4.2 The Stochastic Linear Programming Approach

general case we would have nonlinear programs. Economic applications and comparisons with CCP models are also reported (Sengupta and Gruver, 1969).

However, the general problem of developing a reliability measure for a stochastic LP system is this: If the decision-maker selects a specific set of parameter values $(\bar{A}, \bar{b}, \bar{c})$ and solves the LP model of Eq. (1) with those values, what is the reliability of such a solution, if any, and what are the appropriate measures of reliability of this LP system? If it is possible to develop a satisfactory measure of system reliability of the LP system, it would be of great value in designing and improving LP systems with a given level of system reliability.

One of the simplest ways to develop an operational measure of system reliability (R) for the LP system is to consider a simpler version of the CCP model in which only the elements b_i $(i = 1, 2, \cdots, m)$ of the resource vector b are random and mutually independent statistically. In this case a reliability measure R_i for the ith constraint may be equated to the probability that the ith constraint is not violated, that is,

$$R_i = \text{prob}\,(b_i \geq a'_i x) = 1 - F(a'_i x), \qquad \text{where } F = cdf. \tag{14}$$

If the random elements b_i's are statistically independent and it is assumed that the system reliability R for the LP model as a whole must be an index combining the component reliabilities R_i, then one possible measure of system reliability is

$$R = \prod_{i=1}^{m} R_i. \tag{15}$$

This measure of system reliability is based on three main assumptions:

(a) The LP system consisting of m chance constraints is assumed reliable if and only if each ith constraint is feasible and reliable up to the tolerance level u_i $(i = 1, \ldots, m)$.

(b) The system reliability measure R in Eq. (15) is bounded in the interval $[0, 1]$ so that $\max_i R_i \geq \min_i R_i \geq R$. This implies that the system reliability R cannot have a value higher than its least reliable component.

(c) The system reliability measure R defined in Eq. (15) is based on the probability of simultaneous satisfaction of all m chance-constraints each with component reliability R_i. This is based on the assumption that the failure of any one constraint may lead to infeasibility of solutions. Another measure R^0 based on the probability of simultaneous violation of all m chance constraints would be

$$R^0 = 1 - \prod_{i=1}^{m} (1 - R_i).$$

Once we agree to define our system reliability measure R by Eq. (15), perhaps the simplest way to incorporate this measure in a stochastic LP model is to hypothesize a utility function $U = U(R, z)$ of the decision maker with reliability R and profits $z = c'x$ as two arguments satisfying the usual conditions

$$\partial U/\partial z > 0, \quad \partial U/\partial R > 0, \quad dR/dz < 0,$$

which imply risk aversion and guarantee in cases the U function is concave in its arguments the attainment of its maximum value. One possible choice of such a utility function is

$$U = w_1 \sum_{j=1}^{n} c_j x_j + w_2 \log R = w_1(c'x) + w_2 \sum_i \log u_i,$$

where w_1, w_2 are fixed nonnegative weights. In this case (where only b is assumed random) the transformed programming problem which incorporates the system reliabilities may be formulated as follows.

$$\begin{aligned} \text{maximize} \quad & U(x, u) = w_1 c'x + w_2 \sum_{i=1}^{m} \log u_i, \\ \text{subject to} \quad & x \geq 0, \quad 0 \leq u_i \leq 1, \\ \text{and} \quad & R_i(a_i'x) - u_i \geq 0, \quad i = 1, 2, \ldots, m, \\ \text{where} \quad & R_i(a_i'x) = 1 - F_i(a_i'x). \end{aligned} \quad (16)$$

The case of different distribution assumptions for the cumulative variates $F_i(a_i'x)$, its use in decomposed LP models and subsystems and its extension to other random parameters (A, c) have been investigated (Sengupta and Portillo-Campbell, 1970c).

4.2.5. Distribution Problems in SLP Models

We consider in this section very briefly the following aspects of distribution under the SLP framework: (a) The relations between the passive and the active approaches to SLP, (b) the specification of wider class of risk attitudes, and (c) the need for developing nonparametric and distribution-free procedures useful for decision-making.

(A) Technical Methods of SLP

Consider the LP model of Eq. (1) once again.

$$\text{maximize } z = c'x, \quad \text{subject to } Ax \leq b, \quad x \geq 0. \quad (17)$$

The dual of this problem is

4.2 The Stochastic Linear Programming Approach

$$\text{minimize } z^* = b'w, \quad \text{subject to } A'w \geqq c, \ w \geqq 0. \tag{18}$$

In SLP we assume a joint probability distribution

$$P(A, b, c) \tag{19}$$

of the elements of matrix A and the vectors b and c. Problems connected with the mathematical expectation of the objective function of Eq. (17) under assumption of Eq. (19) have been studied by methods of two-stage programming under uncertainty. Tintner distinguishes two approaches to SLP problems:

(a) The passive approach: Here we derive (or approximate numerically) the distribution of the objective function of Eq. (17) under assumption of Eq. (19). Then for all possible variations of the parameters (elements of the matrix A and the vectors b and c), conditions of simple linear programming are assumed fulfilled. This approach may be used to compare, for instance, two different production situations or planning in two different countries.

(b) The active approach: Here the problem is to maximize the objective function of Eq. (17) under the conditions

$$AX \leqq BU, \quad x \geqq 0.$$

Here U is an allocation matrix $(m \cdot n)$: $[u_{ij}]$ with feasibility conditions

$$u_{ij} \geq 0 \quad \text{and} \quad \sum_{j=1}^{n} u_{ij} = 1.$$

Also, X is a diagonal square matrix with the elements of the vector x in the diagonal; B is a diagonal square matrix with the elements of the vector b in the diagonal.

The problem is again the derivation of the distribution of the objective function. But now the distribution will depend upon the elements of U (allocation matrix). We may have, for example, one production problem and consider the probability distribution of profits under various allocations of resources. Or in economic planning, we might consider the effect of the allocation of investment to various industries upon the objective function of the planner.

A selection k is a choice of m elements of the vector x, the corresponding rows of the matrix A in Eq. (17) and the corresponding element of the vector b. A selection is feasible, if the feasibility conditions of Eq. (17) are fulfilled. It is optimal, if the objective function is a maximum.

We now state without proof a number of theorems on stochastic linear programming which might be of interest to workers in the field of operations research, management science, production economics, and economic planning.

Proofs of the theorems and examples of applications will be found in the references quoted.

(B) *Weak Theorems on Stochastic Linear Programming* (Tintner et al., 1963)

Theorem 1 Let $z^{(k)}(x)$ be the objective function of Eq. (17) for a selection k of a passive stochastic programming problem and let $V(k)$ be the region in the parameter space (i. e., the space of the matrix A, the vectors b and c), where $z^{(k)}(x)$ is feasible and optimal. If v is a point in V, there is a selection k^* for the dual problem of Eq. (18) with corresponding objective function $z^{*(k^*)}(v)$ such that $z^{*(k^*)}(v) = z^{(k)}(v)$ for all x in a certain neighborhood $N_e(v)$.

Theorem 2 In the active case, assume that the elements of U can only assume a number of discrete values. Designate a given combination of these values by g. Also, k is a given selection. Designate the objective function for g and k by $z_g^{(k)}(v)$, where v is a feasible parameter point. Assume now that $z_g^{(k)}(v) \; z_{g'}^{(k')}(v)$ where k' is another selection and g' another combination of values of elements of the matrix U. Then there is a neighborhood $N_e(v)$ such that for all x in $N_e(v)$: $z_g^{(k)}(x) \geq z_{g'}^{(k')}(x)$.

Theorem 3 Assume again like in Theorem 2 that $z_g^{(k)}(v) \geq z_{g'}^{(k')}(v)$. Now consider a selection k^* for the dual. Then there is a neighborhood $N_e(v)$ about the region of feasibility in the parameter space, such that $z_h^{*(k^*)}(x) \geq z_{h'}^{*(k^{*\prime})}(x)$; here h' is another combination of values of the elements of the matrix U; $k^{*\prime}$ is another selection of the dual. However, the selection h for the dual is not necessarily optimal.

Theorem 4 Consider a point v in the parameter space, which is feasible. Let $z_1(v)$ be the maximum value of the objective function of Eq. (17) in the passive case, and $z_2(v)$ the value of the objective function in the active case. Further, restrict the variation of the elements of the matrix U to discrete values. Let $z_3(v)$ be the value of the objective function for any other values. The we have

$$z_1(v) \geq z_2(v) \geq z_3(v).$$

Corollary 4.1 Under the assumption that random variables in the probability function of Eq. (19) are mutually independent, and that each random variable is replaced by its mathematical expectation, denote the objective functions for the passive case, the continuously active and the discrete active case by $\bar{z}_1(v)$, $\bar{z}_2(v)$, $\bar{z}_3(v)$ respectively, we have

$$\bar{z}_1(v) \geq \bar{z}_2(v) \geq \bar{z}_3(v).$$

Corollary 4.2 Assume that conditions of Theorem 4 are fulfilled and none of the expected values are infinite. Denoting by E mathematical expectation, we have:

4.2 The Stochastic Linear Programming Approach

$$Ez_1(v) \geq Ez_2(v) \geq Ez_3(v).$$

Corollary 4.3 Denote by $*$ the dual problems to the problems in question. we have:

$$z_3^*(v^*) \geq z_2^*(v^*) \geq z_1^*(v),$$

where v^* is the point dual to v.

Theorem 5 Assume now in the active approach that the elements of the matrix A and the vector b are symmetrically and mutually independently distributed.

Then it is approximately true for any feasible x that

$$Ez(x) \geq zx(E),$$

where $x(E)$ denotes the elements of the vector x if all random variables are replaced by their expected values and we neglect cross terms. Also approximately

$$Ex \geq x(E).$$

Corollary 5.1 Let now z' be the maximum objective function under conditions of the theorem. We have

$$Ez'(x) \geq z'(x(E)), \qquad Ex' \geq x'(E),$$

where x' is a vector x corresponding to the optimum.

Corollary 5.2 Denote by z^* the objective function of the dual, then we have

$$Ez^*(x) \geq z^*(E(x))$$

both for all feasible and optimal x.

Corollary 5.3 If the additive random components of the matrix A and the vector b converge to zero in probability, $E(x)$ converges in probability to $x(E)$ and $Ez'(x)$ converges in probability to $z'(x(E))$.

A few words may be added at this stage about the problems of computation of the distribution of optimal profits. Consider the passive approach aspect of the LP model of Eq. (17), when the set $S = (A, b, c)$ has random elements.

Let y be a m-component column vector of slack variables and I the identity matrix of order m. Then we replace the m constraints by

$$Ax + Iy = b, \qquad x \geqq 0, \quad y \geqq 0.$$

Defining the following vectors and matrices

$$\bar{c} = (c, 0), \quad \bar{x} = (x, y) \quad \bar{A} = (A, I) \quad \bar{c} = (m+n) \cdot 1,$$
$$\bar{x} = (m+n) \cdot 1, \quad \bar{A} = m \cdot (m+n),$$

the original LP problem becomes

$$\text{maximize } z = \bar{c}'\bar{x}$$

under the conditions $\bar{A}\bar{x} = b$, $\bar{x} \geqq 0$.

Assume that the rank of the matrix $\bar{A}$ is m. Then we can solve the problem by the method of selection, that is, by combinatorial method of complete selection. There are then at most

$$K = \binom{n+m}{m} = (n+m)!/(m!\,n!)$$

possible selections, not all of which may satisfy feasibility (i. e., nonnegativity) conditions. For each selection we choose m elements of the vectors $\bar{c}$ and $\bar{x}$ and the corresponding m columns of $\bar{A}$. Let us denote selection k ($k = 1, 2, \ldots$, up to at most K) by a superscript. Then we have

$$z^{(k)} = (\bar{c}^{(k)})'\,\bar{x}^{(k)}, \quad \bar{A}^{(k)}\bar{x}^{(k)} = b.$$

If the solution $\bar{x}^{(k)}$ for selection k are nonnegative, we call the selection feasible. Taking only the feasible selections out of K possible selections, we compute

$$\bar{z}^* = (\max_{k}(z^{(k)}|\text{selection } k \text{ feasible}).$$

Now consider the probability distribution of Eq. (19), where the parameter space is defined to be the total range of variation of the random elements of the set (A, b, c). Let S_k be the region in the parameter space where selection k is feasible (i. e., $\bar{x}^{(k)} \geqq 0$). Let T_k be the region where z_k attains its maximum. Further, the intersection V_k

$$V_k = S_k \cap T_k$$

of the two regions S_k and T_k defines the region where selection k is both feasible and maximal (i. e., optimal). By considering the probability of all points in each region V_k, we can compute the probability distribution of the maximum $\bar{z}^*$, if necessary, by numerical and approximate methods; this distribution when nondegenerate could be utilized to specify rules of optimal decision-making under risk. For instance, if $Q(\bar{z}^*)$ denotes the probability distribution of the

4.2 The Stochastic Linear Programming Approach

maximum z^*, which is derived from the probability distribution of Eq. (19), then this provides several useful implications of alternative feasible decision rules that are candidates for the optimal decision rule, namely,

(a) The first four moments of the distribution $Q(\bar{z}^*)$ would indicate the probability of attaining any specific value of the objective function (z_0) [i. e., prob $(\bar{z}^* \leq z_0)$] so that alternative types of deteministic equivalents could be evaluated.

(b) The confidence intervals and other probabilistic limits for $\bar{z}^*$ could also be derived so that, for any particular z_0, the chance-constrained value α_0 where

$$\text{prob } (\bar{z}^* \leq z_0) \geq \alpha_0$$

could be computed, and this may provide indices of sensitivity when properly interpreted.

(c) Further, it is possible sometimes to derive the bounds and inequalities for any optimal solution vector $\bar{x}^*$ corresponding to $\bar{z}^*$ by statistical methods, both parametric and nonparametric.

(d) In some cases, when the coefficients in the objective function only are random, the variance of the optimal $\bar{z}^*$ in relation to the expected value (if too high) may indicate the need for adopting a more cautious policy (like the selection of second best and third best solutions etc.) based on extreme points other than the optimal extreme point. Algorithms for locating several extreme points are also available now (Balinski, 1961; Bracken and Soland, 1966).

(C) Active Approach

We have already mentioned that the active approach defines a particular type of decomposition of the SLP problem through resource allocation matrices. The implications of this aspect are twofold. First, it imparts flexibility in the policy space because the distribution of optimal profits conditional on the specific choice of the allocation matrix could be analyzed. In a sense it is a type of truncation in the distribution space of optimal profits. The dual variables corresponding to the decomposed sub-models of the active approach provide valuable measures of the implicit cost of pre-assigning a specific allocation matrix. Second, the relative contribution of individual activities to the mean or the variance of total optimal profits may be readily assessed in this approach.

The decomposition aspect of the active approach and its application may now be briefiy illustrated. Consider a deterministic LP model

$$\max z = c'x, \quad \text{subject to } Ax \leq b, \quad x \geq 0 \tag{20}$$

and define as before a new set of decision variables u_{ij} where

$$b_i = \sum_{j=1}^{n} b_i u_{ij}, \quad u_{ij} \geq 0, \quad \sum_j u_{ij} = 1.$$

By using these decision variables the active approach specifies a separable problem for each x_j as

$$\max c_j x_j$$

under the restrictions

$$a_{ij} x_j \leq b_i u_{ij},$$

$$u_{ij} \geq 0, \quad \sum_{j=1}^{n} u_{ij} = 1, \quad x_j \geq 0, \qquad (i = 1, \ldots, m; \; j = 1, \ldots, n) \quad (21)$$

Now consider the jth separable linear program of Eq. (21). For a given $j = 1, \ldots, n$ let $W_j = (x_{j1}, x_{j2}, \ldots, x_{jn_j})$ be the set of all extreme points of the set defined by the constraints of the system (21), where each feasible x_j (and hence the elements x_{jn_j} of the set W_j) is a function of the decision variables u_{ij}. We assume for simplicity that there are r_j extreme points for each jth sub-program. Further define

$$P_{jr} = a_j x_{jr} \quad \text{and} \quad c_{jr} = c_j x_{jr}, \qquad (22)$$

for $r = 1, 2, \ldots, r_j$, $j = 1, \ldots, n$, and $a_j = j$th column of A, where each x_{jr} may also be denoted by $x_{jr}(u_{ij})$ as a function of the allocation ratios u_{ij}. By using a set of nonnegative weights s_{jr} for combining the separable sub-programs, one can construct an equivalent linear program as

$$\text{maximize} \sum_j \sum_r c_{jr} s_{jr}$$

under the restrictions

$$\sum_j \sum_r P_{jr} s_{jr} \leq b, \qquad (23)$$

$$\sum_r s_{jr} = 1, \quad s_{jr} \geq 0,$$

all $j = 1, 2, \ldots, n$ and $r = 1, 2, \ldots, r_j$.

Now since any point $x_j = x_j(u_{ij})$ of the feasible set of the jth sub-problem of Eq. (21) can be expressed as a convex combination of its extreme points, therefore $x_j = \sum_r x_{jr} s_{jr}$, where s_{jr} is defined in Eq. (23). This result may be used to prove the following proposition. If there exist numbers (s_{jr}) solving the combined problem of Eq. (23), then the same numbers also solve the original linear program of Eq. (20).

Now let us define provisional dual prices: y: $(m \cdot 1)$ and $\bar{y}$: $(n \cdot 1)$ corresponding to the restrictions of problem (23) and satisfying,

$$y' P_{jr} + \bar{y}_j = c_{jr},$$

that is,

4.2 The Stochastic Linear Programming Approach

$$y'a_j x_{jr}(u_{ij}) + \bar{y}_j = c_{jr}, \quad (a_j = j\text{th column of } A) \quad (24)$$

Using these quantities, the jth sub-problem of Eq. (21) with a modified objective function may be formulated as

$$\text{maximize } (c_j - y'a_j)x_j, \quad (25)$$

under the restrictions of Eqs. (21) and (24)

Now by applying the decomposition algorithm of Dantzig and Wolfe (1960), it can be easily shown that the iterative calculations implies by problems of Eqs. (25) and (23) would terminate in a finite number of steps, thus solving the original linear program of Eq. (20).

Two comments about the active approach should be added. First, it uses the method of reshuffling the resource vector through allocation ratios, much like the method of two-level planning of Kornai and Liptak (1965), but nonetheless sets up the sub-programs of Eq. (25) with revised profit coefficients, much like the decomposition algorithm developed by Dantzig and Wolfe. When the random variations in parameters are further admitted, the active approach of decomposition serves to decompose the overall SLP problem into components. The probabilistic aspects of convergence of the iterative process implied by Eqs. (25) and (23) require a more intensive investigation, before the implications of terminating the iteration at any arbitrary step (hence getting only a suboptimal solution) can be fully evaluated. Second, the active approach is developed essentially to show that the statistical distribution of the optimal objective function [when the program parameters (A, b, c) have random components] conditional on the choice of the allocation matrix (u_{ij}) have interesting features which could be utilized for instance, in minimizing the probability of a specified (disaster) level of profits. In particular, the allocation ratios (u_{ij}) could sometimes be so determined that the optimal risk-minimizing decision rule is sequentially improved according as the different moments of the distribution of the optimal objective function could be specified with additional precision with more samples or more information.

A few words may now be added about the specification of risk in SLP framework. We have seen before that in the CCP approach this specification is done through pre-assigning tolerance measures. In the utility function approach risk aversion parameters are specifically introduced. The following aspects may be referred to in the SLP framework.

(a) The active approach and its dual formulation in the SLP framework provide interesting insights into the risk characteristics associated with the distribution of optimal profits. Suppose $U^{(1)}$ and $U^{(2)}$ are two specific choices of the allocation matrix in the active approach, and the conditional disttibution of optimal profits induced by the parameter variations in (A, b, c) are denoted by $F(\text{opt } z \mid U^{(1)}) = F^{(1)}$ and $F(\text{opt } z \mid U^{(2)}) = F^{(2)}$, respectively. The rules for ordering these cumulative distribution functions $F^{(1)}$, $F^{(2)}$ and so on (either in the entire domain or subregion of the space of parameter variations) specify one kind of risk attitudes. On the other hand, if it is meaningful to define a specific $\bar{U}$ at which the expected profits $E(\text{opt } z \mid \bar{U})$ is maximum for $F(\text{opt } z \mid \bar{U})$ belonging to a certain class of distributions, then one could define loss functions dependent on the derivations of $U^{(1)}$, $U^{(2)}$, and so on from $\bar{U}$ and consequently in the distribution space. This is basically related to the Bayesian problem of choosing a terminal action for a given prior distribution of the parameters. This has been studied for SLP models (Bracken and Soland, 1966) for the case of the vector c following the unknown mean vector of a multivariate normal stochastic process. Problems of defining in some sense a best estimate of the vector c have also been discussed (Brogden, 1955).

(b) Another useful operational characterization is to find out where there exists an interval I_0 in the random space for which the following measure

$$\log(1 - F^{(i)}(t)), \quad i = 1, 2, \ldots, N$$

is either concave or convex in t, $t \in I_0$. In the former (latter) case the conditional distributions would display increasing (decreasing) risk aversion in the sense that the hazard rate $\lambda_i(t)$ defined by

$$\lambda_i(t) = \frac{dF^{(i)}(t)/dt}{1 - F^{(i)}(t)} \quad i = 1, 2, \quad t \in I_0$$

is increasing (decreasing). Other measures of risk aversion have also been proposed (Pratt, 1964). Several types of probabilistic bounds and inequalities (Sengupta, 1969) could be used, once it is known whether we have increasing or decreasing hazard rate cases in distributions (van Zwet, 1964).

(c) Another way of characterization would be to define extremes (or other order statistics) of sample values $(a_{ijt}, b_{it}, c_{jt})$ when they are known to be drawn from a given parent population. For example consider the case that only the vector c with independent elements c_j are normal and the set $(c_{jt}, t = 1, 2, \ldots, T)$ specifies samples of size T drawn independently at random from this population, then we may like to compare for instance the distributions of

$$z^0 = \max_x \sum_{j=1}^{n} (\max_t c_{jt}) x_j$$

with $Ax \leq b$, $x \geq 0$ and

$$z_0 = \max_x \sum_{j=1}^{n} (\min_t c_{jt}) x_j$$

with $Ax \leq b$, $x \geq 0$. Such comparisons may be defined also for given allocation matrices. Also the case of other order statistics besides the extremes may be considered. The distance between these distributions defined in some sense (Matusita, 1966) would provide interesting insights about the deviations from normality.

4.2.6. Use of Distribution-Free Methods

For practical applications the greatest need perhaps is to develop deterministic transfomations of a given SLP problem such that they are more or less robust (i, e., relatively distribution-free). This field is in our opinion wide open, although some work has been done by what has been called the safety-first approach based on Tchebycheff-type probabilistic inequalities. Another approach is povided by methods of distance functions related to the sample and population space. For instance, if we interpret that the sample values of the random variable profit ($z = c'x$) define a sampling distribution which is an approximation to the unknown population distribution of z, then the degree of reliability of a sampling approximation can be measured, in one way, at least by the probability of the maximum discrepancy between the two cumulative distributions, the sampling distribution and the population distribution. And indeed there are Kolmogorov–Smirnov limit theorems (Feller, 1948: Sengupta, 1969a) for specifying such probabilities of maximum discrepancy and the results are distribution-free. To see how these results can be utilized, assume that $z_1, z_2, \ldots, z_N$ are mutually independent random variables with a common cumulative distribution function $F(z)$. Let $z_1^*, z_2^*, \ldots, z_N^*$ be the same set of random variables rearranged in an increasing order of magnitude. Then one can define the empirical sampling distribution of the sample $(z_1, \ldots, z_N)$ by the step-function $F_N(z)$ as

$$F_N(z) = \begin{cases} 0, & \text{for } z < z_1^*, \\ k/N, & \text{for } z_k^* \leq z < z_{k+1}^*; k = 1, \ldots, N-1, \\ 1, & \text{for } z \geq z_N^*. \end{cases}$$

In other words, the sampling approximation $F_N(z)$ is such that $N \cdot F_N(z)$ equals the number of variables which do not exceed z. Since as the sample size N tends to infinity, the sample tends to be co-extensive with the whole population, it is intuitively expected that $F_N(z)$ should tend to $F(z)$ in the limit. Now Kolmogorov and others have shown that the maximum absolute deviation $D_N = \sup |F_N(z) - F(z)|$ is a random variable whose distribution is independent of the specific form of the population distribution $F(z)$, provided $F(z)$ is continuous. The limiting distribution of D_N is known and numerical tables (Owen, 1962) are available. Now, if we consider the differences $(F_N(z) - F(z))$ only, rather than their absolute values and define

$$D_N^* = \sup (F_N(z) - F(z)),$$

then the limiting distribution of D_N^* takes the following simple form

$$\operatorname*{prob}_{N \to \infty} (D_N^* \leq t) \to 1 - \exp(-2t^2 N), \qquad t > 0. \tag{26}$$

Note the similarity of this form with the utility function of Eq. (4) used in Freund's risk-programming model mentioned before. Now different choices of t determine different types of deterministic transformation of the SLP problem. For instance, if t represents the disaster level (or any level the chance of which the decision-maker wants to avoid) then the safety first approach dictates the minimization of the right-hand side of Eq. (26) as an approximation, which again is equivalent to minimizing $(2Nt^2)$ (i. e., minimizing t^2). Now if t is equated to $S(z)/M(z)$ where $S(z)$ and $M(z)$ are estimates of population standard deviation and mean of random profits z, then the following nonlinear fractional functional programming problem results:

minimize $S(z)/M(z)$,

under the restrictions

$Ax \leq b, \qquad x \geq 0.$

To indicate another type of choice of t, assume that a tolerance measure v ($0 < v < 1$) is preassigned by the decision-maker as in chance-constrained programming, that is,

$$\operatorname{prob}(D_N^* \leq t) = v, \qquad 0 < v < 1.$$

This relation can then be called a reliability equation, since it expresses the reliability level of achieving the probability $\operatorname{prob}(D_N^* \leq t)$. Given v, one possible choice of t is

$$t = (r - M(z))/S(z), \qquad r \geq M(z),$$

where r is any pre-assigned level of profits on or above the location parameter $M(z)$ (e. g., the mean of z) and $S(z)$ denotes any scale parameter (e. g., standard deviation of z).

4.2.7. STOCHASTIC APPROACH IN NONLINEAR PROGRAMMING

The stochastic approach in nonlinear programming has been developed mainly through the recent advances in research work in the field of optimal control theory and dynamic programming. Stochastic considerations in optimal control theory are particularly analyzed in the following areas: (a) Analysis of stability

4.3 Economic Applications of Stochastic Programming

characteristics (Kushner, 1967) of optimal control, (b) the methods of analyzing sensitivity of optimal and suboptimal solutions (Rekasius, 1964: Bellman, 1961), (c) computational methods based on conjugate gradient and other penalty function approaches (Balakrishnan and Neustadt, 1964), and (d) the differential game analogs of optimal control problems. Also the effects of singularity in control space, truncation of the random horizon, and the error analysis based on the statistical distribution of characteristic roots associated with linear and nonlinear differential equations arising in optimal control specification are of great importance in the theory of optimal control. We would not survey the stochastic approach in the fields of optimal control and dynamic programming. Instead we would mention very briefly some characteristics of the stochastic approach arising in nonlinear programming.

The analysis of stochastic aspects of (static) nonlinear programming problems has several facets, of which the following appear to be very important from an operational viewpoint.

(a) The use of (i) probabilistic bounds based on Jensen's inequalities and (ii) the bounds established in the mathematical theory of reliability.

(b) The CCP and SLP approaches have been analyzed for the case of linear integer programming problems (e. g., network problems and job assignment problems).

(c) Simulation analysis based on sample drawings from normal, rectangular and binomial populations has been applied to selected types of nonlinear programming problems. Characterization of a nonlinear program in terms of a stochastic game and analyzing the conditions for the existence of optimal solutions have been reported.

(d) Methods of sensitivity analysis (Boot, 1964; Fiacco and McCormick, 1964) around a particular point of the random parameter space but limited to a local neighborhood have been studied for quadratic programming and some other types of nonlinear programming. However, the distribution properties of the partial sensitivity measures are in most cases unknown.

(e) The decomposition methods applied to nonlinear programming have sometimes been used to analyze the sequential convergence of submodel optima to the overall optima. The stochastic aspects of this convergence are basically related to the multidimensional stochastic process known as the Brown–Monroe process (Wetherill, 1966). Some dynamic programming work based on transition probabilities (Sengupta and Fox, 1969) appear to be very relevant here.

4.3 Economic Applications of Stochastic Programming

The economic applications considered in this section are used for illustrating specific types of probabilistic programming (e. g., stochastic, chance-constrained, and reliability programming). Applications to economic models are divided into three groups. For example,

(a) economic growth and planning models,
(b) models of production and resource allocation,
(c) production and inventory decisions of an enterprise.

It should be emphasized that the empirical applications are illustrative in nature; they are utilized essentially to suggest (i) the types of characterization of stochastic considerations in some important operational economic models, (ii) the difficulties of computation, and (iii) the need for various approximations and simple decision rules. For large scale models, the problems of computation may prove enormous in the stochastic case, if the experiences with decomposition algorithms (Sengupta and Fox, 1969) offer any guide whatsoever. A partial solution to this dilemma may be obtained by requiring only suboptimal rather than fully optimal solutions and analyzing the sensitivity of suboptimal solutions. This gives an added support to the various approximation and search techniques (Balakrishnan and Neustadt, 1964; Fiacco and McCormick, 1964).

4.3.1. APPLICATION TO ECONOMIC PLANNING AND GROWTH

The applications of stochastic programming to economic planning problems would be discussed here in terms of specifications and other operational characterizations.

We consider first a static stochastic linear programming model where the input coefficients are stochastic. The cases where the net price coefficients in the linear objective function or the vector of available resources are random are far simpler compared to the case of stochastic input coefficients. The former two cases have been frequently analyzed and applied in resource allocation models under risk (Kataoka, 1963; Sengupta, 1968b). Our first simple example is drawn from the planning statistics in India (Tintner and Sengupta, 1963), where the two activities x_1 and x_2 denote the two components of national income (z), namely, investment goods and consumption goods. In this static model, national income is maximized, subject to the resource restrictions set by capital and labor. The two capital input coefficients were assumed random, since these coefficients were varying for our sample observations much more than the labor coefficients. We had 16 sample observations on the two (aggregative) capital coefficients (i. e., a_{11} and a_{12}) with mean values 3.664476 and 2.685166. Taking these mean values the LP model is

$$\text{maximize } z = x_1 + x_2$$

under the conditions

$$3.664476 x_1 + 2.685166 x_2 \leq 104 \text{ (capital)},$$

4.3 Economic Applications of Stochastic Programming

$$0.000250 x_1 + 0.000413 x_2 \leq 0.014 \text{ (labor)},$$
$$x_1 \geq 0, \quad x_2 \geq 0.$$

Note that the model could be interpreted in a comparative-static framework by assuming that the activities $x_j = x_j(T) - x_j(0)$, $j = 1, 2$ are in fact deviations from known initial values $x_j(0) \geq 0$ and we intend to solve for the next "period" (T) optimal values only. As in comparative-static analysis this approach neglects to consider the particular time paths by which optimal $x_j(T)$ values are attained. Now the capital constraint can be written as

$$a_{11} x_1 + a_{12} x_2 \leq b_1, \quad \text{where } b_1 = 104,$$

where a_{11}, a_{12} are stochastic variables following some probability distributions. Consider how the various methods of probabilistic programming would incorporate the stochastic considerations into the LP model. The chance-constrained programming (CCP) approach would interpret that there would be a finite probability of violating the above resource constraint for some values of the two random variables and would therefore preassign a tolerance measure (e. g., $u = 95\%$) so that the stochastic constraint becomes

$$\text{prob}(a_{11} x_1 + a_{12} x_2 \leq b_1) \geq u.$$

If a_{11} and a_{12} are normally distributed with means $\bar{a}_{11}$, $\bar{a}_{12}$ then this chance-constraint can be reduced to

$$b_1 \geq \bar{a}_{11} x_1 + \bar{a}_{12} x_2 + \sigma F^{-1}(u)$$

where σ denotes the standard deviation of $(a_{11} x_1 + a_{12} x_2)$, that is,

$$\sigma^2 = x_1^2 (\text{var } a_{11}) + x_2^2 (\text{var } a_{12}) + 2 x_1 x_2 \text{ cov}(a_{11}, a_{12})$$

and $F^{-1}(w)$ denotes the inverse of the cumulative normal distribution of a $N(0, 1)$ variate. In the reliability programming approach the tolerance measure, u is not pre-assigned but solved for optimally by adjoining to the original objective function another facet specifying a measure of system reliability (R) as a function of the tolerance measures for different constraints.

Note however that if the capital coefficients are not normally distributed, the above deviation of the CCP approach is not possible and it turns out that our sample observations suggest the empirical distributions to be Beta distributions which have a nonnegative domain. The stochastic linear programming (SLP) approach would utilize the random variations in capital coefficients in two ways. First, by considering some or all of the sample observations to compute the probability distribution of optimal national income. Second, by using selected points from the empirical distribution of the capital coefficients. These two steps may be very simply illustrated. According to the passive approach of SLP, we use the 16 pairs of sample values of the capital coefficients for the two sectors and compute the 16 values of the maximal objective function z^*, the mean and variance of which turn out to be 35.8243 and 108.3267. The empirical probability density function $f(z^*)$ of z^* is estimated to be a Beta-distribution as follows

$$f(z^*) = \frac{573.1890}{(36.1691)} 0.7864 \frac{\Gamma(1.7865)}{\Gamma(0.6453)\Gamma(1.1412)} (z^*)^{-0.3547} (36.1691 - z^*)^{0.1412}.$$

The estimation is based on the method of moments combined with maximum likelihood (Sengupta, 1966). Using this distribution we can make various probabilistic statements. For example, the statistical distribution of z^* has a frequency function with an infinite ordinate and a J-shaped curve for which moments higher than the second are very important and similarly we may compute prob $(z^* \geq z_0)$, where z_0 may be a prescribed level of real national income. The prescribed income level may be given a safety first interpretation (Roy 1952; Sengupta, 1969).

Also, it is possible to consider a direct method of approximation (as distinct from the earlier indirect method) by which we fit empirical probability density functions to the random coefficients (e. g., the two capital coefficients). These empirical frequency functions for the two coefficients a_{11}, a_{12} turn out to be Beta distributions as

$$f(a_{11}) = 6.0930 \left(1 + \frac{a_{11}}{1.6903}\right)^{-0.3887} \left(1 - \frac{a_{11}}{2.5645}\right)^{-0.5897},$$

$$f(a_{12}) = 47.8227 \left(1 + \frac{a_{12}}{7.3612}\right)^{2.1789} \left(1 - \frac{a_{12}}{0.7676}\right)^{0.2272}.$$

We assume mutual independence of the random variables and for each probability distribution the total range except for the 5 per cent upper and lower tails is divided into four equal parts and the cumulative probabilities evaluated for each part. Thus, we derive Table 1 for an approximation of the joint probability distribution of a_{11} and a_{12}.

TABLE 1. Joint Probabilities of a_{11} and a_{12}

			a_{12}		
		0.6109	1.6219	2.6330	3.6441
a_{11}	0.0010	0.0647	0.0713	0.0266	0.0154
	2.3238	0.1125	0.1239	0.0462	0.0268
	4.6466	0.1090	0.1200	0.0447	0.0260
	6.9694	0.0594	0.0654	0.0244	0.0142

4.3 Economic Applications of Stochastic Programming

On the basis of the 16 paired values of a_{11}, a_{12} as the selected points in our parameter space, we derive the maximum value z^* for each sample in Table 2.

TABLE 2. Values of the Optimal z^*

		a_{12}			
		0.6109	1.6219	2.6330	3.6441
a_{11}	0.0010	33.898	33.898	33.898	37.384
	2.3238	50.722	48.315	41.871	28.540
	4.6466	41.585	39.177	35.805	28.540
	6.9694	38.879	37.129	34.981	28.540

By using Table 1 we may easily derive the following sample moments for the distribution of the optimal objective function under the passive approach: the mathematical expectation is 38.056, the variance 114.29, and third and fourth moment about the mathematical expectation -2425.28 and 20661300.98. In order to compare the two methods of approximating the income destribution, we may now test the hypothesis that the two methods of obtaining the sample estimates from populations with continuous cumulative distribution functions $G(r)$ and $H(s)$ are such that $G(t) \neq H(t)$. As a measue of divergence from the null hypothesis given by

$$\max |G(t) - H(t)|$$

we use the Kolmogorov–Smirnov statistic used in nonparametric theory (Feller, 1948; Birnbaum, 1952)

$$D_{m,n} = \max \left| (S_m(t) - T_n(t)) \left(\frac{mn}{m+n} \right)^{-1/2} \right|, \tag{1}$$

where $S_m(t)$ and $T_n(t)$ are the sample or observed cumulative distribution functions (i.e., necessarily step functions). For example, $S_m(t) = (1/m)$ (number of observed values $\leq t$) and $T_n(t) = (1/n)$ (number of observed values $\leq t$) for the optimal values of the objective function for the 16 admissible selections under the direct and indirect approximating procedures, respectively. The ordered values of the optimal objective function under the two approximating procedures are given in Table 3, where it may be seen that the optimal values are with few exceptions below the expected value until the 14th value for the direct method and the 13th value for the indirect method. Hence, putting $m = 14$ and $n = 13$ in Eq. (1) we get the observed value of $D_{m,n}$ equal to at least 0.51, whereas the 99 per cent values of the limiting distribution of $D_{m,n}$, tabulated by Smirnov (1948) is about 1.63 and the 95 per cent value is 1.36. Hence, the observed sample data do not disprove the null hypothesis tested, that the two approximating

TABLE 3. Ordered Maximal Values of z^* of the Primal Stochastic Problem for $K = 16$ Admissible Selections.

Direct Method	Indirect Method
28.5396	22.3600
28.5396	23.8160
28.5396	24.0240
33.8983	25.2720
33.8983	27.7680
33.8983	29.9520
34.9809	32.0320
35.8047	33.9035
37.1294	34.1168
37.3840	34.6492
38.8794	38.8144
39.1774	42.1190
41.5847	45.7532
41.8712	50.3270
48.3147	52.2819
50.7219	56.0000
mean = 38.0558	mean = 35.8243

empirical distributions came from the same population with a continuous cumulative distribution function. We have also used the Kolomogorov statistics

$$D_m = \max_{0 \leq t \leq 16} |S_m(t) - G(t)|,$$
$$D_n = \max_{0 \leq t \leq 16} |T_n(t) - G(t)|, \quad (2)$$

for our observed sample values under the two methods of approximation to test the null hypothesis that our sample values are independent. On the basis of the distribution of the statistics in Eqs. (2) for finite sample size tabulated by Birnbaum (1952) it was found that our sample data do not disprove the null hypothesis at the 95 per cent level of significance. In regard to the application of the nonparametric statistics to our problem, it has to be noted, however, that these are based on the assumption that the cumulative population distribution functions are continuous. Although this assumption is not very restrictive, yet it may not hold, especially in cases of multivariate distributions. For other applications see Sengupta et al. (1963), Sengupta (1969a).

In the SLP framework there is a second method known as the active approach, where we introduce additional decision variables (e. g., allocation of resources to activities and decomposition of the overall LP model) and analyze how the distribution of optimal income and its various characteristics (e. g., mean, variance) would change as these decision variables are parametrically

4.3 Economic Applications of Stochastic Programming

varied. As a simple illustration of the active approach we consider a dynamic two-sector planning model which may be considered in many ways a dynamic extension of the earlier static model. This is a quite conventional planning model (Mahalanobis, 1955; Lange, 1960; Sengupta and Tintner, 1963) having close resemblance with other two-sector growth models of the Harrod–Domar variety (Hicks, 1965; Kaldor, 1961). Let I_t, C_t, and Y_t be investment, consumption and real national income in year t, λ_i is the proportion of investment allocated to the investment goods sector and $\lambda_c = 1 - \lambda_i$ is the remaining proportion allocated to the consumption goods sector. With β_i and β_c as the marginal output-capital coefficients for the two sectors the dynamic programming formulation of the two-sector model in the deterministic case is as follows:

$$\text{maximize } Y_T = C_T + I_T$$

subject to the restrictions

$$I_t \leq I_{t-1} + \lambda_i \beta_i I_{t-1},$$
$$C_t \leq C_{t-1} + \lambda_c \beta_c I_{t-1},$$
$$I_t \geq 0, \quad C_t \geq C_0 > 0, \tag{3}$$
$$\sum_{t=0}^{T} I_t \leq I_s, \quad \lambda_i + \lambda_c = 1,$$

for all $t = 1, 2, \ldots, T$.

Here C_0 is consumption in period zero and I_s the total investment available over the given planning period T. In our empirical application to Indian planning we assume $t = 1, 2, 3, 4$ and maximize the objective Y_4 by choosing C_1, C_2, C_3, C_4 and I_1, I_2, I_3, and I_4. We have used the following data from the planning statistics (third plan) in India: $I_0 = 14.40$, $I_s = 99.0$, $C_0 = 121.70$. All figures are in billions of rupees in constant 1952–1953 prices. Further, the expected values of the output-investment coefficients $\bar{\beta}_c = 0.706$, $\bar{\beta}_i = 0.335$ and their estimated variances $\sigma_c^2 = 0.4582$, $\sigma_i^2 = 0.0319$ are obtained form estimates for the period 1946–1954 which slightly differ from the earlier static model. For the fixed allocation ratio $\lambda_i = 1 - \lambda_c = 1/3$ equal to that used in the Mahalanobis model, we obtain by the simplex method the following optimal solutions:

$I_1 = 16.01$, $\quad C_1 = 128.48$, $\quad I_2 = 17.81$, $\quad C_2 = 136.02$,
$I_3 = 19.80$, $\quad C_3 = 144.00$, $\quad I_4 = 22.02$, $\quad C_4 = 153.72$,
$Y_4 = 175.74$.

Now in the stochastic version of the above model the random variations in the coefficients β_i, β_c are to be considered. Again if we follow the passive approach we may apply the direct method of fitting empirical distribution functions to these coefficients. In this case the empirical density functions turn out to be Gamma frequency functions:

$$P(\beta_i) = (10.508)^{3.520} e^{-10.5088\beta_i} \cdot (\beta_i)^{2.520}/\Gamma(3.520)$$

$$P(\beta_c) = (1.541)^{1.088} e^{-1.541\beta_c} \cdot (\beta_c)^{0.088}/\Gamma(1.088)$$

The estimation method is based on a combination of the method of moments and the maximum likelihood (e. g., the former is used to identify the form of the distribution and the latter for approximation). These approximate estimates are based on 16 sample observations which were statistically tested and found to contain no trend. Assuming independence of the two frequency functions, we derive an approximation to the bivariate distribution of the random parameters β_i and β_c by using the same method illustrated in Table 1. Finally, we derive numerically the approximate distribution of the optimal objective function Y_4 on the basis of the 16 points selected in the parameter space in Table 4.

TABLE 4. Values of Optimal Y_4 When $\lambda_i = 1 - \lambda_c = 1/3$

β_i	β_c			
	0.0444	0.7138	1.3832	2.0526
0.1041	139.95	167.07	194.19	221.31
0.2934	144.54	174.33	204.11	233.90
0.4827	149.89	182.48	215.06	247.72
0.6721	151.68	183.44	215.21	250.30

To compute the moments of the distribution of Y_4 we use the approximate joint probabilities of β_i and β_c and derive as follows: the expected value of Y_4 denoted as $E(Y_4)$ is 180.10, the variance 851.88, third and fourth moments about the mean: 10,912.82 and 173,629.54. Since the optimal value of Y_4 in the non-stochastic case (denoted by $Y_4(E)$) when the mean values of β_i and β_c were used, turned out to be 175.74, it is easily checked that

$$E(Y_4) \geqq Y_4(E)$$

in our empirical problem where E = expectation. Again, one could derive the approximate empirical distribution function of Y_4 on the basis of the sample coefficient values given in Table 4. We note only that this passive approach could be utilized for at least two different purposes. In the first place, it can be used for comparing two method of planning, depending on different random distributions of β_i and β_c. Secondly, it can be used as a check or bound based

4.3 Economic Applications of Stochastic Programming

on probabilities against a purely deterministic result derived from a nonstochastic model. For instance, on the basis of expectation and variance of Y_4 mentioned before, we can state the following distribution-free confidence interval for the sample mean.

$$\text{prob}\left[\{E(Y_4) - \mu(Y_4)\}^2 \leq \frac{V(Y_4)}{n-1} + k\sigma^2\left(\frac{2}{n(n-1)}\right)^{1/2}\right] > 1 - k^{-2},$$

where k is any positive constant and the optimal Y_4 values are a random sample of n observations from a population of mean $\mu(Y_4)$ and variance σ^2. The sample mean and variance are denoted by $E(Y_4)$ and $V(Y_4)$, respectively. This inequality is on the average much more efficient than Tchebycheff's inequality for the mean (Guttman, 1948). Similarly, a confidence interval based on the Kolmogorov statistic can be constructed by using the numerical table for finite sample sizes constructed by Birnbaum, (1952).

Now for the active approach of stochastic linear programming we may define a risk preference functional, depending on income for different years $t = 1, 2, 3, 4$ and the decision variables λ_i, λ_c defined before. This would need, however, the specification of discount rates for future consumption and national income and also some index characterizing the risk aversion or otherwise, of the national policy-maker. This would in general introduce nonlinearity into the problem. Restricting ourselves to the linear case we consider the simplest type of active approach and specify the alternative distributions of the optimal objective function Y_4 for various pre-assigned values of the decision variables λ_i, λ_c. The empirical results are given in Table 5.

TABLE 5. Alternative Implications of the Active Approach

Characteristics of the distribution of Y_4	Policy variables		
	$\lambda_i = 1/3$ $\lambda_c = 2/3$	$\lambda_i = 1/2$ $\lambda_c = 1/2$	$\lambda_i = 2/3$ $\lambda_c = 1/3$
Expectation	180.10	174.20	166.46
Variance	851.88	519.50	247.23
Skewness (μ_3^2/μ_2^3)	0.1926	0.1648	0.1131
Kurtosis (μ_4/μ_2)	2.3903	2.3254	2.3674
Lower 5% probability level	138.94	142.10	143.20
Mode	160.83	157.31	159.89
Coefficient of variation (%)	16.20	13.08	9.44

It appears on the basis of mean values of the terminal income level Y_4 that the policy of allocating one-third of capital to new investment (i. e., $\lambda_i = 1/3$). which incidentally was the allocation decided in the deterministic Mahalanobis

(1955) planning model for India, produces the highest average terminal income. A closer inspection shows however that the variance of optimal terminal income increases with decreasing λ_i. This is partly due to the fact that the output-investment coefficient β_c for the consumption goods sector (which includes agricultural output) has much higher variance than that of the other coefficient β_i; also the selection of λ_i in the model is decided once for all in the sense that it is independent of time. The characterization of a cautious policy which takes into account the risk of variability of optimal terminal income Y_4 along with is average value depends among other things on the planner's risk preference functional, the reliability of the approximate distribution of terminal income and perhaps the particular class of alternative feasible policies considered by the planner. For instance, in the class of policies ($\lambda_i = 1/3, 1/2, 2/3$) considered in Table 5 the planning authority has the choice of a cautious policy in the allocation $\lambda_i = 2/3$ which under the conditions and assumptions of the model attains the optimal level Y_4 at 95% probability level; in this case it has also the lowest variance among the three feasible policies.

Several types of generalization of the above type of dynamic planning models are conceivable; a few of them may be briefly mentioned here. First, note that the objective function could be more generally specified and the allocation ratio $\lambda_i(t) = 1 - \lambda_c(t)$ varying over time during the planning horizon. A formulation in line with the current models of optimal growth and capital accumulation (Cass, 1965; Uzawa, 1966; Shell, 1967) may be suggested as follows:

(a) $$\text{maximize } E\left(\sum_{t=0}^{T} r^t u(c_t)\right),$$

subject to the stochastic constraints

(b) $$ni_t \leq (1 + \lambda_i(t)\beta_i)i_{t-1}, \qquad (4)$$

(c) $$nc_t \leq c_{t-1} + (1 - \lambda_i(t))\beta_c\, i_{t-1},$$

(d) $$i_t \geq 0, \quad c_t \geq c_0 > 0, \quad \sum_{t=0}^{T} i_t \leq i_s,$$

for all $t = 1, 2, \ldots, T$, where the lower case letters (c_t, i_t, i_s, etc.) denote per capita quantities (e. g., $c_t = C_t/L_t$) and labor (L) is assumed to grow exogenously as $L_t = L_0(1 + n)^t$; The planner is assumed to have an instantaneous utility function $u(c_t)$ strictly concave and nondecreasing and r^t is the exogenous discounting function ($0 < r < 1$) for the planning horizon T, and E is the expectation operator. The following special cases are operationally characterized in growth planning framework.

(a) The case with $T = \infty$ in Eq. (4a) and equalities in Eqs. (4b) and (4c) in a one-sector framework has been analyzed by Phelps (1962), Levhari and Srinivasan (1969).

4.3 Economic Applications of Stochastic Programming

(b) The implications of a quadratic objective function in Eq. (4a) in a short run horizon has been investigated by Sengupta (1970) with respect to stability, optimality and recursive characteristics (Sengupta 1966).

(c) The case of restrictions Eqs. (4b) and (4c) as equalities and the objective function of Eqs. (4a) in the form $E(c_T + i_T)$ have been analyzed and compared with cases involving inequality restrictions (Tintner and Sengupta, 1964). Applications to two-sector planning models for India and the United Arab Republic are available (Sengupta et al., 1963; Tintner and Farghali, 1967; Sengupta and Thintner, 1965).

(d) The implications of the planning horizon T becoming random have been analyzed (Yaari, 1965; Phelps, 1962; Mirrlees, 1965) in a slightly different context (e. g., the lifetime allocation problem of an individual's wealth).

(e) A chance-constrained interpretation of the constraints and the multi-period investment allocation problem have been characterized in its distribution (Sengupta and Portillo-Campbell, 1970b) and decision-rule aspects (Naslund and Whinston, 1962; Charnes and Kirby, 1967).

A second type of generalization of the stochastic analog of model in Eq. (3) would be to compare various feedback policies (Box and Jenkins, 1968; Fox et al., 1966) which in effect restrict attention to a certain class of decision rules and investigate their stability properties. The sensitivity of optimal and feedback policies under conditions when some interaction between the prior and posterior distributions of the random state variables C_t or I_t is admitted has been investigated in simulation (Zellner and Geisel, 1968) and numerical frameworks (Boyce, 1968).

Now we consider another application to a static production-allocation problem in stochastic linear programming in order to emphasize the importance of studying the effects of non-normality of distributions. In statistical theory and methods the robustness of various statistics and testing procedures are judged in term of their sensitivity to deviations from normality (Geary, 1947; Bradley, 1968) and in small sample data analysis problems the myth of normality has been seriously questioned (Tukey, 1962). The robustness analysis for solutions in the SLP framework is quite a new field, and hence our empirical results may have some interest here. Consider an LP model with (A, b, c) having random elements and assume that we want to characterize the set of basic feasible solutions apart from the optimum solution and the corresponding values of the objective function. For a fixed q in the sample space we may characterize the finite set of all basic feasible solutions by the subscript $k = 1, 2, \ldots, K_q$ by applying the complete description method of linear programming. Note that K_q may vary from sample to sample for different values of q. We denote by $F_q^{(k)}(z)$ the value of the objective function for a fixed q in the admissible sample space $q = 1, \ldots, Q$ and a particular $k = 1, 2, \ldots, K_q$. Now define

$$F_q^{(i)} = \max_k \{F_q^{(k)}(z) \mid k = 1, 2, \ldots, K_q\},$$

$$F_q^{(j)} = \max_k \{F_q^{(k)}(z) \mid k = 1, 2, \ldots, K_q \text{ and } k \neq i\},$$

$$F_q^{(p)} = \max_k \{F_q^{(k)}(z) \mid k = 1, \ldots, K_q \text{ and } k \neq i, k \neq j\}.$$

We assume without loss of generality that our basic feasible solutions are so defined that $F_q^{(i)}$, $F_q^{(j)}$, and $F_q^{(p)}$ are strictly positive for all admissible q and that by construction

$$F_q^{(i)} > F_q^{(j)} > F_q^{(p)} > 0 \tag{5}$$

since the weak inequalities

$$F_q^{(i)} \geq F_q^{(j)} \geq F_q^{(p)} > 0$$

can be easily reduced to strict inequalities by defining that each of the indices i, j, and p may contain more than one point (i. e., more than one selection), provided they give rise to the same value of the objective function. For example, if there are three points (i. e., three basic feasible selections) in the sequence $k = 1, 2, \ldots, K_q$ for a fixed sample q, which give rise to the identical maximum value $F_q^{(i)}$, then the superscript i contains these three points, so that in the definition of truncated maxima $F_q^{(j)}$, the condition $k \neq i$ has to be interpreted accordingly with suitable modifications. From now on we will designate $F_q^{(i)}$ the regular maximum [i. e., truncated maximum of zero order (or the best solution)], $F_q^{(j)}$ as the truncated maximum of first order (or, the second best solution), and $F_q^{(p)}$ as the truncated maximum of second order (or, the third best solution), and assume that the sample space is such that these three maximum values are generated for each admissible k satisfying the conditions of an ordinary LP problem with finite solutions.

Now consider the three truncated maxima $F_q^{(i)}$, $F_q^{(j)}$ and $F_q^{(p)}$ of orders zero, one or two respectively and let us define over the admissible sample space (i. e., admissible in the sense that the feasibility of solutions is maintained) $q = 1, 2, \ldots, Q$ the expected value of $F_q^{(s)}$ by E_s and its variance by V_s, where $s = i, j,$ or p. Then we have by construction

$$E_i > E_j > E_p > 0$$

provided the expected values are defined. Further, since for each admissible sample q the inequality of Eq. (5) holds for three strictly positive real numbers, Therefore, by the Archimedean property, there must exist variables $h_q(ij) > 1$ and $h_q(jp) > 1$ such that

4.3 Economic Applications of Stochastic Programming

$$F_q^{(i)} = h_q(ij)F_q(ij) \quad \text{and} \quad F_q^{(j)} = h_q(jp)F_q^{(p)}. \tag{6}$$

Without loss of generality, assume that we are comparing only the pair $F_q^{(i)}$ and $F_q^{(j)}$ and for simplicity we use the notation h for $h_q(ij)$, f_1 for $F_q^{(i)}$, and f_2 for $F_q^{(j)}$ and the corresponding expected values are denoted by $Eh = \bar{h}$, $Ef_1 = \bar{f}_1$, $Ef_2 = \bar{f}_2$ and the corresponding variances by V_h, V_1, and V_2, respectively.

With this notation Eq. (6) can be rewritten as

$$f_1 = hf_2. \tag{7}$$

Now we can state the following result.

Theorem 1 Let variances V_h, V_1 and V_2 defined above be bounded and nonzero. Further, let the random variable h defined in Eq. (7) be such that the following function

$$g = g(h, f_2) = h^2 f_2^2$$

is analytic around the mean values $\bar{h}$ and $\bar{f}_2$, for all possible random variations in the admissible sample space. Then under very mild conditions, the following strict inequality

$$V_1 > V_2$$

holds (i. e., the truncated maximum of zero order has higher variance than that of order one).

Proof: We define the standardized variates d_1, d_2, and d_h corresponding to f_1, f_2, and h respectively,

$$d_1 = (f_1 - \bar{f}_1)/\bar{f}_1, \quad d_2 = (f_2 - \bar{f}_2)/\bar{f}_2, \quad d_h = (h - \bar{h})/\bar{h}.$$

Then, from the usual formula for the variance of a product we get

$$V_1 = \mathrm{var}\, f_1 = \mathrm{var}\,(hf_2)$$
$$= (\bar{h}\bar{f}_2)^2 [E\{(d_h + 1)^2 (d_2 + 1)^2\} - \{E(hf_2)/(\bar{h}\bar{f}_2)\}^2],$$

where E denotes the expectation operator. Now consider the function $g = g(h, f_2) = h^2 f_2^2$ which is assumed to be analytic around the mean values $\bar{h}, \bar{f}_2$. By applying Taylor series expansion we get

$$g(h, f_2) = g(\bar{h}, \bar{f}_2) + \left\{(h - \bar{h})\frac{\partial}{\partial h} + (f_2 - \bar{f}_2)\frac{\partial}{\partial f_2}\right\} g(\bar{h}, \bar{f}_2)$$
$$+ \left\{\frac{(h - \bar{h})^2}{2}\frac{\partial^2}{\partial h^2} + \frac{(f_2 - \bar{f}_2)^2}{2}\frac{\partial^2}{\partial f_2^2}\right\} g(\bar{h}, \bar{f}_2)$$
$$+ o(h, f_2),$$

where the term $o(h, f_2)$ is negligible in the sense that its expected value is zero, Hence, taking expectations on both sides, we get

$$Eg(h, f_2) = (\bar{h}\bar{f}_2)^2 + (\bar{h})^2 V_2 + (\bar{f}_2)^2 V_h.$$

By using these results and denoting by $D = (d_h + 1)^2(d_2 + 1)^2$ it is easy to see that

$$ED = 1 - (1/\bar{f}_2)^2 + V_2(\bar{h})^2/V_2(\bar{f}_2)^2,$$

which is a strictly positive quantity, since $\bar{f}_2 > 1$. Again since h and $\bar{f}_2$ are strictly positive for all admissible sample values having nonzero variances, therefore, we can write

$$0 < E(hf_2) < \bar{h}\bar{f}_2$$

Hence, by the Archimedean property there must exist a real constant θ lying in the interval $0 < \theta < 1$ such that

$$E(hf_2) = \theta \bar{h}\bar{f}_2.$$

Hence we may easily derive

$$V_1 - V_2 = (\bar{h}\bar{f}_2)^2 \left[1 - \frac{1}{(\bar{f}_2)^2} + \frac{V_h}{(\bar{h})^2} - \theta \right] + (\bar{h}^2 - 1)V_2.$$

Since $\bar{h} > 1$ and $1/\bar{f}_2 < 1$, it is easily seen that under very mild conditions we would have $V_1 - V_2 > 0$ (i. e., $V_1 > V_2$). For example, it is easy to note that a sufficient (though not necessary) condition is that

$$V_h/\bar{h}^2 \geq \theta$$

Another such condition is

$$\theta + (1/\bar{f}_2)^2 \leq 1,$$

which is very likely to be fulfilled because both θ and $(1/\bar{f}_2)^2$ would be very small.

Corollary If under the conditions of Theorem 1 the statistical distribution of h is independent of f_2, then the strict inequality $V_1 > V_2$ necessarily holds.

This theorem is useful in situations when the coefficients in the objective function alone are probabilistic, since it may indicate their conditions when the first best solution may have higher variance (i. e., higher instability) than the second best. Also it may indicate the degree of sensitivity of the optimal solu-

4.3 Economic Applications of Stochastic Programming

tion compared to other basic feasible solutions in terms of the objective function (Sengupta et al., 1965. Sengupta and Kumar, 1964).

To illustrate the applicability of our theorem to a production planning problem, we have used the annual input-output data of a single farm enterprise in Hancock County (Iowa) for the years 1928–52. These data which were previously analyzed by Babbar (1955) and Tintner (1955) show quite interestingly the role of short-term variations in weather and other factors in agricultural production and resource use. We have considered only three activities x_1 growing corn, x_2 raising flax, and x_3 growing oats. The net prices in dollars of corn, flax, and oats are taken from Babbar's study to be $c_1 = 1.56$, $c_2 = 3.81$, and $c_3 = 0.84$, respectively. The resources land, labor, and capital are given by $b_1 = 148$, $b_2 = 234$, and $b_3 = 1800$, respectively. The vector c of net prices and b of resources are taken as fixed while the coefficient matrix $A = (a_{ij})$ is assumed to be stochastic with 25 sample values for each of the years 1928–52. Labelling the truncated maxima $F_q^{(i)}$, $F_q^{(j)}$, and $F_q^{(p)}$ of orders zero, one, and two as the maximand, truncated maximand, and maxi-minimand respectively, Table 6 specifies the first four moments of the distribution of the sample values of the maximand, the truncated maximand, and the maxi-minimand, where k indicates the sample size.

Here m_1 denotes the arithmetic mean and m_2, m_3, and m_4 are the second, third, and fourth moments about the mean, respectively. It is apparent from Table 6 that in terms of expected value, the maximand exceeds the truncated maximand at least by three times or more, whereas in terms of variance it exceeds it by at least seven times or more.

TABLE 6

The Distribution of Optimal Solutions in the Passive Approach

Type	Sample size k	Unit	Maximand	Truncated maximand	Maxi-minimand
m_1	10	original	4,368.58	1,293.30	286.5000
	18	original	4,775.04	1,552.84	300.0800
	25	original	4,955.16	1,633.38	314.6100
m_2	10	10^3	1,606.02	76.0520	4.523000
	18	10^3	1,249.16	145.2931	6.072000
	25	10^3	1,077.94	150.4412	8.244000
m_3	10	10^9	−1.03226	0.00205	0.0001834
	18	10^9	−1.44766	−0.00950	0.0003111
	25	10^9	−1.44564	−0.01203	0.0005693
m_4	10	10^{12}	7.05285	0.01141	0.0000585
	18	10^{12}	5.82904	0.04144	0.0001083
	25	10^{12}	5.40201	0.04986	0.0001964

It is also interesting to note that for different sample sizes the approximately estimated distribution turned out in each case to be a Beta distribution of the form[1]

$$y = y_0(1 + x/a_1)^{b_1}(1 - x/a_2)^{b_2}, \qquad -a_1 \leq x \leq a_2, \qquad (8)$$

with origin at the mode where y_0 is a constant and a_1, a_2, and b_1, b_2, are the parameters to be estimated. Table 7 summarizes the estimates of these parameters for different sample sizes.

TABLE 7

ESTIMATES OF THE PARAMETERS a_1, a_2, m_1, m_2 OF EQ. (8)

Type	Sample size k	a_1	a_2	b_1	b_2	Mode
Maximand	10	562.85	87.83	2.1667	0.3381	5,203.16000
	18	7,080.19	7,509.85	3.5721	0.1068	5,964.5087
	25	7,277.82	−170.22	4.910	−0.0116	6,166.5539
Truncated	10	1,058.10	643.62	3.57720	2.17590	1,242.08
Maximand	18	396.63	5,317.31	0.42896	0.16118	1,802.26
	25	1,118.60	640.76	1.30150	0.74550	1,751.47
Maxi-	10	328.94	79.39	3.6687	0.8854	248.42
minimand	18	393.36	571.75	3.6233	0.7450	349.15
	25	435.36	21.94	1.9306	0.0973	417.45

By an appropriate change of unit, Eq. (8) can be put in the form of a standard Beta distribution as

$$y = f(x) = [\Gamma(a + b)/\Gamma(a)\Gamma(b)]x^{a-1}(1 - x)^{b-1},$$

where x has the finite range $0 \leq x \leq 1$ and the parameters a and b are given by $a = b_1 + 1$ and $b = b_2 + 1$. The log likelihood, say $L(a, b)$, of the sample is then given by

$$[L(a, b)/k] = \ln \Gamma(a + b) - \ln \Gamma(a) - \ln \Gamma(b)$$
$$+ (a - 1)\overline{\ln x} + (b - 1)\overline{\ln(1 - x)},$$

where k is the number of the sample and $\overline{\ln Q}$ is expected value of $\ln Q$. The maximum likelihood equations obtained, by equating the partials of $L(a, b)$ to zero, are

---

[1] The method of estimation utilized is a combination of the Pearsonian method of moments and the maximum likelihood method. In particular, the iterative method of scoring of the maximum likelihood technique is applied to improve the efficiency of the initial trial estimates (see in particular, Sengupta and Kumar, 1964).

4.3 Economic Applications of Stochastic Programming

$$\psi(a+b) - \psi(a) = -\overline{\ln x},$$
$$\psi(a+b) - \psi(b) = -\overline{\ln(1-x)},$$

where the Digamma function $\psi(s) = d\ln\Gamma(s)/ds$ is approximated by

$$\psi(s) = \ln s - (2s)^{-1} - (12s^2)^{-1} + 0.1(120s^4)^{-1} + o(1/s^6). \tag{9}$$

The asymptotic variances and covariances of the estimates could be obtained through the inverse of the so-called-information matrix. In particular, the variances var $\hat{a}$ and var $\hat{b}$ of the estimates a and $\hat{b}$ are given approximately by

$$\text{var } \hat{a} = 2(\hat{a})^2 K^{-1}[1 + (\rho/2)(\hat{a} + \hat{b} + 1)^{-1}]$$
$$\text{var } \hat{b} = 2(\hat{b})^2 K^{-1}[1 + (\rho/2)(\hat{a} + \hat{b} + 1)^{-1}]$$
$$\text{cov }(\hat{a}, b) = 2(\hat{a}\,\hat{b})K^{-1}[1 - (\rho/2)(\hat{a} + \hat{b} + 1)^{-1})],$$

where $\rho = \hat{b}/\hat{a}$, assuming $\hat{b} > \hat{a}$ and K = sample size. By following this procedure, we have in Table 8 the asymptotic variances of a and b. It is apparent that, in terms of asymptotic efficiency, our estimates of the parameters of the distribution, which are obtained by approximating the right-hand side of Eq. (9) by retaining its first two terms and then applying the iterative method of scoring, are quite good.

TABLE 8

Asymptotic Variances of Estimates of $\hat{a}$ and $\hat{b}$

Type	Sample size k	var $\hat{a}$	var $\hat{b}$	car $(\hat{a}, \hat{b})$
Maximand	10	2.0826	0.3718	0.8149
	18	2.3649	0.1386	0.5520
	25	2.8819	0.0790	0.4770
Truncated maximand	10	4.3438	2.0914	2.8007
	18	0.2561	0.1691	0.1606
	25	0.4556	0.2620	0.2972
Maxi-minimand	10	4.4759	0.7299	1.7134
	18	2.4359	0.3470	0.8734
	25	0.7127	0.0999	0.2477

It may be easily verified that the mean and variance computed from the estimated statistical distributions of the maximand, the truncated maximand, and the maxi-minimand satisfy the inequality proved in Theorem 1.

Two comments are in order. First, the above results in the theory of second best are applicable only for "wait-and-see" or passive situations. An active approach could be developed however by introducing additional decision variables.

For instance, the allocation ratios u_{ij} and then the stochastic optimization problem would be dependent on the selection of optimal allocation ratios (Sengupta et al., 1963). Secondly, in terms of deviations from normality particularly under small sample conditions the maximand, the truncated maximand, and the maxi-minimand have different stability characteristics (i. e., degrees of convergence to asymptotic normality), although each of them follow the Beta distribution with third and fourth moments playing important roles (Sengupta and Sanyal, 1970). Applications to dynamic investment planning models emphasizing the implications of unequal variances between sectors have been reported in the literature (Frisch, 1964; Sengupta, 1963; Tintner and Sengupta, 1964).

4.3.2 Application of CCP Models to Production

This section considers application of CCP methods to problems of resource allocation and production. In the field of operations research the CCP methods have been applied in various situations; some of the important situations are (a) in characterizing the various certainty equivalents (Charnes and Cooper, 1963), (b) in building into the constraints appropriate safety margins (van de Panne and Popp, 1963; Sengupta, 1969), (c) in deriving linear and other decision rules for multi-period investment allocation and capital budgeting problems (Naslund, 1966; Naslund and Whinston, 1962; Charnes and Kirby, 1967; Hillier, 1967), and (d) in reliability programming for CCP tolerance measures (Sengupta 1969, 1969b; Sengupta and Gruver, 1969).

We consider here a simple application of CCP methods to develop a set of optimal tolerance measures instead of pre-assigning them. This also shows indirectly the implicit cost of preassigning the tolerance measures rather then optimizing them. Perhaps one of the simplest ways to develop an operational measure of reliability for the LP system is perhaps to consider a simplified version of the CCP approach under the assumption that only the elements b_i of the resource vector b are random in the constraints: $Ax \leq b$, $x \geq 0$ and mutually independent statistically (but the net price vector c and the coefficient matrix A are fixed constants). In this chance-constrained LP approach a tolerance level in terms of probability measures, one for each probabilistic constraint is pre-assigned by the decision maker and this set of tolerance measures is supposed to indicate the limit up to which constraint violations are permitted. This view of linear chance-constraints allows the interpretation of a LP model as a *system*, where each probabilistic constraint can be viewed as a *system component*, each with its reliability (i.e., its tolerance measure). The system reliability for the LP model would be defined in this view in terms of the reliability of the individual components. For instance, consider the LP system

$$\max z = c'x, \quad a_i'x \leq b_i, \quad x \geq 0; \quad i = 1, \ldots, m, \qquad (10)$$

where a_i denotes the ith column of matrix A, x is a column vector with n elements, prime denotes transpose and it is assumed that each b_i is random with a nonnegative domain and mutually independent such that the cumulative distribution function

$$F(a'_i x) = \text{prob}(b_i \leq a'_i x) = 1 - u_i, \qquad 0 < u_i < 1 \tag{11}$$

is specified (i. e., either pre-assigned or assumed known or estimable). A reliability measure R_i for the ith constraint may then be set up in terms of the probability that the ith constraint violated

$$R_i = u_i = \text{prob}(b_i \geq a'_i x) = 1 - F(a'_i x). \tag{12}$$

If the random elements b_i's are statistically independent and it is assumed that the system reliability R for the LP model as a whole must be an index combining the component reliabilities R_i, then one possible measure of system reliability is

$$R = \prod_{i=1}^{m} R_i = \prod_{i=1}^{m} u_i. \tag{13}$$

This measure R of system reliability is based on three main assumptions.

(i) The LP system consisting of m chance constraints is assumed reliable, if and only if each ith constraint is feasible and reliable up to the tolerance level $u_i = (i = 1, \ldots, m)$

(ii) The system reliability measure R in Eq. (13) is bounded in the interval [0, 1] as also the component reliabilities R_i or u_i; hence

$$1 \geq \max_i R_i \geq R = \prod_{i=1}^{m} R_i \leq \min_i R_i \geq 0,$$

This implies that the system reliability cannot have a value higher than that of its least reliable component, for example, any component having zero reliability automatically makes the system reliability take on a zero value.

(iii) The system reliability meased on the probability of *simultaneous nonviolation* of all m chance-constraints, each with its reliability level u_i ($i = 1, \ldots, m$).

If we restrict ourselves for the present to the measure R of system reliability defined in Eq. (13) and (12), then one simple way of incorporating the system reliability into the original LP problem with linear chance-constraints is to associate a monotonic pay-off function for achieving the level R of system reliability and thereby hypothesize, as in the portfolio analysis model of Markowitz (1959) a utility function $U = U(R, z)$ of the decision maker with reliability

(R) and profits ($z = c'x$) as two arguments satisfying the usual condtions,

$$\partial U/\partial z > 0, \qquad \partial U/\partial R > 0, \qquad dR/dz < 0. \tag{14}$$

which imply risk aversion and guarantee in cases the U function is concave in its arguments the attainment of its maximum value. One possible choice of the utility function satisfying the conditions (14) is

$$U = w_1 \sum_{j=1}^{n} c_j x_j + w_2 \sum_{i=1}^{m} \log u_i = w_1(c'x) + w_2 \log R, \tag{15}$$

where w_1, w_2 are scalar nonnegative constant weights assumed known. In this case the transformed programming problem which incorporates the system reliability may be presented as follows:

$$\text{maximize } U = w_1 c'x + w_2 \sum_{i=1}^{m} \log u_i \tag{16}$$

$$R_i(a_i'x) - u_i = 0, \qquad 0 \leq u_i \leq 1, \qquad x \geq 0, \qquad i = 1, \ldots, m,$$

where the component reliability $R_i = R_i(a_i'x)$ is defined by $\text{prob}(b_i \geq a_i'x)$, where b_i is assumed random in a nonnegative domain. Note that if we interpret the linear constraints in Fq. (11) as in chance-constrained linear programming (i. e., $R_i(a_i'x) \geq u_i$), then the constraints of the above nonlinear program would appear as follows

$$R_i(a_i'x) - u_i \geq 0, \qquad i = 1, \ldots, m, \qquad 0 \leq u_i \leq 1, \qquad x \geq 0. \tag{17}$$

This constraint (17) may be further simplified by introducing assumptions about the specific probability distribution of the random variables b_i. For instance, if each b_i followed a two-parameter exponential density $p(b_i) = \lambda_i \exp\{\lambda_i(b_i - \theta_i)\}$; $b_i \geq \theta_i \geq 0$ with parameters $\lambda_i > 0$ and $\theta_i > 0$, then the reliability constraint of Eq. (17) becomes

$$\exp\{-\lambda_i(a_i'x - \theta_i)\} - u_i \geq 0, \qquad i = 1, \ldots, m, \qquad u_i, x \geq 0, \qquad u_i \leq 1.$$

Since u_i is nonnegative and for all practical purposes greater zero, this can also be written as

$$a_i'x - \theta_i \leq -\log u_i/\lambda_i; \qquad i = 1, \ldots, m.$$

Note that in this case λ_i and θ_i can be replaced by their best estimates and on defining $x_{n+i} = -\log u_i$, the transformed programming problem can be reduced to a LP problem for choosing nonnegative activities $x_j \geq 0$ which

4.3 Economic Applications of Stochastic Programming

$$\text{maximize } U = w_1 \sum_{j=1}^{n} c_j x_j - w_2 \sum_{i=1}^{m} x_{n+i}$$

under the constraints

$$\lambda_i \sum_{j=1}^{n} a_{ij} x_j - x_{n+i} \leq \lambda_i \theta_i, \qquad i = 1, 2, \ldots, m. \tag{18}$$

$$x_j \geq 0, \qquad j = 1, \ldots, n+m$$

Three points about this program are to be noted. First, note that a set of optimal u_i's is here solved for along with the optimal activities x_j ($j = 1, \ldots, n$). This is different from the chance-constrained LP approach, where the set of tolerance measures u_i's is to be pre-assigned by the decision maker. Second, the shadow prices (i.e., optimal dual variables) associated with the optimal values of $x_{n+i} = -\log u_i$ in the dual LP problem corresponding to Eq. (18) provide important information about the profitable direction of change of each resource b_i; in particular if a specific set of tolerance measures u_i is pre-assigned by the decision maker exogenously, then its implicit cost can be compared with that given by the optimal shadow prices (specific to x_{n+i}'s) which are optimal in the LP problem dual to Eq. (18). Third, note there is no component reliability associated with the profit function $z = c'x$, since the introduction of utility function U and the definition of system reliability adopted here preclude this necessity. However, if it is possible to associate a reliability measure u_0 with the profit function $z = c'x$, for example,

$$\text{prob}(z = c'x \geq m'x) = u_0, \qquad m = (m_j), \qquad j = 1, \ldots, n, \tag{19}$$

where profit levels are assumed random and the level $z_0 = m'x$ (e.g., this may be the expected profit) of profits is so determined that Eq. (19) holds, then in one version, the transformed programming model may be specified mostly in terms of reliability measures, for example,

$$\text{maximize } U = \sum_{i=0}^{m} \log u_i,$$

under the conditions

$$R(z_0) = R(m'x) \geq u_0,$$
$$R_i(a_i'x) - u_i \geq 0, \qquad i = 1, \ldots, m,$$
$$x \geq 0, \qquad 1 \geq u_i \geq 0, \qquad i = 0, 1, \ldots, m.$$

Note that in this case the reliability measure $R(z_0)$ is conditional on the specified level z_0 of profits; also it depends on the form of the probability distribution of profits.

Now consider a second alternative approach of incorporating system reliability measures into the original LP model involving chance constraints. In this approach we do not introduce any separate utility function satisfying conditions of Eq. (14). Instead we consider the LP model in Eq. (1) under the assumption that the parameters (c_j, a_{ij}, b_i) of the set c, A, b have multivariate distributions with finite means and covariances; also we restrict ourselves to the following linear class of decision rules as in the approach of chance-constrained linear programming:

$$x = Db, \qquad D: n \cdot m, \qquad b: m \cdot 1, \tag{20}$$

where the elements of matrix D are the decision variables to be optimally determined. The reliability measures u_i $(i = 1, \ldots, m)$ are now introduced as

$$\text{prob}(a_i' x \leq b_i) = 1 - \text{prob}(b_i \leq a_i' x) \geq u_i, \qquad i = 1, \ldots, m, \tag{21}$$

where the probability measures u_i are assumed pre-assigned by the decision maker. This decision-rule aspect of CCP models in the context of reliability analysis has been discussed by a number of authors (Naslund, 1967; Sengupta, 1969b). Here we restrict ourselves to the so-called zero-order class of decision rules which assume that the elements b_i of the vector b alone are independent and normally distributed with finite means $m_i = Eb_i$ and standard deviations $s_i = \sqrt{\text{var }} b_i$; $i = 1, 2, \ldots, m$. By the assumption of independence of the distribution of b_i in respect of the decision vector x, the CCP model could still be solved with pre-assigned tolerance measures and the optimal solutions of the decision variables x would still be useful in a passive sense in what has been sometimes termed "wait and see situations". The active approach specified by the linear class of decision rules in Eq. (20) on the other hand is more applicable for "here and now" situations.

Taking the passive approach version (i. e., zero-order decision rules) of the CCP model, the chance-constrained inequalities of Eq. (21) may be transformed as follows:

$$a_i' x \leq m_i + s_i k_i, \qquad k_i = \phi^{-1}(1 - u_i), \qquad i = 1, \ldots, m \tag{22}$$

where $\phi^{-1}(w)$ is the inverse of the cumulative distribution of a unit normal variate. For fixed u_i ($0 < u_i < 1$) the constant $k_i = \phi^{-1}(1 - u_i)$ can be read off from normal distribution tables and $k_i \leq 0$ for $u_i \geq 0.50$. The transformed LP model now becomes

$$\text{maximize } z = c'x \tag{23}$$

under the constraints (6c) and $x \geq 0$

4.3 Economic Applications of Stochastic Programming

Since u_i's are pre-assigned constants, this problem can be alternatively written as follows in a form comparable to the linear reliability model in Eq. (18) mentioned before

$$\text{maximize } U(z) = w_1(c'x) + w_2 \sum_{i=1}^{m} (\log u_i), \qquad (24)$$

under the constraints

$x \geq 0$ and Eq. (22) and $u_i \geq 0.50$ (pre-assigned).

The two LP problems in Eqs. (18) and (24) can be directly compared in several respects and in principle the actual levels of u_i at the optimal basic solution can be computed, although the computation required in Eq. (24) may be more involved. Note that the reliability programming (RP) model in Eq. (18) can be used to analyze the effects on the optimal solution due to departures from normality in the CCP model in Eq. (24). For instance, assume that the RP model in Eq. (18) has been solved and a set of optimal u_i's determined and let these optimal u_i's be in fact the tolerance measures pre-assigned in the CCP model in Eq. (24). If the optimal profits and the shadow prices associated with tolerance measures u_i are not much affected, then the CCP model based on normality assumtions may be taken to be robust in the sense of insensitivity to slight departures from normality. Ideally of course departures from normality would have to be tested by considering other distributions besides the exponential. However, since the exponential provides a limiting distribution under certain conditions for the class of nonnegative random variables, the comparison of the two models in Eqs. (18) and (24) would provide very important indications.

Now we consider the following RP models for applying the LP models in Eq. (18) with or without reliability optimization.

Problem 1A: The LP model in Eq. (18) with $w_1 = w_2 = 1.0$ under the additional resitictions

$$x_{n+i} \leq -\log u_i^* \ (i = 1, 2, \ldots, m); \ u_i^* \text{ pre-assigned}$$

(log is natural log, i.e., ln)

Problem 1B: the LP model in Eq. (18) with $w_1 = w_2 = 0.50$ under the additional restrictions

$$\sum_{i=1}^{m} x_{n+i} \leq -\log R^* = -\sum_{i=1}^{m} \log u_i^*; \ u_i^* \text{ pre-assigned}$$

Problem 2A: the LP model in Eq. (18) with $w_1 = 1.0$, $w_2 = 0$ and $x_{n+i} = -\log u_i$ pre-assigned $(i = 1, 2, \ldots, m)$

Problem 2B: the LP model in Eq. (23) under the assumption that each b_i is normally independently distributed with mean m_i and standard deviation s_i.

The coefficient matrix $A = a_{ij}$ used for the above problems were as follows

	Corn x_1	Oats x_2	Soybeans x_3	Flax x_4	Wheat x_5	
	0.31772	0.27870	0.70812	0.96956	1.00365	Capital
	0.02255	0.07523	0.05485	0.21186	0.42324	July labor
$A =$	0.00000	0.08370	0.00000	0.30910	0.08650	Aug. labor
	0.05253	0.00000	0.11681	0.00000	0.00000	May labor
	0.02274	0.02770	0.05862	0.09242	0.09081	Land

The net prices (c_j) for the five crops are $c_1 = 1.0025$ (corn), $c_2 = 0.6455$ (oats), $c_3 = 2.3690$ (soybeans), $c_4 = 2.9145$ (flax), and $c_5 = 1.7408$ (wheat). These estimates are derived from quarterly data which were previously utilized in related studies (Tintner, 1955; Sengupta and Portillo–Campbell, 1970) on stochastic programming. Resource availabilities other than land were assumed to be chance-constrained and the following means (m_i) and standard deviations (s_i) were utilized:

$$\text{capital: } m_1 = 1800, \quad s_1 = 60.00000,$$
$$\text{July labor: } m_2 = 234, \quad s_2 = 21.63331,$$
$$\text{August labor: } m_3 = 234, \quad s_3 = 21.63331,$$
$$\text{May labor: } m_4 = 182, \quad s_4 = 19.07878.$$

Although the above values of standard deviations were developed for illustrative purposes only, they have some basic similarity with the empirical studies of Babbar (1955) and Tintner (1955) on stochastic linear programming for a typical farming enterprise in Hancock County (Iowa). When a normal distribution is assumed for the elements of vector b, the above parameters are sufficient to fully determine the distribution function. When a two-parameter exponential is assumed as in Eq. (18) the following estimates of the parameters λ_i and θ_i are used

$$\lambda_1^{-1} = s_1 = 60.00000, \quad \lambda_2^{-1} = s_2 = 21.63331,$$
$$\lambda_3^{-1} = 21.63331, \quad \lambda_4^{-1} = 19.07878,$$
$$\theta_1 = m_1 - s_1 = 1740.0, \quad \theta_2 = m_2 - s_2 = 212.36669,$$
$$\theta_3 = m_3 - s_3 = 212.36669 \quad \text{and} \quad \theta_4 = 162.92122.$$

The results of numerical application of these problems are reported in the following tables. In these tables the term "component reliability" means the

TABLE 1. Optimal Solutions and Shadow Prices of CCP and RP Models of Maximization

Activity		OPT. (RP) 1A(2 par.exp.)	0.95 CCP 2A(2 exp.)	CCP 2B(normal)	Component Reliability OPT. (RP) 1A(2 exp.)	0.90 CCP 1A(2 exp.)	CCP 2B(normal)	0.75 OPT. (RP) 1A(2 exp.)	0.50 OPT. (RP) 1A(2 exp.)
Corn	(1)	0	0	0	0	0	0	0	884.75684
Oats	(2)	0	0	0	0	0	0	0	0
Soybeans	(3)	1403.13133	1403.13132	1289.42137	1411.96268	1411.96267	1348.75987	1441.74132	1110.08817
Flax	(4)	644.36189	644.36188	602.71221	647.59663	647.59665	624.44667	658.50396	679.07083
Wheat	(5)	0	0	0	0	0	0	0	0
$-\ln u_1$	(6)	0			0	0		0	0
$-\ln u_2$	(7)	0.05129			0.10536			0.28768	0.69315
$-\ln u_3$	(8)	0			0			0	0
$-\ln u_4$	(9)	0.05129			0.10536			0.28768	0.69315
Shadow price, Row									
Capital	(1)	0	0	0	0	0	0	0	0
July labor	(2)	13.75578	13.75578	13.75578	13.75578	13.75578	13.75578	13.75578	5.58483
Aug. labor	(3)	0	0	0	0	0	0	0	0
May labor	(4)	13.82155	13.82155	13.82155	13.82155	13.82155	13.82155	13.82155	8.26561
$-\ln u_1$	(5)	0			0			0	0
$-\ln u_2$	(6)	296.58310			296.58310			296.58310	119.81839
$-\ln u_3$	(7)	0			0			0	0
$-\ln u_4$	(8)	262.69833			262.69833			262.69833	156.69767
Land	(9)	0	0	0	0	0	0	0	18.71659
$U = c'x + \Sigma \ln u_i$		5201.77938	5201.88192	4811.12340	5232.01973	5232.23051	5015.03705	5333.98792	5494.39742
$c'x = z$		5201.88196	5201.88192	4811.12340	5232.23045	5232.23051	5015.03705	5334.56328	5495.78372
System reliability									
R		0.9024	$\geq$0.814506	$\geq$0.814506	0.8100	$\geq$0.65610	$\geq$0.65610	0.5625	0.2500
Capital u_1		1.00	0.95	0.95	1.00	0.90	0.90	1.00	1.00
July labor u_2		0.95	0.95	0.95	0.90	0.90	0.90	0.75	0.50
Aug. labor u_3		1.00	0.95	0.95	1.00	0.90	0.90	1.00	1.00
May labor u_4		0.95	0.95	0.95	0.90	0.90	0.90	0.75	0.50

TABLE 2. OPTIMAL SOLUTION AND SHADOW PRICES OF CCP AND RP MODELS OF MAXIMIZATION

Activity		Component reliability			
		0.50 CCP 2A (2 par, exp.)	0.30 OPT.(RP) 1A (2 exp.)	0.30 CCP 2A(2 exp.)	$R^* \geq 0.25$ OPT.(RP) 1B (2 par. exp.)
Corn	(1)	884.75724	2810.49120	2810.49170	230.81043
Oats	(2)	0	0	0	0
Soybeans	(3)	1110.08803	327.50960	327.50938	1517.38431
Flax	(4)	679.07082	701.59820	701.59821	581.70940
Wheat	(5)	0	0	0	0
$-\ln u_1$	(6)		1.08512		0
$-\ln u_2$	(7)		1.20397		0
$-\ln u_3$	(8)		0.20789		0
$-\ln u_4$	(9)		1.20397		1.38630
Shadow price, Row					
Capital	(1)	0	0.01667	0 (slack)	0
July labor	(2)	5.58483	5.45627	5.58483	5.58483
Aug. labor	(3)	0	0.04623	0	0
May labor	(4)	8.26561	8.24236	8.26561 (slack)	8.26561
$-\ln u_1 \leq$	(5)		0		0
$-\ln u_2 \leq$	(6)		117.03713		0
$-\ln u_3 \leq$	(7)		0		0
$-\ln u_4 \leq$	(8)		156.25409		156.69767
land	(9)	0	18.68189	18.71659	18.71659
$U = c'x + \Sigma \ln u_i$		—	5634.35435		5519.96029
$c'x = z$		5495.78376	5638.05530	5638.05533	5521.34659
System reliability		≥ 0.0625	0.0241	≥ 0.0081	0.2500
u_1		0.50	0.338	0.30	1.00
u_2		0.50	0.300	0.30	1.00
u_3		0.50	0.813	0.30	1.00
u_4		0.50	0.300	0.30	0.25

minimum value which each tolerance measure u_i could attain. For Problem 1A the following alternative values of $u_i^* = 0.95, 0.90, 0.75, 0.50, 0.30$ (i. e., $-\log 0.95 = 0.05129$ through $-\log 0.30 = 1.20397$) were pre-assigned, whereas for Problem 1B, the pre-assigned value of R^* is 0.25 (i. e., $-\log R^* = -\Sigma \log u_i^* = 0.38629$). For problem 2A the pre-assigned values of u_i are 0.95, 0.90, 0.50, and 0.30, whereas for Problem 2B only two pre-assigned levels $u_i = 0.95$ or 0.90 are considered for all $i = 1, 2, 3, 4$ which means that $k_i = $ either -1.2816 or -1.6449 in Eq. (22) for $i = 1, 2, 3, 4$.

In Tables 1 and 2 the first nine rows give the optimal activity levels, the next nine rows the shadow prices, the two following rows give the optimal value of the objective function with and without the sum $\Sigma - \ln u_i$ respectively. Note that the value of the term $\Sigma - \ln u_i$ is very small relative to the $c'x = z$ value

4.3 Economic Applications of Stochastic Programming

of profits in each case and can be neglected for most problems. The last four rows give the preassigned values of u_i's in the CCP cases and the solution u_i's (i. e., anti-log of x_{n+i} at the optimal solution) in the optimal RP [i.e., opt·(RP)] cases. Note that for the CCP cases these u_i's are actually the lower bound for the set of feasible tolerance measures and the effective (i. e., binding) set of u_i's can be obtained by further calculation as will be shown below. The system reliability (R) is shown in the tables as the product of u_i's (i. e., $\sum_{i=1}^{u} u_i$).

Several interesting points are worth mentioning in Tables 1 and 2. First, the optimal levels of crop-activities are very close for CCP and Opt. (RP) models for a fixed two-parameter exponential distribution assumed in both cases, even when the component reliability level is varied from 0.95 to 0.30. This similarity is basically due to the fact that the optimal basis remains the same in the two cases of CCP and Opt·(RP) for a fixed level of component reliability. Second, the optimal levels of crop-activities and also profits are lower for the normal distribution case (CCP) compared to the two-parameter exponential case (CCP). This suggests that the normal approximation for non-normal underlying distributions (e. g., a two-parameter exponential) may be very pessimistic under certain range of the component reliability. Third, a comparison of the optimal solutions of the model 1B (2 par. exp.) Opt·(RP) with those of the other models show very markedly the scope of reliability optimization permitted through our analysis. The trade-off between system reliability and optimal profits is apparent. For instance, the last column of Table 2 shows a profit level of \$5521.35 with a reliability sum $\Sigma u_i = 3.25$, which is higher by about \$289.11 than that of models in Problem 2A (2par. exp.) CCP at a component reliability of 0.90 with a reliability sum $u_i = 3.60$. Note also that it is of some interest to analyze the shadow prices associated with the tolerance measures (u_i), which provide a good indicator of the relative cost or profit of increasing or decreasing the level of any u_i around the optimal solution. The use of such shadow prices are reported in Sengupta and Gruver (1969).

Note that the tolerance measures (u_i) are all pre-assigned in the CCP problems but after the problem has been solved, those restraints which are not binding are actually satisfied at a level higher than the pre-assigned and not always at a level of certainty (i. e., $u_i = 1.0$). The level at which u_i is effectively satisfied can however be determined after a little computation. As a first example consider Problem 2A (2 par. exp. CCP) where each u_i is pre-assigned at or above 0.30. If we look at the capital constraint which is not binding we find that 1805.10693 was used at the optimal solution. Set this equal to $-(1/\lambda_1)\ln u_1^0 + \theta_1$ and then solve for u_1^0.

$$u_1^0 = \exp[-\lambda_1(1805.10693 - \theta_1)] = 0.338 \text{ (approx.)}.$$

Note that this value of u_1^0 equals the value obtained for u_1 in Problem 1A at

0.30 component reliability in Table 2. The general formula for the effective $u_i = u_i^0$ is

$$u_i^0 = \exp[-\lambda_i(b_i^0 - \theta_i)], \quad i = 1, \ldots, m, \tag{25}$$

where b_i^0 denotes the amount of resource b_i used at the optimal solution. A similar general formula can be used for Problem 2B (normal, CCP), for example,

$$u_i^0 = 1 - \phi\left(\frac{b_i^0 - m_i}{s_i}\right), \quad i = 1, 2, \ldots, m \tag{26}$$

where b_i^0 is as before, (m_i, s_i) are the mean and standard deviation of the random element b_i and ϕ is the cumulative unit normal distribution. For a specific example consider Problem 2B (CCP, normal) at the 0.95 component reliability level and again the capital constraint. The (optimal) amount of capital used at the optimal solution is 1497.4307. Then from Eq.(26) we get

$$u_i^0 = 1 - \phi\left(\frac{1497.4307 - 1800.0}{60}\right) = 1 - \phi(-5) = 1 \text{ (approx.)}.$$

In this case $\phi(-5)$ is extremely small making u_1^0 approximately equal to 1.0. Similar calculations could be made for finding the levels of effective u_i for each constraint, if it is not binding at an optimal solution.

4.3.3 Application to Production and Inventory Scheduling

One most important problem in production scheduling is to specify the time path of input utilization and output sequence in some optimal fashion. The scheduling model formalized and applied by Holt *et al.*, (1955) specifies the objective of short-run production planning in terms of minimizing the total costs (comprising labor costs, inventory and shortage costs and overtime costs) of meeting demand within the constraints of the given system. Their problem was to determine an optimal linear decision rule for making production and laborforce decisions in successive time periods for a paint factory. With a fixed short-run planning horizon $T(0 \leq t \leq T)$ denote by $W_t =$ the volume of employment, $P_t =$ quantity of output of the single product produced by the factory, $I_t =$ inventory minus backlogs at the end of period t, $S_t =$ forecast of sales and $c_k =$ cost coefficients assumed constant. At any period t the total cost (C_t) comprises the following cost components,

Regular payroll	$[c_1 W_t$
Hiring and layoff	$+ c_2(W_t - W_{t-1})^2$
Overtime	$+ c_3(P_t - c_4 W_t)^2 + c_5 P_t - c_6 W_t$
Inventory and shortage	$+ c_7(I_t - c_8 - c_9 S_t)^2]$,

4.3 Economic Applications of Stochastic Programming

where the inventory level satisfies the relation

$$I_t = I_{t-1} + P_t - S_t \tag{27}$$

Total cost is simply $\sum_{t=1}^{T} C_t$, that is, that obtained by adding the component costs over the planning horizon. The decision rule that minimizes the expected value of the total cost function is obtained by differentiating with respect to each decision variable P_t, W_t, I_t. The resulting decision rules are of the following form

$$P_t = \sum_{i=0}^{T-1} a_i \hat{S}_{t+i} + bW_{t-1} + cI_{t-1} + d,$$

$$W_t = \sum_{i=0}^{T-1} f_i \hat{S}_{t+i} + gW_{t-1} + hI_{t-1} + k,$$

where $\hat{S}_{t+i}$ is the forecast level of demand at $t + i$ months hence made in month t and the coefficients a_i, b, c, f_i, g, h, and so on, depend only on the cost parameters and may therefore be determined once for all so long as the cost parameters do not change. The cost coefficients used in the application by Holt, Modigliani, Muth and Simon to the paint factory case are

$c_1 = 340,$ $c_2 = 64.3,$ $c_3 = 0.20,$ $c_4 = 5.67,$ $c_5 = 51.2,$

$c_6 = 281,$ $c_7 = 0.0825,$ $c_8 = 320,$ and $c_9 = 0.0.$

We retain the other assumptions of the Holt–Modigliani–Muth–Simon model, for example, perfect competition, homogeneous work force and the single-product firm. However, the inventory relation of Eq. (27) is now assumed chance-constrained

$$\text{prob}\,(P_t + I_{t-1} \geq S_t) \geq u_t; \quad t = 1, 2, \ldots, T, \quad 0 \leq u_t \leq 1, \tag{28}$$

where the sales variable S_t is assumed to follow a two-parameter exponential density function for all t

$$f(S) = m \exp(-m(S - S_0)), \quad \infty \geq S \geq S_0 \geq 0. \tag{29}$$

Utilizing Eq. (29) in (28) and after a little rearrangement of terms we can rewrite the chance-constrained inequality in Eq. (28) as follows

$$P_t + I_{t-1} - \lambda y_t \geq S_0, \quad t = 1, 2, \ldots, T, \tag{30}$$

where $\lambda = 1/m = \bar{S} - S_0$, $\bar{S}$ being the expected value of S

$$y_t = -\log(1 - u_t)$$

The use of exponential distribution is motivated by its wide applications in statistical theory of reliability (Barlow and Proschan, 1965) and the fact that it retains the constraint linearity of the transformed model, after the reliability measures are incorporated.

Note that the quadratic scheduling model we have chosen to illustate the application of reliability programming has been used very extensively by Theil (1964, 1967) and others (Box and Jenkins, 1962) in connection with certainty equivalence theorems and loss computaions. The final form of the above model used in reliability programming (Sengupta and Portillo–Campbell, 1970c) is as follows

$$\text{minimize} \sum_{t=1}^{T} (w_1 C_t - w_2 y_t)$$

subject to

$$W_t \geqq W_0, \quad P_t + I_{t-1} - \lambda y_t \geqq S_0,$$
$$y_t \geqq 0.69315, \quad y_t \leqq 6.90776, \quad \text{for} \quad t = 1, 2, \ldots, T,$$

where the total cost C_t is defined before, w_1, w_2 are nonnegative weights and the bounds on y_t are imposed only to make sure that the component reliability measure u_t lies between 0.50 and 0.99 for all t. These *a priori* bounds are pre-assigned, since the measure of system reliability is based here on the probability that not all the constraints can be violated.

TABLE 1. Optimal Solutions of Quadratic Reliability Program

Time t	$w_1 = 0.8$ $w_2 = 0.2$ Work force W_t	$W_0 = 90$ $W_t \geqq 75$ Production P_t	$S = 500$ $S_0 = 450$ Inventory I_t	$I_0 = 320$ Total cost = 158611.04 Reliability measure y_t	Component reliability u_t
1	82	335.378	320.03	4.107	0.983
2	76	303.843	320.03	3.477	0.969
3	75	297.263	320.03	3.346	0.904
4	75	297.263	320.03	3.346	0.904
5	75	297.263	320.03	3.346	0.904
6	75	297.263	320.03	3.346	0.904
	$w_1 = 0.8$ $w_2 = 0.2$	$W_0 = 90$ $W_t \geqq 75$	$\bar{S} = 500$ $S_0 = 300$	$I_0 = 0$ Total cost = 170102.51	
1	83	438.63	320.0	0.693	0.500
2	77	307.06	320.0	1.635	0.805
3	75	297.25	320.0	1.586	0.795
4	75	297.25	320.0	1.586	0.795
5	75	297.25	320.0	1.586	0.795
6	75	297.25	320.0	1.586	0.795

$$\text{System reliability} = 1 - \prod_{t=1}^{T}(1 - u_t).$$

Note that this view of combining component reliabilities is different from the earlier reliability programming model considered in Section 4.3.2.

The numerical results of our quadratic programming calculations are summarized in Table 1 for two specific set of values of $(\bar{S} - S_0)$.

A comparison with the CCP approach with pre-assigned tolerance measures is possible by means of results reproduced in Table 2.

TABLE 2. OPTIMAL SOLUTION OF EXPONENTIAL CCP MODELS WITH PREASSIGNED TOLERANCE MEASURES

Time t	$w_1 = 0.8$ $w_2 = 0.2$ Work force W_t	$W_0 = 90$ $W_t \geq 75$ Production P_t	$\bar{S} = 500$ $S_0 = 300$ Inventory I_t	$I_0 = 0$ Component reliability y_t	 u_t
1	89	899.146	494.957	2.996	0.95
2	81	404.189	513.277	2.996	0.95
3	77	385.870	519.572	2.996	0.95
4	75	379.574	519.572	2.996	0.95
5	75	379.574	519.572	2.996	0·95
6	75	379.574	320.000	2.996	0.95
				Total cost = 401949.88	
1	86	679.423	351.151	1.897	0.85
2	78	328.274	364.017	1.897	0.85
3	75	315.407	364.017	1.897	0.85
4	75	315.407	364.017	1.897	0.85
5	75	315.407	364.017	1.897	0.85
6	75	315.407	320.000	1.897	0.85
				Total cost = 237309.42	

It is apparent that reliability optimization permits considerable reduction in cost in terms of this quadratic programming model; also the optimal decision variables are insensitive to planning periods beyond $T = 2$. A substitution curve between system reliability and system costs can be drawn on the basis of these calculations which would show the rates of trade-off between tolerance measures and component costs. Note that such a framework could be easily modified to analyze the implications of assuming normal and other distributions as possible characterizations of random sales (Sengupta and Portillo–Campbell, 1970c).

4.4 Stochastic Programming and Economic Decisions

This section analyses the characterizations of selected economic decision problems under uncertainty in terms of stochastic programming. Specifically the analysis is restricted here to the following problems: (a) portfolio selection problem under risk, (b) competitive allocation models under uncertainty, and (c) the decision criteria under risk and uncertainty.

4.4.1 Dynamic Portfolio Selection Problems

The selection of an efficient portfolio when returns are random has been one of the most popular applied fields in probabilistic programming. The Markowitz–Tobin mean variance type analysis (Markowitz, 1959, Tobin, 1968), which was essentially developed in the static context of a one-period model has been extended and applied in recent times to multi-period situations. In terms of economic applications the following two aspects are worth mentioning.

(i) Problems of lifetime allocation of wealth between safe and risky assets and the optimal consumption and investment decisions have been treated by Samuelson (1969) and Merton (1969) under discrete and continuous time horizons.

(ii) Problems of investment allocation over time have been analysed in terms of dynamic chance-constrained programming models by Naslund and Whinston (1962), Charnes and Kirby (1967), Naslund (1967), Sengupta and Portillo–Campbell (1970b) and others.

Consider first the discrete-time portfolio allocation model due to Samuelson (1969), where an individual is supposed to maximize a discounted utility function over his lifetime. That is,

(a) maximize $\sum_{t=0}^{T} (1 + \delta)^{-t} U(C_t),$

subject to

(b) $\dfrac{W_{t+1}}{1 + r} = W_t - C_t,$ (1)

(c) $W_0, W_T = C_T$ pre-assigned,

by optimal allocation of wealth (W_t) between consumption (C_t) and investment at an exogenous rate of yield r. So far there is nothing stochastic in the model and if the utility function U is concave, the recursion conditions for a regular interior maximum can be easily obtained by differentiating partially with respect to each W_t, that is,

$$\frac{1 + \delta}{1 + r} U'\left(W_{t-1} - \frac{W_t}{1 + r}\right) = U'\left(W_t - \frac{W_{t+1}}{1 + r}\right). \quad (2)$$

For this model prime denotes differentiation. Now assume there are two assets, one is completely safe and the other risky and at any time t, w_t is the fraction of wealth allocated to the risky asset with $(1 - w_t)$ going into the safe asset. The constraint of Eq. (1b) then becomes

4.4 Stochastic Programming and Economic Decisions

$$C_t = W_t - [(1 - w_t)(1 + r) + w_t Z_t]^{-1} W_{t+1}, \quad (3)$$

where Z_t is a random variable for each t [in general, Z_t will be subject to a stochastic process, as Merton (1969) has used in his general model with more than two assets]. For simplicity, it is assumed that probability distributions of Z_t for different t are independent of time, that is,

$$P(z_0, z_1, \ldots, z_t, \ldots, z_T) = P(z_0) P(z_1) \ldots P(z_T), \quad (4)$$

where prob $(Z_t \leq z_t) = P(z_t)$. The stochastic model now becomes one of maximizing the expected value of the objective function of Eq. (1a) subject to Eqs. (3), (4), and (1c) by choosing the sequence of decision variables $\{C_t, w_t\}$. This type of problem is best characterized and solved by dynamic programming algorithms (Bellman, 1961; Dreyfus, 1964), which essentially derive recursion conditions of optimality through analyzing the T-period decision problem into a series of single-period optimum allocation problems. Thus we begin at the end period T of the stochastic model denoted by

$$J_T(W_0) = \max_{\{C_t, w_t\}} [E \sum_{t=0}^{T} (1 + \delta)^{-t} U(C_t) \text{ subject to Eqs. (3), (4)}], \quad (5)$$

$J_T(W_0)$ denotes the maximal value of expected sum of discounted utilities from a T-stage process starting from the initial wealth W_0. Suppose the initial wealth is W_{T-1} and we have a single period allocation problem, then from Eq. (5) we may write

$$J_1(W_{T-1}) = \max_{\{C_{T-1}, w_{T-1}\}} [U(C_{T-1}) + E(1 + \delta)^{-1} U(W_T)], \quad (6)$$

where $W_T = C_T$ can be explicitly written as Eq. (7) by using Eq. (3)

$$W_T = (W_{T-1} - C_{T-1})[(1 - w_{T-1})(1 + r) + w_{T-1} Z_{T-1}]^{-1}. \quad (7)$$

Note that the right-hand side of Eq. (6) merely says that if we start with initial wealth W_{T-1} in a single period allocation decision then the optimal decision concerns consuming it at $(T - 1)$ with utility $U(C_{T-1})$ or investing it for consumption one period later $(C_T = W_T)$ with a discounted utility having expected value $E(1 + \delta)^{-1} U(W_T)$. Since current consumption C_{T-1} is known once we have made our decision, the expected value operator E operates only on the random variable of the next period. Now by differentiating the expression of Eq. (6) with respect to C_{T-1}, w_{T-1} separately and equating to zero, that is,

$$0 = U'(C_{T-1}) - (1 + \delta)^{-1} EU'(C_T)[(1 - w_{T-1})(1 + r) + w_{T-1} Z_{T-1}], \quad (8)$$

$$0 = EU'(C_T)(W_{T-1} - C_{T-1})(Z_{T-1} - 1 - r), \quad (9)$$

and solving these two equations simutltaneously we get our optimal decisions $(C_{T-1}^*, w_{T-1}^*))$ say as functions of initial wealth W_{T-1} alone. Substituting these optimal values the maximum value $J_1(W_{T-1})$ in Eq. (6) can be explicitly computed.

Once we know $J_1(W_{T-1})$ explicitly we may compute the optimal value one period earlier by assuming that we have started with W_{T-2}. Then

$$J_2(W_{T-2}) = \max_{\{C_{T-2}, w_{T-2}\}} [U(C_{T-2}) + E(1 + \delta)^{-1} J_1(W_{T-1})], \qquad (10)$$

where $W_{T-1} = (W_{T-2} - C_{T-2})[(1 - w_{T-2})(1 + r) + w_{T-2} Z_{T-2}]$. Again differentiating with respect to C_{T-2}, w_{T-2} separately, equating to zero and solving two simultaneous equations,

$$0 = U'(C_{T-2}) - (1 + r)^{-1} E J_1'(W_{T-2})[(1 - w_{T-2})(1 + r) + w_{T-2} Z_{T-2}], \quad (11)$$
$$0 = E J_1'(W_{T-1})(W_{T-2} - C_{T-2})(Z_{T-2} - 1 - r), \qquad (12)$$

we get optimal C_{T-2}^*, w_{T-2}^* and an explicit value of $J_2(W_{T-2})$.

Thus continuing recursively in this manner we may determine optimal decision rules for consumption and portfolio selection at any t in the form

$$C_t^* = f[W_t; Z_{t-1}, \ldots, Z_0] = f_{T-t}(W_t),$$

if Z's are independently distributed, and

$$w_t^* = g[W_t; Z_{t-1}, \ldots, Z_0] = g_{T-t}(W_t),$$

if Z's are independent.

A few comments may be made about this type of application. First, as Samuelson has emphasized that with a little reinterpretation other growth models under uncertainty could be subsumed under this model [e.g., the Ramsay problem analyzed by Mirrlees (1965) with Harrod-neutral technical change as a random variable]. Second, note that the specific distribution characteristics of the random variables $Z_{t-1}, \ldots, Z_0$ are introduced into the model in a very specific way. The generating stochastic processes (Bharucha-Reid, 1960) for these variables may be varied, and if there are inequality restrictions on the decision space the sequence of optimal decisions may have switching and instability apart from the fact that the simultaneous equations like Eqs. (11) and (12) may be very difficult to solve computationally (Sengupta and Fox, 1969). Third, the random variations in return may lead the decision-maker to adopt a safety-first principle which is closely related (see Sengupta, 1969) to the approach of chance-constrained programming. An application of this approach is considered next.

4.4 Stochastic Programming and Economic Decisions

We consider now a second type of application of a dynamic investment allocation model due to Naslund (1967) (see also Naslund and Whinston, 1962, Sengupta and Portillo-Campbell, 1970b). This model poses the individual's decision problem as one of optimally determining whether to invest his savings in the stock market or to retain it in cash; in each period two types of constraints are considered by the individual: First, a risk constraint which in effect sets a tolerance limit on losses beyond a certain level. Second, a capital constraint which probabilistically stipulates that invested capital should be below a limit which varies according to accumulated capital gains. Mathematically the decision problem is to

(a) maximize $E\left[\sum_{i=1}^{T} x_i \frac{p_i - p_{i-1}}{p_{i-1}}\right]$

subject to

(b) $\text{prob}\left[x_i \frac{p_i - p_{i-1}}{p_{i-1}} \geq L_i\right] \geq \alpha_i, \quad 0 < \alpha_i \leq 1$ \hfill (13)

(c) $\text{prob}\left[x_i \leq K_i + \sum_{j=2}^{i} x_{j-1} \frac{p_{j-1} - d_{j-2}}{p_{j-2}}\right] \geq \eta_i, \quad 0 < \eta_i \leq 1,$

where
- x_i = the accumulated amount of savings invested in stock in period $i = 1, 2, \ldots, T$
- p_i = the price of stock or group of stocks in period $i = 1, \ldots, T$
- L_i = the maximum loss that the person is willing to take in period $i = 1, 2, \ldots, T$
- α_i = the risk or tolerance pre-assigned for year $i = 1, \ldots, T$
- K_i = the accumulated capital that the individual can use for investment either in cash or stocks apart from the returns on earlier investments
- η_i = the risk or tolerance pre-assigned for the capital constraint in period $i = 1, \ldots, T$

Following Osborne (1959) it is assumed that the random variable $(p_i - p_{i-1})/p_{i-1}$ is approximately normally distributed with mean μ and variance σ^2. In solving the above model, consideration is given to the fact that certain data which are presently unavailable (e.g., the stock prices) will, to varying degrees, be available in the future, when actual decisions are made. For this reason Naslund does not solve directly for the optimal x_i's but a linear decision rule of the following form

$$x_i = \sum_{j=2}^{i} \beta_j \left(1 + \frac{p_{j-1} - p_{j-2}}{p_{j-2}}\right) + \gamma_i; \quad i = 1, 2, \ldots, T \quad (14)$$

is introduced in the model replacing the x_i's and the unknown constants β_j and

γ_i are optimally solved for. Naslund's interpretation of the linear decision rule of Eq. (14) is as follows:

In period zero each x_i is a random variable being a linear function of random variables with a known distribution determined by the β_j's, γ_i's. We can say that we have selected for each period a randomized decision function which is optimal for the class of probability distributions which can be formed from linear combinations of the given random variables. Since x_k in period k is a function of known past prices at that time the decision rule can be interpreted as the demand of a particular person to hold funds invested in the stock market (Naslund, 1967, pp. 86-87).

Assuming a three-period ($T = 3$) horizon the optimization problem finally reduces to the following

$$\max E[\gamma_1 \Delta p_0 + \{\beta_2(1 + \Delta p_0) + \gamma_2\}\Delta p_0 + \{\beta_3(1 + \Delta p_1) + \gamma_3\}\Delta p_2]$$

subject to

$$\text{prob } [\gamma_1 \Delta p_0 \geq L_1] \geq \alpha_1,$$
$$\text{prob } [\{\beta_2(1 + \Delta p_0) + \gamma_2\}\Delta p_1 \geq L_2] \geq \alpha_2,$$
$$\text{prob } [\{\beta_3(1 + \Delta p_1) + \gamma_3\}\Delta p_2 \geq L_3] \geq \alpha_3,$$
$$\gamma_1 \leq K_1,$$
$$\text{prob } [\beta_2(1 + \Delta p_0) + \gamma_2 \leq K_2 + \gamma_1 \Delta p_0] \geq \eta_2,$$
$$\text{prob } [\beta_3(1 + \Delta p_1) + \gamma_3 \leq K_3 + \gamma_1 \Delta p_0 + \beta_2 \Delta p_1(1 + \Delta p_0) + \gamma_2 \Delta p_1] \geq \eta_3,$$

where Δp_k denotes $(p_{k+1} - p_k)/p_k$ for $k = 0, 1$.

A numerical illustration of this model has been discussed by Naslund (1967) and a comparison of this solution with the general optimal solution which is not limited to the linear decision rule of Eq. (14) has been made by Sengupta and Portillo-Campbell (1970b).

A few remarks may be made concerning this type of CCP formulation in a dynamic context. First, in a general case x_i would be a vector for each i, so that investment in different kinds of stocks may be considered; also the sensitivity of the optimal objective function due to marginal variations in the parameters μ and σ of the distribution of $p_{i+1} - p_i)/p_i$ may be investigated through the dual problems. This would be of some help in pre-assigning the tolerance measures. Second, the formal generalization is not difficult for the case when the Δp_k variables are subject to stochastic processes, except that the explicit computation problems become very nonlinear and complicated (Sengupta, 1966). Third, the uses of such dynamic CCP models in problems related to portfolio selection have been investigated in economic literature, for instance, the problem of incidence of taxation on risk-bearing (Naslund 1968), the problem of building Pareto-class distributions into the random variations (Mandelbrot, 1963: Fama, 1965) of the model, and so on.

4.4.2 COMPETITIVE MODELS UNDER UNCERTAINTY

The economic theory of general equilibrium under a competitive market framework has been generalized in recent times in several directions (Debreu, 1959; Koopmans and Bausch, 1959). These include the detailed specification of mathematical conditions required for consumer choice, producer behavior, and market interactions under various restrictions and regularity conditions (Arrow, 1964; Morishima, 1964). However, from an operational standpoint the framework of competitive equilibrium, which under certain conditions is known to lead to an optimum has been attacked on two fundamental grounds. First, the role of uncertainty is introduced to show that competitive equilibrium of a "certain" environment may not always exist due to the absence of markets for exchanging contingent claims on future goods, and even when it exists it need not allocate resources efficiently (Stigum, 1969). Second, the attainment of competitive equilibrium where it exists and satisfies optimality conditions may be computationally very expensive and institutionally unrealizable. The latter situations have been characterized by various decomposition techniques and decentralization procedures of two-level planning (Sengupta and Fox, 1969; Dantzig, 1968; Malinvaud, 1967).

Radner (1968) has argued that the existence and optimality of competitive framework in "classical" equilibrium theory (as exemplified by Arrow–Debreu models) face two great obstacles due to the fact that economic decision makers do not have unlimited computational capacity for choice among strategies and that they have uncertain information about each other's behavior. The computational limitations and the externality due to uncertain information (Marschak, 1964) about each other's behavior are aggravated by the fact that markets for exchanging contingent claims on future goods do not exist in the real world (Stigum 1969); also, if the latter markets for future goods are hypothetically constructed and rules of guidance through "interim" shadow prices are laid down, they need not always be stable and attainable (Hahn and Negishi, 1962; Kurz, 1968). Several basic issues are yet unresolved in this framework and these include among others some of the following aspects: meaning of internal prices in a hypothetically competitive vis-a-vis non-competitive market (Shubik,1962), the interaction between subjective and empirical probability outcomes under situations when decisions can be deferred (Robbins, 1956), and the incidence of aggregations and coalitions for allowing organizational slacks and tolerances in an otherwise unstable competitive framework.

In a more specific partial equilibrium context of the theory of a single-product firm in a competitive industry, Tisdell (1968) has considered the case of static "uncertainty in prices" and his simple quadratic models show that increased price uncertainty does reduce the expected profit for the competitive industry as a whole under a wide range of conditions. However, price uncertainty is defined here in a very specific way as $E(p - \hat{p})^2$, where E is expectation, p the actual price of the product, and $\hat{p}$ its shadow (or optimal) price. Two basic difficulties of such formulation have to be noted. First, the meaning of shadow prices used to define static price uncertainty becomes imprecise and confused, unless it is related either to entrepreneurial expectations and environments or to the random processes generating them. The treatment of inventories in various dynamic models of firm behavior (Sengupta and Fox 1969) shows how important it is to characterize the different sources of random variations before one could hypothesize the existence of "shadow prices." Second, the dynamic extension of such a theory is beset with great obstacles, since it does not consider the information network required to sustain the apparently optimal framework of shadow prices; also it neglects to consider any imperfections in the rigid framework of an otherwise competitive firm-industry model. Note that uncertainty in the sense of Radner (1968) and Stigum (1969) is completely neglected here.

4.4.3 DECISION CRITERIA UNDER UNCERTAINTY

We have so far discussed stochastic phenomena in the form of probability distributions (or families of probability distributions). From an operational or practical viewpoint this characterization is perhaps the most important and useful. However, it is important to realize that there exists a facet of uncertainly which cannot be characterized by the ordinary notions of probability distributions. There are deep philosophical problems (Tintner, 1949) concerning this type of uncertainty and its relations with inference and decisions, which we do not propose to discuss here. Instead we like to mention very briefly three important aspects of uncertainty which may have some operational implications for economic decision models. They are the following:

(a) Treatment of uncertainty in utility theory of consumer behavior,
(b) Decision rules under incomplete specification of probability distribution of outcomes,
(c) The role of simulation in problems of stochastic control.

The theory of consumer behavior as it has developed in economic theory displays a certain type of dichotomy so far as uncertainty is concerned. This dichotomy has been analyzed recently in some detail by van Praag (1968) to show the importance of cardinality as a tool yet to be fully realized. The dichotomy exists because, on the one hand, there is the von Neumann–Morgenstern approach based on expected utilities (i.e., $U(x) = E[U(x; S)]$ where E is expectation, x is the commodity space, and S is the random state of nature) which leads to a cardinal utility function $U(x)$ over the budget space, and, on the other hand, we have the ordinal theory initiated by Pareto and developed by a number of economists (e.g., Arrow, Debreu, Samuelson) which deals with uncertainty by defining each pair $(x; s)$ where $s \in S$ as a separate commodity and postulating a preference ordering on this space of commodities. However,

4.4 Stochastic Programming and Economic Decisions

the dichotomy appears to be more apparent than real, if contingent claims on future goods are admitted as commodities, the states of nature are subject to a probability distribution of subjective, objective or mixed character and a set of axioms are introduced. For example, Borch (1968) has shown that if the preference ordering over a set of prospects satisfies three simple axioms concerning the existence of certainty equivalents, monotonicity, and boundedness, then this would lead to a cardinal utility function which is bounded and unique up to a positive linear transformation. All there has to be done in this framework is to introduce "commodities" like contingent claims. However, it is not at all clear how, in view of the computational limitations mentioned in the preceeding section, such reconciliation is going to be helpful (Chipman, 1960; Fishburn, 1964).

To view uncertainty in terms of the parameters of subjective probability of objective distribution of random states of nature, raises a number of fundamental issues in statistical theory of probability and inference (Kendall, 1949; Savage, 1962; Birnbaum, 1962), which we would not discuss. Instead, we like to emphasize those attempts which seek to characterize decision rules under uncertainty in situations where the probability distribution of states of nature is only incompletely known (this incomplete knowledge may be due to computational limitations, environment, or information networks). First, there is a set of decision criteria developed in the theory of games against nature (e.g., Wald's minimax rule, Savage's minimum regret, Hurwicz's pessimism-optimism index, etc.) which are based in a sense on some form of extremes. From an empirical standpoint, the performance of these alternative criteria in the face of new information concerning the incompletely known distribution of states of nature requires to be evaluated. This is very similar to the development in the nonparametric theory of statistics, where tests which are relatively distribution-free and robust in some sense (Sengupta, 1969; Bradley, 1968) are emphasized. Some evaluation in this framework of the above decision criteria in the context of stochastic linear programming are available in the literature (Sengupta, 1968; Sengupta and Sanyal, 1970). Second, there is an important contribution by Shackle (1949, 1961), where the decision criterion is based on some measures of best and worst outcomes called "focus gain" and "focus loss" which are determined by the individual's attitudes towards possible payoffs and his degrees of potential surprise. Here the extreme outcomes are weighted in a sense by the individual's attitude and degrees of potential surprise. In this theory potential surprise is a measure of possibility, not probability and hence the statistical theory of extreme values do not easily apply. It seems, however, that the performance of this criterion can be operationally tested only through concrete specification of potential surprise possibly as vector functions (many-valued anticipations) coupled with a subjective probability distribution of the individuals attitude towards new information. The approach of order-statistics based on more than two extremes has scope here. Third, there is the safety-first principle (Roy, 1952) by which the chance of disaster is minimized in the face of no knowledge about the probability distribution of disaster. In most economic situations, of course, some information would be available and it would increase sequentially. The problem is to utilize this sequential information process for arriving at decision rules which tend to improve. Some work in the context of stochastic programming are available in this framework (Sengupta and Portillo–Campbell, 1970b).

Simulation methods provide another indication of the relative sensitivity of decision rules under uncertainty, when the latter is specified in terms of deviations from a class of distributions (e.g. normal) or in the acceptance of suboptimal and therefore pessimistic outcomes. In problems of stochastic control and programming such methods are frequently adopted to analyze the stability of different kinds of the various types of optimal controls. In macrodynamic policy models in economics the simulation method are widely applied to feedback, adaptive and optimal policies. However, the evaluation and comparison of alternative simulation approaches (Orcutt and others, 1961; Hufschmidt, 1963) particularly in applications to economic systems (Duesenberry, *et al.*, 1960; Naylor *et al.*, 1968) need much more intensive analyses than those attempted so far.

REFERENCES AND BIBLIOGRAPHY

Ackoff, R. L., ed. (1963). "Progress in Operations Research," Vol. 1. Wiley, New York.
Adelman, I., and Thorbecke, E., ed. (1966). "The Theory and Design of Economic Development." Johns Hopkins Press, Baltimore, Maryland.
Aitchison, J., and Brown, J. A. C. (1957). "The Log-Normal Distribution." Cambridge Univ. Press, London and New York.
Aizerman, M. A. (1963). "Theory of Automatic Control" (*English Transl*. by R. Feinstein). Pergamon, Oxford.
Allen, R. G. D. (1959). "Mathematical Economics," 2nd Revised ed. Macmillan, New York.
Allen, R. G. D. (1967). "Macro-economic Theory: A Mathematical Treatment." St. Martin's Press, New York.
Alper, P. (1967). Introduction of control concepts in educational planning. "Mathematical Models in Educational Planning," pp. 259-273. OECD, Paris.
Alper, P., and Smith, C. (1967). An application of control theory to a problem in educational planning. *IEEE Trans. Automatic Control* **12**, 176-178.
Anderson O. N. (1929). Die Korrelationsrechnung in der Konjunkturforschung" Schroeder, Bonn.
Anderson R. L. (1942). Distribution of the serial correlation coefficient. *Ann. Math. Statist.* **13**, 1-13.
Anderson, T. W. (1959). On asymptotic distributions of estimates of parameters of stochastic difference equations. *Ann. Math. Statist.* **30**, 676-687.
Anderson, T. W. (1962). The choice of the degree of a polynomial regression as a multiple decision problem. *Ann. Math. Statist.* **33**, 255-265.
Anderson T. W. (1964). Some approaches to the statistical analysis of time series. *Austral. J. Statist.* **6**, 1-11.
Anderson, T. W., and Rubin, H. (1949). Estimation of the parameters of a single equation in a complete system of stochastic equations. *Ann. Math. Statist.* **20**, 46-63.
Ando, A., Fisher, F., and Simon, H. (1963). "Essays on the Structure of Social Science." MIT Press, Cambridge, Massachusetts.
Antosiewicz, H. A. (1963). Linear control systems. *Arch. Ration. Mech. Anal.* **12**, 313-324.
Aoki, M. (1967). "Optimization of Stochastic Systems." Academic Press, New York.
Aris, R. (1963). "Discrete Dynamic Programming." Ginn, New York.

References and Bibliography

Armitage, P., and Smith, C. (1967). The development of computable models of the British educational system and their possible uses. "Mathematical Models in Educational Planning." OECD, Paris.

Arrow, K. J. (1951). Alternative approaches to the theory of choice in risk-taking situations. *Econometrica* **19**, 404–437.

Arrow, K. J. (1951). "Social Choice and Individual Values." Wiley, New York.

Arrow, K. J. (1952). An extension of the basic theories of classical welfare economics. *2nd Berkeley Symp. Math. Stat. Probability* **2**, 507–32.

Arrow, K. J. (1962). The economic implications of learning by doing. *Rev. Econ. Stud.* **29**, 155–173.

Arrow, K. J. (1964). Control in large organizations. *Management Sci.* **10**, No. 3.

Arrow, K. J. (1964). The role of securities in the optimal allocation of risk bearing. *Rev. Econ. Stud.* **31**, 91–96.

Arrow, K. J. (1965). Aspects of the theory of risk-bearing. *Yrjo Jahnsson Lecture Series*, Helsinki.

Arrow, K. J. (1968). Applications of control theory to economic growth, in Dantzig, G. B. and Veinott, Jr., A. F. eds. Vol. 12.

Arrow, K. J., and Enthoven, A. C. (1961). Quasi-concave programming. *Econometrica* **29**, 779–800.

Arrow, K. J., and Hurwicz, L. (1959). Decentralization and computation in resource allocation, *In* "Essays in Economics and Econometrics" (R. W. Pfouts, ed.). Univ. of North Carolina Press, Chapel Hill, North Carolina.

Arrow, K. J., Hurwicz, L., and Uzawa, H., eds. (1958). "Studies in Linear and Nonlinear Programming." Stanford Univ. Press, Stanford, California.

Arrow, K. J., Karlin, S., and Scarf, H., eds. (1958). "Studies in the Mathematical Theory of Inventory and Production." Stanford Univ. Press, Stanford, California.

Arrow, K. J. and Kurz, M. (1970). "Public investment, the rate of return and optimal fiscal policy." Johns Hopkins Press, Baltimore, Maryland.

Arrow, K. J., and Harris, T., and Marschak, J. (1952). Optimal inventory policy. *Econometrica* **19**, 250–272.

Aseltine, J. A. (1958). "Transform Methods in Linear System Analysis." McGraw-Hill, New York.

Athans, M., and Falb, P. L. (1966). "Optimal Control: An Introduction to the Theory and its Applications." McGraw-Hill, New York.

Aumann, R. J. (1964). Markets with a continuum of traders. *Econometrica* **32**, 39–50.

Babbar, M. M. (1955). Distribution of solutions of a set of linear equations with applications to linear programming. *J. Amer. Stat. Assoc.* **50**, 155–64.

Bailey, N. T. J. (1964). "The Elements of Stochastic Processes." Wiley, New York.

Balakrishnan, A. V., and Neustadt, L. W., eds. (1964). "Computing Methods in Optimization Problems." Academic Press, New York.

Balakrishnan, A. V., and Neustadt, L. W., eds. (1967). "Mathematical Theory of Control." Academic Press, New York.

Balinski, M. L. (1961). An algorithm for finding all vertices of convex polyhedral sets. *SIAM J.* **9**, 72–88.

Bardhan, P. K. (1967). Optimum foreign borrowing, in Shell, K. ed.

Barlow, R. E., and Proschan, F. (1965). "Mathematical theory of reliability." Wiley, New York.

Barnett, S. (1962). Stability of the solution to a linear programming problem. *Operational Res. Quart.* **13**, No. 3.

Bartlett, M. S. (1935). Some aspects of the time correlation problem in regard to tests of significance. *J. Roy. Statist. Soc.* **98**, 536–543.

Bartlett, M. S. (1961). "An Introduction to Stochastic Processes." Cambridge Univ. Press, London and New York.

Bartlett, M. S. (1962). "Stochastic population models." Methuen, London.

Basmann, R. L. (1957). A generalized calssical method of linear estimation of coefficients in a structural equation. *Econometrica* **125,** 77–83.

Basmann, R. L. (1960). On finite sample distribution of generalized classical linear identifiability test statistics. *J. Amer. Statist. Assoc.* **55,** 650-659.

Baumol, W. (1959). "Economic Dyanmics," 2nd ed. Macmillan, New York.

Baumol, W. (1965). "Economic Theory and Operations Research," 2nd ed. Prentice-Hall, Englewood Cliffs, New Jersey.

Beale, E. M. L. (1961). The use of quadratic programming in stochastic linear programming. Rand Corp. Rep. P-2404-1, August 15.

Beckmann, M. J. (1968). "Dynamic Programming of Economic Decisions." Springer-Verlag, New York.

Bellman, R. (1957). "Dynamic Programming." Princeton Univ. Press, Princeton, New Jersey.

Bellman, R. (1961). "Adaptive Control Processes: A Guided Tour." Princeton Univ. Press, Princeton, New Jersey.

Bellman, R. (1962). On the determination of optimal trajectories, in Leitman, G. ed.

Bellman, R., and Dreyfus, S. E. (1962). "Applied Dynamic Programming." Princeton Univ. Press, Princeton, New Jersey.

Bellman, R., and Kalaba, R. (1965). "Dynamic Programming and Modern Control Theory." Academic Press, New York.

Beltrami, E. J. (1970). "An algorithmic approach to nonlinear analysis and optimization." Academic Press, New York.

Bereanu, B. (1966). On stochastic linear programming: the Laplace transform of the distribution of the optimum and applications. *J. Math. Anal. Appl.* **15,** 280-94.

Bertram, J. E., and Sarachik, P. E. (1959). Stability of circuits with randomly time-varying parameters. *IRE, PGIT-5 Special Suppl.* p. 260.

Bharucha-Reid, A. T. (1960). "Elements of the Theory of Markov Processes and Applications." McGraw Hill, New York.

Birnbaum, Z. W. (1952). Numerical tabulation of the distribution of Kolmogorov's statistic for finite sample size. *J. Amer. Stat. Assoc.* **47,** 425-41.

Birnbaum, Z. W. (1953). Distribution-free tests of fit for continuous distribution functions. *Ann. Math. Statist.* **24,** 1-8.

Birnbaum, Z. W. (1962). On the foundations of statistical inference. *J. Amer. Statist. Assoc.* **57,** 269-326.

Blaug, M. (1962). "Economic Theory in Retrospect." Irwin, Homewood, Illinois.

Blumen, L. M., Kogan, P. J., and McCarthy, W. (1955). The industrial mobility of labor as a probability process. *Cornell Stud. Ind. Labor Relations.* **7.**

Boot, J. C. G. (1964). "Quadratic Programming." North-Holland Publ., Amsterdam.

Borch, K. (1968). "The Economics of Uncertainty." Princeton Univ. Press, Princeton, New Jersey.

Borch, K. (1969). A note on uncertainty and indifference curves. *Rev. Econ. Studies.* **36,** 1-4.

Bose, S. (1968). Optimal growth and investment allocation. *Rev. Econ. Stud.* **35,** 465-480.

Box, G. E. P., and Hunter, J. S. (1954). A confidence region for the solution of a set of simultaneous equations with an application to experimental design. *Biometrika* **41,** 190-199.

Box, G. E. P. and Jenkins, G. M. (1962). Some statistical aspects of adaptive optimization and control. *J. Roy. Statist. Soc. Ser. B* **24,** 297-343.

Box, G. E. P., and Jenkins, G. M. (1968). Discrete models for feedback and feedforward control. "The Future of Statistics." (D. G. Watts, ed.) Academic Press, New York.

Box, G. E. P. and Jenkins, G. M. (1970). "Time series analysis." Holden-Day, San Francisco, California.

Boyce, W. E. (1968). Random eigenvalue problems. "Probabilistic Methods in Applied Mathematics." (A. T. Bharucha-Reid, ed.), Academic Press, New York.

Bracken, J., and McCormick, G. P. (1968). "Selected Applications of Nonlinear Programming." Wiley, New York.

Bracken, J., and Soland, R. M. (1966). Statistical decision analysis of stochastic linear programming problems. *Naval Res. Logist. Quart.* **13**, 205–226.

Bradley, J. V. (1968). "Distribution-free Statistical Tests." Prentice-Hall, Englewood Cliffs, New Jersey.

Brockett, R. W. (1966). The status of stability theory for deterministic systems. *IEEE Trans. Automatic Control* **11**, 596–606.

Brogden, H. E. (1955). Least squares estimates and optimal classification. *Psychometrika* **20**, 249–242.

Bruno, M. (1967). Optimal accumulation in discrete capital models. Shell, K. ed.

Bruno, M., Burmeister, E., and Sheshinski, E. (1966). The nature and significance of the reswitching of techniques. *Quart. J. Econ.* **80**, 526–554.

Buchanan, L. (1968). Optimal control of macroeconomic systems. Unpublished Ph. D. thesis. Univ. of California, Los Angeles.

Buchanan, L., and Stubberud, A. R. (1969). Problems in optimal control of macroeconomic systems. "Computing Methods in Optimization Problems: Lecture Notes in Operations Research and Mathematical Economics," Vol. 14. Springer-Verlag, Berlin.

Bucy, R. S. (1965). Non-linear filtering. *IEEE Trans. Automatic Control* p. 198.

Burmeister, E., and Dobell, A. R. (1970). "Mathematical Theories of Economic Growth." Macmillan, New York.

Cairncross, A. K. (1953). "Home and Foreign Investment." Cambridge Univ. Press, London and New York.

Caratheodory, C. (1935). "Variationsrechnung.". Teubner, Leipzig.

Carnap, R. (1950). "Logical Foundations of Probability." Univ. of Chicago Press, Chicago, Illinois.

Cass, D. (1965). Optimum growth in an aggregative model of capital accumulation. *Rev. Econ. Stud.* **32**, 223–240.

Cass, D., and Stiglitz, J. E. (1969). The implications of alternative saving and expectations hypotheses for choices of technique and patterns of growth. *J. Political Econ.* **77**, 586–627.

Central Planning Bureau (1956). Scope and Method of Central Planning of the Netherlands. The Hague.

Central Planning Bureau (1961). Central Economic Plan 1961. The Hague.

Chakravarty, S. (1962). The existence of an optimum savings program. *Econometrica* **30**, 178–187.

Chakravarty, S. (1965). Optimal program of capital accumulation in a multi-sector economy. *Econometrica* **33**, 557–70.

Chakravarty, S. (1966). Optimal savings with finite planning horizon: a reply. *Int. Econ. Rev.* **7**, 119–123.

Chakravarty, S. (1969). "Capital and Development Planning." MIT Press, Cambridge, Massachusetts.

Chamberlin, E. H. (1948). "The Theory of Monopolistic Competition," 6th ed. Harvard Univ. Press, Cambridge, Massachusetts.

Champernowne, D. G. (1953). A model of income distribution. *Econ. J.* **63,** 318–51.
Champernowne, D. G. (1969). "Uncertainty and estimation in economics." Vol. 1–3. Holden-Day, San Francisco, California.
Charnes, A., and Cooper, W. W. (1959). Chance constrained programming. *Management Sci.* **6,** pp. 73–79.
Charnes, A., and Cooper, W. W. (1962). Normal deviates and chance constraints. *J. Amer. Statist. Assoc.* **57,** 134–43.
Charnes, A., and Cooper, W. W. (1963). Deterministic equivalents for optimizing and satisficing under chance constraints. *Operations Res.* **11,** 18–39.
Charnes, A., and Kirby, J. L. (1967). Some special P-models in chance constrained programming. *Management Sci.* **14,** 183–195.
Charnes, A., Cooper, W. W., and Ferguson, R. O. (1955). Optimal estimation of executive compensation by linear programming. *Management Sci.* **1,** 138–151.
Chenery, H. B. (1961). Comparative advantage and development policy. *Amer. Econ. Rev.* **51,** 18–51.
Chenery, H. B. (1963). The use of interindustry analysis in development programming. "Structural Interdependence and Economic Development" (T. Barna, ed.). St. Martin's Press, New York.
Chenery, H. B. (1966). Comparative advantage and development policy. American Economic Association, and Royal Economic Society. *Surveys Econ. Theory* **2,** 125–155.
Chenery, H. B., and Bruno, M. (1962). Development alternatives in an open economy: the case of Israel. *Econ. J.* **72,** 79–103.
Chenery, H. B., and MacEwan, A. (1966). Optimal patterns of growth and aid: the case of Pakistan. Adelman, I, and Thorbecke, E. eds.
Chenery, H. B., and Uzawa, H. (1958). Nonlinear programming in economic development. "Studies in Linear and Non-Linear Programming" (K. J. Arrow, L. Hurwicz, and H. Uzawa, eds.). Stanford Univ. Press, Stanford, California.
Chernoff, H. (1968). Optimal stochastic control. Dantzig, G. B. and Veinott, Jr., A. F. eds. Vol. 12.
Chipman, J. S. (1960). The foundations of utility. *Econometrica* **28,** 193–224.
Christ, C. (1966). "Econometric Models and Methods." Wiley, New York.
Churchman, C. W., Ackoff, R. L., and Arnoff, E. L. (1957). "Introduction to Operations Research." Wiley, New York.
Cochrane, D., and Orcutt, G. (1949). Application of least squares regression to relationships containing autocorrelated error terms. *J. Amer. Statist. Assoc.* **44,** 32–61.
Connors, M., and Teichrow, D. (1967). "Optimal Control of Dynamic Operations Research Models." Int. Textbook Co. Scranton, Pennsylvania.
Cootner, P. H. ed. (1964). "The Random Character of Stock Market Prices." MIT Press, Cambridge, Massachusetts.
Courtillot, M. (1962). On varying all the parameters in a linear programming problem. *Operations Res.* **10.**
Cox, D. R., and Miller, H. D. (1965). "The Theory of Stochastic Processes." Wiley, New York.
Dantzig, G. B. (1948). Programming in a Linear Structure. Comptroller, USAF, Washington, D. C.
Dantzig, G. B. (1955). Linear programming under uncertainty. *Management Sci.* **1,** 197–206.
Dantzig, G. B. (1957). On the status of multi-stage linear programming problems. *Bull. Int. Stat. Inst.* **36,** Part 3, 303–20.
Dantzig, G. B. (1963). "Linear Programming and Extensions." Princeton Univ. Press, Princeton, New Jersey.

Dantzig, G. B. (1968). Large-scale linear programming. "Mathematics of the Decision Sciences," Part 1, Lectures in Applied Mathematics, Vol. 11. *Amer. Math. Soc.* Rhode Island.

Dantzig, G. B., and Madansky, A. (1961). On the solution of two-stage linear programs under uncertainty. *Prod. 4th Berkeley Symp. Math. Stat. Probobility*, Vol. 1. Univ, of California Press, Berkeley, California.

Dantzig, G. B., and Veinott, A. F., Jr., eds. (1968). "Mathematics of the Decision Sciences," Vol. 11-12. Amer. Math. Soc. Providence, Rhode Island.

Dantzig, G. B., and Wolfe, P. (1960). The decomposition principle for linear programs. *Operations Res.* **8,** 101-111.

Davis, H. T. (1941). "The Analysis of Economic Time Series," Principia, Bloomington, Indiana.

Day, R. H. (1963). "Recursive Programming and Production Response." North Holland Publ., Amsterdam.

Day, R. H. (1963). On aggregating linear programming models of production. *J. Farm Econ.* **45,** 797-813.

D'Azzo, J. J., and Houpis, C. H. (1966). "Feedback Control: Systems Analysis and Synthesis," 2nd ed. McGraw Hill, New York.

Debreu, G. (1951). The coefficient of resource utilization. *Econometrica* **19,** 273-92.

Debreu, G. (1954). Representation of a preference ordering by a numerical function. "Decision Processes" (R. M. Thrall, C. H. Coombs, and R. L. Davis, eds), pp. 159-165. Wiley, New York.

Debreu, G. (1959). "Theory of Value." Wiley, New York.

Derman, C. (1970). "Finite state Markovian decision processes." Academic Press, New York.

Dhrymes, P. (1964). On the theory of the monopolistic multi-product firm under uncertainty. *Int. Econ. Rev.* **5,** No. 3.

Dhrymes, P. (1970). "Econometrics: Statistical foundations and applications." Harper and Row, New York.

Dilberto, S. P. (1967). The Pontryagin maximum principle. Leitman, G. ed.

Dixit, A. K. (1968). Optimal development in the labor-surplus economy. *Rev. Econ. Stud.* **35,** 23-34.

Domar, E. D. (1946). Capital expansion, rate of growth and employment. *Econometrica* **14,** 137-147.

Dorfman, R., Samuelson, P., and Solow, R. M. (1958). "Linear Programming and Economic Analysis." McGraw-Hill, New York.

Dostupov, B. G. (1957). Approximate determination of the output coordinate probability characteristics for nonlinear automatic control systems. *Automat. Remote Control* **18,** 1045-1056.

Dreyfus, S. E. (1964). Some types of optimal control of stochastic systems. *SIAM J. Control.* **2,** 120-134.

Dreyfus, S. E. (1965). "Dynamic Programming and the Calculus of Variations." Academic Press, New York.

Dreyfus, S. E. (1968). Introduction to stochastic optimization and control. Karreman, H. F. ed.

Duesenberry, J. S. (1949). "Income, Savings and the Theory ot Consumer Behavior." Harvard University Press, Cambridge, Massachusetts.

Duesenberry, J. S., Eckstein, O., and Fromm, G. (1960). A simulation of the U. S. economy in recession. *Econometrica* **28,** 749-809.

Duesenberry, J. S., Fromm, G., Klein, L. R., and Kuh, E. ed. (1965). "The Brookings Model of the United States." Rand McNally, Chicago, Illinois.

Duesenberry, J. S., Klein, L. R., and others, eds. (1965). "The Brookings Quarterly Econometric Model of the United States." North-Holland Publ., Amsterdam.

Duffin, R. J., Peterson, E. L., and Zener, C. (1967). "Geometric Programming: Theory and Application." Wiley, New York.

Durbin, J., and Watson, G. S. (1950, 1951). Testing for serial correlation in least squares regression. *Biometrika* December and June.

Dvoretsky, A., Kiefer, J., and Wolfowitz, I. (1952). The inventory problem I. Case of known distributions of demand. *Econometrica* **20**, 450-466.

Eckaus, R. S., and Parikh, K. S. (1966). Planning for Growth: Multisectoral Inter-temporal Models Applied to India. Center for International Studies, MIT, Cambridge, Massachusetts.

Ellman, M. (1969). Aggregation as a cause of inconsistent plans. *Economica* February, 69-74.

Elmaghraby, S. (1959). An approach to linear programming under uncertainty. *Operations Res.* **7**, 208-216.

Elmaghraby, S. (1960). Allocation under uncertainty when the demand has continuous d.f. *Management Sci.* Vol. **6**, 270-274.

Evans, G. C. (1930). "Mathematical Introduction to Economics." McGraw Hill, New York.

Fama, E. F. (1965). Portfolio analysis in a stable Paretian market. *Management Sci.* **11**, 404-419.

Fama, E. F. (1971). Risk return and equilibrium. *J. Polit. Econ.* **79**, 30-55.

Fan, L. T., and Wang, C. S. (1964). "The Discrete Maximum Principle." Wiley, New York.

Farrar, D. E. (1962). "The Investment Decision under Uncertainty." Prentice Hall, Englewood Cliffs, New Jersey.

Fei, J. C. H., and Ranis, G. (1966). Agrarianism, dualism and economic development. Adelman, I. and Thorbecke, E. eds.

Feldbaum, A. A. (1960). Dual-control theory. Oldenburger, R. ed.

Feldbaum, A. A. (1965). "Optimal Control Systems." Academic Press, New York.

Feldstein, M. S. (1969). Mean variance analysis in the theory of liquidity preference and portfolio selection. *Rev. Econ. Stud.* **36**, 5-12.

Feller, W. (1948). On the Kolmogorov-Smirnov limit theorems for empirical distributions. *Ann. Math. Statist.* **19**, 177-189.

Feller, W. (1966, 1968). "An Introduction to Probability Theory and its Applications." Vol. 1-2. Wiley, New York.

Fellner, W. (1965). "Probability and Profit: A Study of Economic Behavior along Bayesian Lines." Irwin, Homewood, Illinois.

Fels, E., and Tintner, G. (1967). Methodik der Wirtschaftswissenschaft. "Handbuch der Arbeitsmethoden." Oldenbourg, Munich.

Fiacco, A. V., and McCormick, G. P. (1964). The sequential unconstrained minimization technique for nonlinear programming: a primal dual method. *Management Sci.* **10**, 360-366.

Fiacco, A. V., and McCormick, G. P. (1968). "Nonlinear Programming: Sequential Unconstrained Minimization Techniques." Wiley, New York.

Fishburn, P. C. (1964). "Decision and value theory." Wiley, New York.

Fishburn, P. C. (1968). Utility theory. *Management Sci.* **14**, 335-378.

Fisher, F. (1963). See Ando *et al.* (1963).

Fisher, F. (1965). Identifiability criteria in nonlinear systems: a further note. *Econometrica* January.

Fisher, F. (1966). Dynamic structure and estimation in economywide econometric models. Duesenberry, J. S. and others eds. The Brookings Quarterly Econometric Model of the United States.

Fisher, W. (1969). "Clustering and Aggregation in Economics." Johns Hopkins Press, Maryland.

Fisk, P. R. (1967). "Stochastically Dependent Equations." Hafner Publ., New York.

Fisz, M. (1963). "Probability Theory and Mathematical Statistics," 3rd ed. Wiley, New York.

Florentin, J. J. (1961). Optimal control of continuous time, Markov, stochastic systems. *J. Electron. Control* 10, 473-488.

Foley, D. F., Shell, K., and Sidrauski, M. (1969). Optimal fiscal and monetary policy and economic growth. *J. Political Econ.* 77, 698-719.

Fox, K. A. (1963). Spatial price equilibrium and process analysis in the food and agricultural sector. Cowles Commission Monograph on "Studies in Process Analysis: Economywide Production Capabilities." Wiley, New York.

Fox, K. A., Sengupta, J. K., and Thorbecke, E. (1966). "The Theory of Quantitative Economic Policy with Applications to Economic Growth and Stabilization." North Holland Publ., Amsterdam and Rand McNally, Chicago. Illinois.

Frank, M., and Wolfe, P. (1956). An algorithm for quadratic programming. *Naval Res. Logist. Quart.* 3, 95-110.

Freund, R. F. (1956). The introduction of risk into a programming model. *Econometrica* 24, 253-63.

Frisch, R. (1955). The mathematical structure of decision models: the Oslo sub-model. *Metroeconomica* 7, 111-136.

Frisch, R. (1960). "Planning for India." Asia Publ. House, Calcutta, India.

Frisch, R. (1961). A reconsideration of Domar's theory of economic growth. *Econometrica* 29,

Frisch, R. (1964). Parametric solution and programming of the Hicksian model. Rao, C. R. ed. "Essays on Econometrics and Planning" (C. R. Rao, ed.). Statistical Publ. Soc., Caluctta, India.

Frisch, R. (1965). "Theory of Production." Rand McNally, Chicago.

Frisch, R., and Holme, H. (1935). The characteristic solutions of a mixed difference and differential equation occurring in economic dynamics. *Econometrica* 3, 225-239.

Frisch, R. and Waugh, F. V. (1933). Partial time regressions as compared with individual trends. *Econometrica* 1, 387-401.

Fromm, G. and Taubman, P. (1968). "Policy simulations with an econometric model." North-Holland Publ., Amsterdam,

Funk, P. (1962). "Variationsrechnung und ihre Anwendung in Physik und Technik." Springer-Verlag, Berlin.

Galbraith, J. K. (1962). "Economic Development in Perspective." Harvard, Cambridge, Massachusehs.

Galbraith, J. K. (1967). "The New Industrial State." Houghton, Boston, Massachusetts.

Gale, D. (1967). On optimal development in a multi-sector economy. *Rev. Econ. Stud.* 34, 1-18.

Geary, R. C. (1947). Testing for normality. *Biometrika* 34, 209-242.

Georgescu-Roegen, N. (1966). "Analytical Economics." Harvard Univ. Press, Cambridge, Massachusetts.

Geyer, H., and Oppelt, W., eds. (1957). "Volkswirtschaftliche Regelvorgaienge." Oldenbourg, Munich.
Ghellinck, G. T., and Eppen, G. D. (1967). Linear programming solutions for separable Markovian decision problems. *Management Sci.* **13,** 371-394.
Gibrat, V. (1931). *Les inegalites economiques.* Paris.
Gilbert, E. G. (1963). Controllability and observability in multivariable control systems. *SIAM J. Control* **1,** 128-151.
Gnedenko, B. V., and Kolmogorov, A. N. (1968). "Limit Distributions for sums of Independent Random Variables." Addison-Wesley, Reading, Massachusetts.
Goldberger, A. S. (1964). "Econometric Theory." Wiley, New York:
Goldman, S. M. (1968). Optimal growth and continual planning revision. *Rev. Econ. Studi.* **35,** 145-154.
Goodwin, R. M. (1961). Optimal growth path for an underdeveloped economy. *Econ. J.* **71,** 756-774.
Goodwin, R. M. (1964). Stabilizing the exchange rate. *Rev. Econ. Statis.* **46,** 160-162.
Granger, C. W. J., and Morgenstern, 0. (1963). Spectral analysis of New York stock market data. *Kyklos* **16,** 1-27.
Granger, C. W. J., and Hatanaka, M. (1964). "Spectral Analysis of Economic Time Series." Princeton Univ. Press, Princeton, New Jersey.
Graves, R. L., and Wolfe, P., ed. (1963). "Recent Advances in Mathematical Programming." McGraw-Hill, New York.
Grenander, V., and Rosenblatt, M. (1957). "Statistical Analysis of Stationary Time Series." Wiley, New York. 6
Gumbel, E. J. (1958). "The Statistics of Extremes." Wiley Ner York.
Gupta, A. K. (1952). Estimation of the mean and standard deviation of a normal population from a censored sample. *Biometrika* **39,** 260-273.
Guttman, L. (1948). A distribution-free confidence interval for the mean. *Ann Math. Statis* **19,** 410-13.
Haavelmo, T. (1944). The probability approach in econometrics. *Econometrica Suppl.* **12,**
Haavelmo, T. (1954). "A Study in the Theory of Economic Evolution." North Holland Publ., Amsterdam.
Hader, J., and Russell, W. R. (1969). Rules for ordering uncertain prospects. *Amer. Econ. Rev.* March, pp. 25-34
Hadley, G. (1962) "Linear programming." Addison-wesley, Reading, Massachuschs.
Hadley, G. (1964) "Nonlinear and Dynamic Programming." Addison-Wesley, Reading, Massachusetts.
Hahn, F. (1966). Equilibrium dynamics with heterogeneous capital goods. *Quart. J. Econ.* **80,** 633-646.
Hahn, F. (1969). On money and growth. *J. Money Credit Banking* **1,** 172-187.
Hahn, F., and Matthews, R. c. O. (1964. E. J.; also 1966). The theory of economic growth: a survey. American Economic Association, Royal Economic Society: "Surveys of Economic Theory," Vol. 2, pp. 1-124. Macmillan, London.
Hahn, F. H., and Negishi, T. (1962). A Theorem on non-tatonnement stability. *Econometrica* **30,** 463-469.
Halkin, H. (1966). A maximum principle of the Pontryagin type for systems described by nonlinear difference equations. *SIAM J. Control* **4,** 90-111.
Hamada, K. (1969). Optimal capital accumulation by an economy facing an international capital market. *J. Political Econ.* **77,** 684-697.
Hannan, E. J. (1960). "Time Series Analysis." Wiley, New York.

Hannan, E. J. (1963). The estimation of seasonal variations in economic time series. **58**, 31–44.
Hannan, E. J. (1970), "Multiple Time Series." Wiley, New York.
Hanson, M. A. (1960). Errors and stochastic variations in linear programming. *Aus. J. Statist.* **2**, No. 2.
Hanson, M.,A. (1961). A duality theorem in nonlinear programming with nonlinear constraints. *Aus. J. Statist.* **3**, 64–72.
Hanssmann, F. (1962). "Operations Research in Production and Inventory Control." Wiley New York.
Hardt, J. P., Hoffenberg, M., Kaplan, N., and Levine, H. S., eds. (1970). "Mathematics and Computers in Soviet Economic Planning." Yale Univ. Press, New Haven, Connecticut.
Harrod, R. F. (1952). "Towards a Dynamic Economics." Macmillan, New York.
Harsanyi, J. C. (1966). A general theory of rational behavior in game situations. *Econometrica* **34**, 613–634.
Hart, B. I. (1942). Tabulation of the probabilities for the ratio of the mean square successive difference to the variance. *Ann. Math. Statist.* **13**, 207–14.
Hart, P. E., Mills, G., and Whitaker, J. K., eds. (1964). "Econometric Analysis for National Economic Planning." Colston Papers No. 16, London.
Hartman, P., and Winter, A. (1954). On non-oscillatory linear differential equations with monotone coefficients. *Amer. J. Math.* **76**, 207–219.
Hass, J. E. (1968). Transfer pricing in a decentralized firm. *Management Sci. Ser. B* **14**, 310–328.
Hayek, F. A. (1952). "The Counter Revolution of Science." Free Press, Glencoe, Illinois.
Heady, E., and Candler, W. (1958). "Linear Programming Methods." Iowa State Univ. Press. Ames, Iowa.
Heal, G. M. (1969). Planning without prices. *Rev. Econ. Stud.* **36**, 347–362.
Hestenes, M. R. (1966). "Calculus of Variations and Optimal Control Theory." Wiley, New York.
Hickman, B. G., ed. (1965). "Quantitative Planning of Economic Policy." Brookings, Washington, D. C.
Hicks, J. R. (1946). "Value and Capital," 2nd ed. Oxford Univ. Press (Clarendon), London and New York.
Hicks, J. R. (1965)· "Capital and Growth." Oxford Univ. Press, New York and London.
Hillier, F. S. (1967). Chance-constrained programming with zero-one or bounded continuous decision variables. *Managemeni Sci.* **14**, 34–57.
Ho, Y., and Brentani, P. B. (1963). On computing optimal control with inequality constraints. *SIAM J. Control* **1**, 319–347.
Ho, Y. (1965). Differential games and optimal control theory. *Proc. N. E. C.* **21**, 613–615.
Ho, Y., and Newbold, P. M. (1967). A descent algorithm for constrained stochastic extrema. *J. Optimization Theory Appl.* **1**, 215–231.
Hoffmann, W. G. (1955). "British Industry: 1700–1950. Oxford Univ. Press, London and New York.
Hoggatt, A. C. and Balderston, F. eds. (1963). "Symposium on Simulation Models." South-Western Publ., Chicago, Illinois.
Hohenbalken, B., von and Tintner, G. (1962). Econometric models for the OEEC member countries, the U. S. and Canada. *Weltwirtschaftliches Arch.* **89**.
Holt, C. C. (1962). Linear decision rules for economic stabilization and growth. *Quart. J. Econ.* **76**, 20–45.
Holt, C. C., Modigliani, F. and Simon, H. (1955). A linear decisson rule for production and employment scheduling. *Management Sci.* **2**, 1–30.

References and Bibliography

Holt, C. C. Modigliani, F., Muth, J. F., and Simon, H. A. (1960). "Planning, Production, Inventories, and Work Force." Prentice Hall, Englewood, Cliffs, New Jersey.

Holtzman, J. M. (1966). Convexity and the maximum principle for discrete systems. *IEEE Trans. Automatic Control* **11**, 30-35.

Holtzman, J. M., and Halkin, H. (1966). Directional canvexity and the maximum principle for discrete systems. *SIAM J. Control* **4**, 263-275.

Hood, W. C., and Koopmans, T. C., eds. (1953). "Studies in Econometric Method." Wiley, New York.

Hotelling, H. (1927). Differential equations subject to error, and population estimates. *J. Amer.) Statist. Assoc.* **22**, 288-314.

Hotelling, H. (1933). Analysis of a complex of statistical variables into principal components. *J. Educ. Phychol.* **24**, 417.

Howard, R. A. (1960). "Dynamic Programming and Markov Processes." Wiley New York.

Hufschmidt, M. M. (1963). Simulating the behavior of a multi-unit multi-purpose water resource system. "Symposium on Simulation Models" (A. C. Hoggatt and F. Balderston, eds.), South-Western Publ., Chicago, Illinois.

Ichimura, S. (1960). Dynamic input-output and linear programming models. U. N. Experts. "Programming Techniques for Economic Development with special reference to Asia and the Far East." U. N. Publ., Bangkok.

Inagaki, M. (1963). The golden utility path. Netherlands Economics Institute Rotterdam.

Intriligator, M. D. (1971). "Mathematical optimization and economic theory." Prentice Hall, Englewood Cliffs, New Jersey.

Isaacs, R. (1965). "Differential Games." Wiley, New York.

Isard, W.(1951). Interregional and regional input-output analysis: a model of space economy. *Rev. Econ. Statist.* **33**, 318-28.

Ito, K. (1951). On stochastic differential equations. *Mea. Amer. Math. Soc.* **4**, 1-51.

Jenkins, G. M. and Watts, D. G. (1968). "Spectral analysis and its applications." Holden-Day, San Francisco, California.

Johansen, L. (1960). "A Multisectoral Study of Growth." North Holland Publ. Amsterdam.

Johansen, L. (1967). Some theoretical properties of a two-sector model of optimal growth. *Rev. Econ. Stud.* **34**, 125-142.

Johnson, C. D., and Gibson, J. E. (1963). Singular solutions in problems of optimal control. *IEEE Trans. Autmatic Control* **8**, 4-15.

Johnston, J. (1963). "Econometric Methods." McGraw Hill, New York.

Jorgenson, D. W. (1960). A dual stability theorem. *Econometrica* **28**, 892-899.

Jorsenson, D. W. (1961). The development of a dual economy. *Econ. J.* **71**, pp. 309-334.

Jorgenson, D. W. (1966). Rational distributed lag functions. *Econometrica* **34**, 134-149.

Kalaba, R. (1962. Computations and considerations for some deterministic and adaptive control processes. Leitman, G. ed.

Kaldor, N. (1961). Capital accumulation and economic growth. "The Theory of Capital," pp. 177-220. Macmillan, London.

Kaldor, N., and Mirrlees, J. A. (1962). A new model of economic growth. *Rev. Econ. Stud.* **29**, 174-192.

Kalecki, M. (1935). A macrodynamic theory of business cycles. *Econometrica* **3**, 327-344.

Kalecki, M. (1954). "Theory of Economic Dynamics." Allen and Unwin, London.

Kailath, T., and Frost, P. (1968). Mathematical modeling of stochastic processes. "Stochastic Problems in Control: A Symposium of the American Automatic Control Council." *Amer. Soc. of Mechanical Engineers*, New York.

Kalman, R. E. (1961). On the general theory of control systems. *Proc. Int. Cong. Automatic Control* **1**, 481-492. Butterworths, London.

Kalman, R. E. (1963). Mathematical description of linear dynamical systems. *SIAM J. Control* **1**, 152-192.

Kantorovich, L. V. (1939). "Mathematical Methods of Organization and Planning of Production" (Russian). Univ. of Leningrad, Leningrad.

Kantorovich, L. V. (1959). "Economic Calculation of Optimal Utilization of Resources." Acad. of Sci. of the U. S. S. R.,Moscow.

Kantorovich, L. V. (1963). "Calcul economique et utilisation des resources." Dunod, Paris.

Kantorovich, L. V., and Akilov, G. P. (1964). "Functional Analysis in Normed Spaces." Pergamon Press, Oxford.

Kantorovich, L. V., and Krylov, V. I. (1958). "Approximation Methods of Higher Analysis." Noordhoff.

Karlin, S. (1956). "Mathematical Methods and Theory in Games, Programming and Economics," Vol. I. Addison-Wesley, Reading, Massachusetts.

Karreman, H. F., ed. (1968). "Stochastic Optimization and Control." Wiley, New York.

Kataoka, S. (1963). A stochastic programming model. *Econometrica* **31**, 181-96.

Kelley, H. J. (1964). A transformation approach to singular subarcs in optimal trajectory and control problems. *SIAM J Control* **2**, 234-240.

Kelley, J. E. (1960). The cutting plane method for solving convex programs. *SIAM* **8**, 703-12.

Kemeny, J. G., and Thompson, G. L. (1957). The effect of psychological attitudes on the outcomes of games. *Ann. Math. Stud.* **39**, 273-298.

Kemeny, J. G., Morgenstern, O., Thompson, G. L. (1956). A generalization of the von Neumann model of an expanding economy. *Econometrica* **24**, 115-135.

Kendall, D. (1949). Stochastic processes and population growth. *J. Roy. Statist. Soc. Ser. B* **11**, 230-264.

Kendall, M. G. (1949). On the reconciliation of theories of probability. *Biometrika* **36**, 101-116.

Kendall, M. G., and Stuart, A. (1961, 1963). "The Advanced Theory of Statistics," Vol. 1-3 Hafner Publ., New York.

Keynes, J. M. (1936). The General Theory of Employment, Interest and Money." Harcourt, New York.

Kingman, J. F. C. (1963). On inequalities of the Tchebychev type. *Pro. Cambridge Philosophical Soc.* **59**, 135-146.

Klein, L. R. (1947). The use of econometric models as a guide to economic policy. *Econometrica* **15**, 111-51.

Klein, L. R. (1962). "An Introduction to Econometrics." Prentice-Hall, Englewood Clifls, New Jersey.

Klein L. R. (1953). "Textbook of Econometrics." Harper, New York.

Klein, L. R., and Goldberger, A. S. (1955). "An Econometric Model of the United States 1929-1952." North Holland Publ., Amsterdam.

Klein, L. R. *et al.* (1961). A model of Japanese growth. *Econometrica* **29**, 277-92.

Koopmans, T. (1957). "Three Essays on the State of Economic Science." McGraw-Hill, New York.

Koopmans, T. (1960). Stationary ordinal utility and impatience. *Econometrica* **28**, 287-309.

Koopmans, T. (1965). On the concept of optimal economic growth. "The Econometric Approach to Development Planning." pp. 225-287. North Holland Publ., Amsterdam.

Koopmans, T., ed. (1951). "Activity Analysis of Production and Allocation." Wiley, New York.

Koopmans, T. (1951). Analysis of production as an efficient combination of activities. (T. C. Koopmans, ed.), "Activity Analysis of Production and Allocation." Wiley, New York.

Koopmans, T. (1966). On the concept of optimal economic growth. "The Econometric Approach to Development Planning." North-Holland Publ., Amsterdam and Rand McNally, Chicago, Illinois.

Koopmans, T. (1967). Objectives, constraints and outcomes in optimal growth models. *Econometrica* **35**, 1-15.

Koopmans, T., and Bausch, A. (1959). Selected topics in economics involving mathematical reasoning. *SIAM Rev.* **1**, 79-148.

Kopp, R. E., and McGill, R. (1964). Several trajectory optimization techniques. "Computing Methods in Optimization Problems" (A. V. Balakrishnan and L. W. Neustadt, eds.). Academic Press, New York.

Kornai, J., and Liptak, Th. (1962). A mathematical investigation of some economic effects of profit sharing in socialist firms. *Econometrica* **30**,

Kornai, J., and Liptak, Th. (1965). Two-level planning. *Econometrica* **33**, 141-169.

Kozin, F. (1965). On relations between moment properites and almost sure Lyapunov stability for linear stochastic systems. *J. Math. Anal. Appl.* **10**, 342-353.

Kreindler, E. and Sarachik, P. E. (1964). On the concepts of controllability and observability of linear systems. *IEEE Trans. Automatic Control* **9**, 129-36.

Kreko, B. (1965). "Lehrbuch der linearen Optimierung." V. E. B. Deutscher Verlag der Wissenschaften, Berlin.

Kromphardt, W., Henn, R., and Forstner, K. (1962). "Linear Entscheidungsmodelle. Springer-Verlag, Berlin.

Kuenne, R. E. (1963). "The Theory of General Economic Equilibrium." Princeton Univ. Press, Princeton, New Jersey.

Kuhn, H. W. (1963). Locational problems and mathematical programming. "Colloquium on Applications of Mathematics to Economics" (A. Prekopa, ed.) Hungarian Acad. of Sci. Budapest.

Kuhn, H. W., and Szego, G. P. ed. (1969). "Mathematical Systems: Theory and Economics," Vol. I-II. Springer-Verlag, Berlin.

Kuhn, H. W., and Tucker, A. W. (1951). Nonlinear programming. *Proc. Berkeley Symp. Math. Statist. Probability* Berkeley.

Kulikowski, R. (1969). *See* Kuhn, H. W., and Szego, G. P., ed. (1969).

Kurz, M. (1965). Optimal paths of capital accumulation under the minimum time objective. *Econometrica* **33**, 42-66.

Kurz, M. (1968). The general instability of a class of competitive growth processes. *Rev. Econ. Stud.* **35**, 155-174.

Kushner, H. J. (1963). On stochastic extremum problems: calculus of variations. Lincoln Lab., MIT Rep. JA-2229.

Kushner, H. J. (1965). On the stochastic maximum principle. *J. Math. Anal. Appl.* **11**, 78-92.

Kushner, H. J. (1967). "Stochastic Stability and Control. Academic Press, New York.

Lange, O. (1957). Some observations on input-output analysis. *Sankhya* **18**, 305-36.

Lange, O. (1960). Output-investment ratio and input-output analysis. *Econometrica* **28**, 310-324.

LaSalle, J. P. (1962). Complete stability of a nonlinear control system. *Prod. Nat. Acad. Sci.* **48**, 600-603.

Lasdon, L. S. (1970). "Optimization Theory for Large Systems." MacMillan, New York.

Lasdon, L. S., Mitter, S. K., and others (1967). The conjugate gradient method for optimal control problems. *IEEE Trans. Automatic Control* **12**, 132-138.

Lavi, A., and Vogl, T. P. (1966). "Recent Advances in Optimization Techniques." Wiley, New York.

Lee, E. B., and Markus, L. (1961). Optimal control for nonlinear processes. *Arch. Rational Mech. Anal.* **8**, 36-58.

Lee, E. B., and Markus, L. (1967). "Foundations of Optimal Control." Wiley, New York.
Lefeber, L. (1958). "Allocation in Space." North-Holland Publ., Amsterdam.
Lefkowitz, I. (1965). Multi-level approach applied to control system design. *Joint Automatic Control Conf., Reprint Tech. Papers* **6**, 100–109.
Lehmann, E. L. (1955). Ordered families of distributions. *Ann. Math. Statist.* **26**, 399–419.
Leitman, G., ed. (1962). "Optimization Techniques." Academic Press, New York.
Leitman, G., ed. (1967). "Topics in Optimization." Academic Press, New York.
Leondes, C. T., ed. (1965). "Modern Control System Theory." McGraw-Hill, New York.
Leontief, W. (1963). Multiregional input-output analysis. "Structural Interdependence and Economic Development" (T. Barna, ed.). St. Martin's Press, New York.
Lerner, A. P. (1944). "Economics of Control." Macmillan, New York.
Leser, C. E. V. (1966). "Econometric Techniques and Problems." Hafner, New York.
Lesourne, J. (1960). "Technique économique et gestion industrielle." Dunod, Paris.
Letov, A. M. (1961). "Stability of Nonlinear Controls" (English Trans. from Russian). Princeton Univ. Press, Princeton, New Jersey.
Levhari, D. (1965). A nonsubstitution theorem and switching of techniques. *Quart. J. Econ.* **79**, 98–105.
Levhari, D. and Srinivasan T. N. (1969). Optimal savings under uncertainty. *Rev. Econ. Studies* **36**, 153–163.
Lévy, P. (1925). "Calcul des Probabilités." Gauthier Villars, Paris.
Lintner, J. (1965). Security prices risk and maximal gains from diversification. *J. Finance* **20**, 587–615.
Lord, F. M. (1964). Human judgments as fallible measures. "Human Judgments and Optimality" (M. W. Shelly and G. L. Bryan, eds.). Wiley, New York.
Luce, R. D., and Raiffa, H. (1957). "Games and Decisions." Wiley, New York.
Machol, R. E., and Gray, P., eds. (1962). "Recent developments in Information and Decision Processes." Macmillan, New York.
Madansky, A. (1962). Methods of solution of linear programs under uncertainty. *Operations Rev.* **10**, 463–70.
Madansky, A. (1963). Linear programming under uncertainty. Graves, R. L. and Wolfe, P. eds.
Mahalanobis, P. C. (1955). Approach of operational research to planning in India. *Sankhya* **16**, 3–62.
Mahalanobis, P. C. (1957). Planning in India. *Sankhya* **17**, 4–16.
Malinvaud, E. (1966). "Statistical Methods of Econometrics." Rand McNally, Chicago, Illinois.
Malinvaud, E. (1967). Decentralized procedures for planning. "Activity Analysis in the Theory of Growth and Planning" (E. Malinvaud and M. O. L. Bacharach, eds.). Macmillan, London.
Malinvaud, E. (1970). "Statistical Methods of Econometrics." North-Holland Publ., Amsterdam.
Malinvaud, E., and Bacharach, M. O. L., ed, (1967). "Activity Analysis in the Theory of Growth and Planning." St. Martin's Press, New York.
Mandelbrot, B. (1960). The Pareto-Levy law and the distribution of income. *Int. Econ. Rev.* **1**, 79–106.
Mandelbrot, B. (1961). Stable Pareto random functions and the multiplicative variation of income. *Econometrica* **29**, 517–43.
Mandelbrot, B. (1963). The variation of speculative prices. *J. Business* **36**, 394–419.
Mangasarian, O. L. (1963). Stability criteria for nonlinear ordinary differential equations. *SIAM J. Control* **1**, 311–318.

Mangasarian, O. L. (1964). Nonlinear programming problems with stochastic objective functions. *Management Sci.* **10**, 353–59.

Mangasarian, O. L. (1969). "Nonlinear Programming." McGraw-Hill, New York.

Mann, H. B., and Wald, A. (1943). On the statistical treatment of linear stochastic difference equations. *Econometrica* July.

Manne, A. S. (1961). Capacity expansion and probabilistic growth. *Econometrica.* **29**, 632–649.

Manne, A. S., and Veinott, A. F., Jr. (1967). Optimal plant size with arbitrary increasing time path of demand. "Investment for Capacity Expansion" (A. S. Manne, ed.). MIT Press, Cambridge, Massachusetts.

Marglin, S. A. (1963). "Approaches to Dynamic Investment Planning." North-Holland Publ., Amsterdam.

Marglin, S. A. (1967). The rate of interest and the value of capital with unlimited supplies of labor. Shell, K. ed.

Markowitz, H. (1959). "Portfolio Selection." Wiley, New York.

Marschak, J. (1950). Rational behavior, uncertain prospects and measurable utility. *Econometrica* **18**, 111–141.

Marschak, J. (1964). Actual versus consistent decision behavior. *Behav. Sci.* **9**, 103–110.

Marschak, T. A. (1966). Computation in organizations. Working Paper No. 156. Center for Research in Management Science. Univ. of California, Berkeley.

Marschak, T. A. (1968). Computation in organizations: comparison of price mechanisms and other adjustment processes. "Risk and Uncertainty: Proceedings of a Conference of International Economic Association." Macmillan, London.

Marshall, A. (1948). "Principles of Economics," 8th ed. Macmillan, New, York.

Martos, B., and Kornai, J. (1966). Experiments in Hungary with industry-wide and economy-wide programming. *Math. Optimization Tech.* Proceedings of a Conference. Rome.

Marx, K. (1933). "Capital." Kerr, Chicago, Illinois.

Massé, P. (1959). "Le choix des investissements." Dunod, Paris.

Massé, P. (1962). "Optimal Investment Decisions" (English Transl.). Prentice-Hall, Englewood Cliffs, New Jersey.

Massera, J. L. (1956). Contributions to stability theory. *Ann. Math.* **64**, 182–205.

Matusita, K. (1966). A distance and related statistics in multivariate analysis. "Multivariate Analysis" (P. R. Krishnaiah, ed). Academic Press, New York.

McFadden, D. (1967). The evaluation of development programmes. *Rev. Econ. Stud.* **34**, 25–50.

McFadden, D. (1969). On the controllability of decentralized macroeconomic systems: the assignment problem. Kuhn, H. W. and Szego, G. P. ed.

McGrath, R. J., Rajaraman, V., and Rideout, V. C. (1961). A parameter perturbation adaptive control system. *IRE Trans. Automatic Control* **6**.

Menges, G. (1961). "Oekonometrie." Wiesbaden: Betriebswirtschaftlicher Verlag.

Merriam, C. W. (1960). Use of a mathematical error criterion the in design of adaptive control system. *AIEE Trans.* Part II, **79**, 506–512.

Merton, R. C. (1969). Lifetime portfolio selection under uncertainty: the continuous-time case. *Rev. Econ. Statist.* **51**, 247–257.

Mesarovic, M. D., Macko, D. and Takahara, Y. (1970). "Theory of hierarchical multi-level systems." Academic Press, New York.

Mesarovic, M. D., Pearson, J. D. and others (1965). A multi-level structure for a class of linear dynamic optimization problems. *Joint Automatic Contronl Conf., Preprint Tech. Papers* **6**, 93–99.

Mills, E. S. (1962). "Price, Output and Inventory Policy." Wiley, New York.

Mills, H. D. (1956). Marginal values of matrix games and linear programs. "Liner Inequalities and Related Systems" (H. W. Kuhn, and A. W. Tucker, ed.) Princeton Univ. Press, Princeton, New Jersey.

Milnor, J. (1954). Games against nature. "Decision Processes" (R. M. Thrall, C. H. Coombs, and R. L. Davis, ed.) Wiley, New York.

Mirrlees, J. A. (1965). Optimum allocation under uncertainty. Mimeographed manuscript.

Mirrlees, J. A. (1967). Optimum growth when technology is changing. *Rev. Econ. Stud.* **34**, 95–124.

Montias, J. M. (1967). Soviet optimizing models for multiperiod planning. Hardt, J. P., Hoffenberg, M., Kaplan, N. and Levine, H. S. eds.

Mood, A. M., and Graybill, F. A. (1963). "Introduction to The Theory of Statistics," 2nd ed. Macmillan, New York.

Moran, P. A. P. (1959). "Theory of Storage." Methuen, London.

Morgenstern, O. (1963). "On the Accuracy of Economic Observations." 2nd ed. Princeton Univ. Press, Princeton, New Jersey.

Moriguti, S. (1951). Extremal properties of extreme value distributions. *Ann. Math. Statist.* **22**, 523–536.

Morishima, M. (1964). "Equilibrium, Stability and Growth." Oxford Univ. Press (Clarendon), London and New York.

Morishima, M (1965). Balanced growth and technical progress in a log-linear multisectoral economy. "The Econometric Approach to Development Planning," pp. 529–556. North Holland Publ., Amsterdam.

Morishima, M. (1969). "Theory of Economic Growth." Oxford Univ. Press (Clarendon), London and New York.

Morris, W. T. (1968). "Decentralization in Management Systems." Ohio State Univ. Press, Columbus, Ohio.

Morse, P. M. (1958). Queues, Inventories and Maintenance." Wiley, New York.

Mossin, J. (1966). Equilibrium in a capital asset market. *Econometrica* **34**, 768–783.

Mukherjee, V., Tintner, G., and Narayanan R. (1964). A generalized Poisson process for the explanation of economic development with applications to Indian data. *Arthaniti* **8**, 156–164.

Murphy, R. E. (1965). "Adaptive Processes in Economic Systems." Academic Press, New York.

Naslund, B. (1966). A model of capital budgeting under risk. *J. Business* **39**, 257–271.

Naslund, B. (1967). "Decisions under Risk." Stockholm School of Economics, Stockholm.

Naslund, B. (1968). Some effects of taxes on risk-taking. *Rev. Econ. Studies* **35**, 289–306.

Naslund, B., and Whinston, A. W. (1962). A model of multi-period investment under uncertainty. *Management Sci.* **8**, 184–200.

Naylor, T. H., Balintfy, J. L., Burdick, D. S. and Chu, K. (1966). "Computer Simulation Techniques." Wiley, New York.

Naylor, T. H., Wertz, K., and Wonnacott, T. H. (1968). Methods for evaluating the effects of economic policies using simulation experiments. *Rev. Int. Statist. Inst.* **36**, 184–200.

Naylor, T. H., Wertz, K., and Wonnacott, T. H. (1969). Spectral analysis of data generated by simulation experiments with econometric models. *Econometrica* **37**, 333–352.

Negishi, T. (1962). The stability of a competitive economy. *Econometrica* **30**, 635–669.

Nehari, Z. (1957). Oscillation criteria for second-order linear differential equations. *Trans. Amer. Math. Soc.* **85**, 428–445.

Nemhauser, G. L. (1966). "Introduction to Dynamic Programming." Wiley, New York.

Nerlove, M. (1964). Spectral analysis of seasonal adjustment procedures. *Econometrica* July.

Neustadt, L. W. (1960). Synthesizing time-optimal control systems. *J. Math. Anal. Appl.* **1**, 484–493.

Newman, P. (1968). "Readings in Mathematical Economics," Vol. I–II. Johns Hopkins Press, Baltimore, Maryland.

Newton, R. H. C. (1969). Statistical optimal control. *Int. J. Control* **10**, 303–313.

Nikaido, H. (1964). Monopolistic factor price imputation and linear programming. *Z. Nationalokon.* **24**, 298–301.

Nikaido, H. (1968). "Convex Structures and Economic Theory." "Academic Press, New York.

Nikaido, H. (1970). "Introduction to Sets and Mappings in Modern Economics" (*English Transl.* by K. Sato.) North-Holland Publ., Amsterdam.

O'Carroll, F. M. (1963). Income distribution in relation to economic development. "Economic Planning." *In.* L. J. Zimmerman, ed. pp. 101–144. Monton Publ., the Hague.

Oldenburger, R., ed. (1966). "Optimal and Self-optimizing Control." MIT Press, Cambridge, Massachusetts.

Orcutt, G. (1968). *See* Watts, D. G., ed. (1968).

Orcutt, G., and James, F. S. (1948). Testing the significance of correlation between time series. *Biometrika* **35**, 397–413.

Orcutt, G., Greenberger, M., Korbel, J., and Rivlin, A. M. (1961). "Microanalysis of Scioeconomic Systems." Harper, New York.

Osborne, M. (1959). Brownian motion in the stock market. *Operations Res.* March–April.

Owen, D. B. (1962). "Handbook of Statistical Tables." Addison-Wesley, Reading, Massachusetts.

Ozga, S. A. (1965). "Expectations in Economic Theory." Aldine Publ., Chicago, Illinois.

Paiewonsky, B. (1967). Synthesis of optimal controls. Leitman, G. ed.

Pallu de la Barriere, R. (1967). "Optimal Control Theory." Saunders, Philadelphia, Pennsylvania.

Pareto, V. (1897). "Cours d'Economie Politique." Rouge, Lausanne.

Pareto, V. (1927). "Manuel d'economie politique," 2nd ed. Giard, Paris.

Parzen, E. (1960). "Modern Probability Theory and its Application." Wiley, New York.

Pasinetti, L. L. (1965). A new theoretical approach to the problems of economic growth. "The Econometric Approach to Development Planning." North-Holland Publ., Amsterdam and Rand McNally, Chicago, Illinois.

Pearson, J. D. (1966). Multi-level control systems. "Theory of Self-Adaptive Control Systems" (P. H. Hammond, ed.). Plenum Press, New York.

Pfanzagl, J. (1959). A general theory of meausrement: application to utility. *Naval Res. Logist. Quart.* pp. 283–294.

Pfouts, R. W. (1961). The theory of cost and production in the multi-product firm. *Econometrica.* **29**, 650–658.

Phelps, E. S. (1961). The golden rule of accumulation. *Amer. Econ. Rev.* **51**, 638–642.

Phelps, E. S. (1962). The accumulation of risky capital: a sequential utility analysis. *Econometrica.* **30**, 729–743.

Phelps, E. S. (1966). "Golden Rules of Economic Growth." Norton, New York.

Phillips, A. W. (1954). Stabilization policy in a closed economy. *Econ. J.* **64**, 290–323.

Pillai, K. C., and Ramachandran, K. V. (1954). On the distribution of the ratio of the ith observation in an ordered sample from a normal polulation. *Ann. Math. Statist.* **25**, 565–72.

Plant, J. B. (1968). "Some Iterative Solutions in Optimal Control." MIT Press, Cambridge, Massachusetts.

Pontryagin, L. S. (1961). Optimal regulation processes. *Amer. Math. Soc. Transl. Ser. 2*, **18**, 295-339.
Pontryagin, L. S. (1965). On some differential games. *SIAM J. Control* **3**, 39-52.
Pontryagin, L. S., Boltyanskii, V. G., Gamkrelidze, R. V., and Mishchenko, E. F. (1962). "The Mathematical Theory of Optimal Processes." Wiley (Interscience), New York.
Portes, R. D. (1969). The enterprise under central planning. *Rev. Econ. Stud.* **36**, 197-212.
Prais, S. J., and Houthakker, H. S. (1955). "The Analysis of Family Budgets." Cambridge Univ. Press, London and New York.
Pratt, J. W. (1964). Risk aversion in the small and in the large. *Econometrica* **32**, 122-136.
Pratt, J. W., Raiffa, H., and Schlaifer, R. (1964). The foundations of decision under uncertainty: an elementary exposition. *J. Amer. Statist. Associ.* **59**, 353-375.
Prekopa, A. (1966). On the probability distribution of the optimum of a random linear program. *SIAM J. Control* **4**, 211-22.
Pugachev, V. F. (1963). On an optimization criterion for the economy. Veinstein, A. L. ed.
Quandt, R. (1958). The estimation of the parameters of a linear regression system obeying two separate regimes, *J. Amer. Statis. Assoc.* **53**, 873-880.
Quandt, R. (1960). Tests of the hypothesis that a linear regression system obeys two separate regimes. *J. Amer. Statist. Assoc.* **55**, 324-330.
Quandt, R. E. (1966). Statistical discrimination among alternative hypotheses and some economic regularities. *J. Regional Sci.* **5**, 1-24.
Quenouille, M. H. (1947). A large sample test for goodness of fit of autoregressive schemes. *J. Roy. Statist. Ser. A.* **110**, 123-129.
Quenouille, M. (1957). "The Analysis of Multiple Time Series." Hafner, New York.
Radner, R. (1961). Paths of economic growth that are optimal with regard only to final states, a turnpike theorem. *Rev. Econ. Stud.* **28**, 98-104.
Radner, R. (1962). Team decision problems. *Ann. Math. Stattist.* **33**, 857-881.
Radner, R. (1963). "Notes on the Theory of Economic Planning." Centre of Economic Research, Athens.
Radner, R. (1966). Optimal growth in a linear-logarithmic economy. *Int. Econ. Rev.* **7**, 1-33.
Radner, R. (1967). Efficiency prices for infinite horizon production programmes. *Rev. Econ. Studi.* **34**, 51-66.
Radner, R. (1968). Competition equilibrium under uncertainty. *Econometrica* **36**, 31-58.
Radstrom, H. (1964). A property of stability possessed by certain imputations. *Ann. Math. Stud.* **52**, 513-529.
Ramsay, F. P. (1928). A mathematical theory of savings. *Econ. J.* **38**, 543-559.
Rao, J. N. K., and Tintner, G. (1962). The distribution of the ratio of the variances of the variate difference in the circular case. *Sankhya Ser. A* **24**, 385-394.
Rao, J. N. K, and Tintner, G. (1963). On the variate difference method. *Aust. J. Statist.* **5**, 106-116.
Rekasius, Z. V. (1961). A general performance index for analytical design of control systems. *IRE Trans. Automatic Control* **6**, 217-221.
Rekasius, Z. V. (1964). Suboptimal design of intentionally nonlinear controllers. *IEEE Trans. Automatic Control* **9**, 380-386.
Rhodes, D. D. (1940). Population mathematics III. *J. Roy. Statist. Soc. Ser. A* **103**, 362-387.
Robbins, H. (1956). See Machol, R. E., and Gray, P., ed. (1962).
Robinson, J. (1938). "Economics of Imperfect Competition." Macmillan, London.
Robinson, J. (1962). A neoclassical theorem. *Rev. Econ. Stud.* **29**, 219-226.
Roos, C. F. (1934). "Dynamic Economics." Principia, Bloomington, Indiana.

Rosen, J. B. (1962). Controllable stability and equivalent nonlinear programming problem. "Contributions to Differential Equations." Academic Press, New York.

Rosenfeld, J. L. (1961). Adaptive decision process. Unpublished Ph. D. thesis, MIT.

Rosenfeld, J. L. (1964). Adaptive competitive decision. "Advances in Game Theory." Princeton Univ. Press, Princeton, New Jersey.

Rosenstein-Rodan, P. N. (1943). Problems of industrialization of Eastern and Southeastern Europe. *Econ. J.* **53**, 202–211.

Rostow, W. W. (1960). "The Stages of Economic Growth." Cambridge Univ. Press, London and New York.

Roy, A. D. (1952). Safety first and the holding of assets. *Econometrica* **20**, 431–49.

Rozonoer, L. I. (1959). Pontryagin's maximum principle in the theory of optimum systems. Oldenburger, R. ed.

Ruubel, K. V. (1959). A criterion of control inaccuracy. *Automat. Remote Contr.* **20**, 831–835.

Saaty, T. L. (1959). Coefficient perturbation of a constrained extremum. *Operations Res.* **7**, No. 3.

Saaty, T. L. (1961). "Elements of Queuing Theory." McGraw-Hill, New York.

Saaty, T. L., and Webb, K. W. (1960). Sensitivity and renewals in scheduling aircraft, overhaul. *Proc. and Int. Conf. Operations Res.* London.

Saaty, T. L. and Webb, K. W. (1961). Sensitivity and renewals in scheduling aircraft overhaul. *Proc. Second Intern'l Conf. on Operations Research.* pp. 708–715.

Sakawa, Y. (1966). Optimal control of a certain type of linear distributed parameter systems. *IEEE Trans. Automatic Control* **11**, 35–41.

Samuels, J. C. (1960). On the stability of random systems and the stabilization of deterministic systems with random noise. *J. Acoust. Soc. Amer.* **32**, 594-000.

Samuelson, P. (1959). Efficient paths of capital accumulation in terms of the calculus of variations. "Mathematical Methods in the Social Sciences" (K. J. Arrow, S. Karlin, and P. Suppes, eds.). Stanford Univ. Press, Stanford, Calfornia.

Samuelson, P. (1965). A catenary turnpike theorem involving consumption and the golden rule. *Amer. Econ. Rev.* **55**, 486–496.

Samuelson, P. (1967). Indeterminacy of development in a heterogenous capital model with constant saving propensity. Shell, K. ed.

Samuelson, P. (1969). Lifetime portfolio selection by dynamic stochastic programming. *Rev. Econ. Statist.* **51**, 239–246.

Sandee, J. (1960). "A Demonstration Planning Model for India." Statist. Publ. Soc. Calcutta.

Sarhan, A. E., and Greenberg, B. G., ed. (1962). "Contributions to Order Statistics." Wiley, New York.

Savage, L. J. (1962). "The Foundations of Statistical Inference." Methuen, London.

Schultz, P. R. (1965). Some elements of stochastic approximation theory and its application to a control problem. "Modern Control Systems Theory" (C. T. Leondes, ed.) McGraw-Hill, New York.

Schultz, W. C., and Rideout, V. C. (1961). Control system performance measures: past, present and future. *IRE Trans. Automatic Control* **6**, 22–34.

Schumpeter, J. (1934). "Theory of Economic Development." Harvard Univ. Press, Cambridge, Massachusetts.

Schumpeter, J. (1954). "History of Economic Analysis." Oxford Univ. Press, New York.

Sengupta, J. K. (1963). Models of agriculture and industry in less developed economies. Structural Interdependence and Economic Development" (T. Barna. ed.) St. Martins Press, New York.

Sengupta, J. K. (1964). On the relative stability and optimality of consumption in aggregative growth models. *Economica* **31**, 33–50.

Sengupta, J. K. (1964) Policy criteria for stabilization and growth. *Oxford Econ. Papers* **16**, 407–417.

Sengupta, J. K. (1965a). On the sensitivity of optimal solution under investment planning and programming. *Arthaniti* **8**, 1–23.

Sengupta, J. K. (1965b). Specification and estimation of structural relations in policy models. "Quantitative Planning of Economic Policy" (B. G. Hickman, ed.) Brookings, Washington, D. C.

Sengupta, J. K. (1965c). Cobweb cycles and optimal price stabilization through buffer funds. *Indian Econ. J.* **13**, 351–364.

Sengupta, J. K. (1965d). Income stabilization policy: an elementary empirical analysis. *Econ. Int.* **18**, 2–16.

Sengupta, J. K. (1966a). On the stability of truncated solutions under stochastic linear programming. *Econometrica* **34**, 77–104.

Sengupta, J. K. (1966b). Truncated decision rules and optimal economic growth with a fixed horizon. *Int. Econ. Rev.* **7**, 42–64.

Sengupta, J. K. (1966c). Recursive constraints and stochastic linear programming. *Metrika* **10**.

Sengupta, J. K. (1968). Some observations on the optimal growth path for an underdeveloped economy. *Metroeconomica* **20**, 149–172.

Sengupta, J. K. (1968a). Econometric models of risk programming. *Indian Econ. J. Econmet. Ann.* **15**, 423–441.

Sengupta, J. K. (1968b). Methods of linear programming under risk. (author; mimeographed).

Sengupta, J. K. (1968c). A computable approach to risk programming in linear models. "Papers in Quantitative Economics." Univ. Press of Kansas, Lawrence, Kansas.

Sengupta, J. K. (1969). Distribution problems in stochastic and chance constrained programming. "Economic Models, Estimation and Risk Programming: Essays in honor of G. Tintner" (K. Fox, J. K. Sengupta, and G. V. L. Narasimham, eds.), 391–424. Springer-Verlag, Berlin and New York.

Sengupta, J. K. (1969a). Safety first rules under chance-constrained linear programming. *Operations Res.* **17**, 112–132.

Sengupta, J. K. (1969b). A system reliability approach to linear programming. Accepted for publication in *Unternehmensforschung*.

Sengupta, J. K. (1969c). The extremes of extreme value solutions under risk programming. (author; mimeographed).

Sengupta, J. K. (1969d). Chance-constrained linear programming with chi-square type deviates. (author; mimeographed).

Sengupta, J. K. (1969e). A generalization of some distribution aspects of chance-constrained linear programming. See Sengupta, J. K. (1970e).

Sengupta, J. K. (1969f). A statistical reliability approach to linear programming. Accepted for publication in *Unternehmensforschung*.

Sengupta, J. K. (1970a). Optimal stabilization policy with a quadratic criterion function. *Rev. Econ. Stud.* **37**, 127–145.

Sengupta, J. K. (1970b). Optimum growth under dynamic stochastic programming. (author, mimeographed).

Sengupta, J. K. (1970c). Economics of decomposition and divisionalization under transfer. Accepted for publication in *Z. die Gesamte Staatswissench*.

Sengupta, J. K. (1970d). The stability in optimum growth models. (author, mimeographed).

Sengupta, J. K. (1970e) A generalization of some distribution aspects of chance-constrained linear programming. *Int. Econ. Rev.* **11**, No. 2, 287-304.

Sengupta, J. K. (1970f). Stochastic linear programming with chance constraints. *Int. Econ. Rev.* **11**, No. 1, 101-116.

Sengupta, J. K. (1970g). Application of system reliability measures in chance-constrained programming. (Sent for publication).

Sengupta, J. K. (1970h). Adaptive control in models of optimum economic growth. (Sent for pulication).

Sengupta, J. K. (1970i). On the active approach of stochastic linear programming. *Metrika* **15**, 59-70.

Sengupta, J. K., and Fox, K. A. (1969). "Economic Analysis and Operations Research: Optimization Techniques in Quantitative Economic Models." North-Holland Publ., Amsterdam.

Sengupta, J. K., and Fox, K. A. (1970). A computable approach to optimal growth of an Z. *die Gesamte Staatswissensch.* **126**, 97-125.

Sengupta, J. K., and Gruver, G. (1969). A linear reliability analysis in programming with chance constraints. *Swedish J. Econ.* 221-246.

Sengupta, J. K., and Gruver, G. (1970). Chance-constrained linear programming under truncation and varying sample sizes. Accepted for publication in *Swedish J. Econ.*

Sengupta, J. K., and Kumar, T. (1964). An application of sensitivity analysis to a linear programming problem. *Unternhmungsforschung* **9**, 18-36.

Sengupta, J. K., and Portillo-Campbell, J. H. (1970). A fractile approach to linear programming under risk. *Management Sci. Theory Ser.* **16**, 298-308.

Sengupta, J. K., and Portillo-Campbell, J. H. (1970a). The approach of geometric programming with economic applications. (authors; mimeographed).

Sengupta, J. K., and Portillo-Campbell, J. H. (1970b). Optimal investment allocation through reliability programming (authors; mimeographed).

Sengupta, J. K., and Portillo-Campbell, J. H. (1970c). A reliability programming approach to production planning. (authors; mimeographed).

Sengupta, J. K., and Sanyal, B. C. (1970). Sensitivity analysis methods for a crop-mix problem in linear programming. *Unternehmensforschung* **14**, 2-26.

Sengupta, J. K., and Sanyal, B. C. (1970a). A fractile programming approach with extreme sample estimates of parameters. *Statist. Neerland.* **24**, 51-59.

Sengupta, J. K., and Sanyal, B. C. (1970b). A distributional analysis of linear programs under risk. (Sent for publication).

Sengupta, J. K., and Sen, A. (1969). Models of optimal capacity expansion for the firm: an appraisal. *Metroeconomica* **21**, 1-28.

Sengupta, J. K., and Tintner, G. (1963a). On some economic models of development planning. *Econ. Int.* **16**, 1-19.

Sengupta, J. K., and Tintner, G. (1963b). On some aspects of trend in the aggregative models of economic growth. *Kyklos* **14**, 47-61.

Sengupta, J. K., and Tintner, G. (1964). An approach to a stochastic theory of economic development with applications. "Problems of Economic Dynamics and Planning." Essays in honour of M. Kalecki. PWN Polish Sci. Publ., Warsaw.

Sengupta, J. K., and Tintner, G. (1965). The flexibility and optimality of Domar-type growth models. *Metroeconomica* **17**, 3-16.

Sengupta, J. K., and Tintner, G. (1966). On the stability of solutions under recursive programming. *Unternehmungsforschung* **10**, 1-14.

Sengupta, J. K., and Tintner, G. (1970). The approach of stochastic linear programming: methods and computation. (authors; mimeographed).

Sengupta, J. K., and Walker, D. A. (1964). On the empirical specification of optimal economic policy for growth and stabilization under a macrodynamic model. *The Manchester School*. September, pp. 215–238.

Sengupta, J. K., Tintner, G., and Millham, C. (1963). On some theorems of stochastic linear programming with applications. *Management Sci.* **10**, 143–159.

Sengupta, J. K., Tintner, G., and Morrison, B. (1963). Stochastic linear programming applications to economic models. *Economica* **30**, 262–276.

Sengupta, J. K., Millham, C., and Tintner, G. (1965). On the stability of solutions under error in stochastic linear programming. *Metrika* **9**, 47–60.

Sengupta, S. S. (1967). "Operations Research and Sellers' Competition." Wiley, New York.

Shackle, G. L. S. (1949). "Expectation in Economics." Cambridge Univ. Press, London and New York.

Shackle, G. L. S. (1961). "Decision Order and Time." Cambridge Univ. Press, London and New York.

Shapley, L. S. (1953). (1953). Stochastic games. *Pro. Nat. Acad. Sci.* **39**, 1095–1100.

Shell, K. ed. (1967). "Essays on the Theory of Optimal Economic Growth." MIT Press, Cambridge, Massachusetts.

Shell, K. and Stiglitz, J. E. (1967). The allocation of investment in a dynamic economy. *Quart. J. Econ.* **81**, No. 4, 592–609.

Shell, K. (1969). Applications of Pontryagin's maxmum principle to economics. *In. Mathematical Systems: Theory and Economics.*" Vol. 1. H. W. Kuhn and G. P. Szego, eds. pp. 241–292. Springer-Verlag Berlin and New York.

Shubik, M. (1959) "Strategy and Market Structure." Wiley, New York.

Shubik, M. (1959). Edgeworth market games. *Ann. Math. Stud.* **40**, 267–278.

Shubik, M, (1962). Incentives, decentralized control, the assignment of joint costs and internal pricing. *Management* **8**.

Sidrauski, M. (1969). Rational choice and patterns of growth. *J. Political Econ.* **77**, 575–585.

Simon, H. A. (1955). On a class of skew distribution functions. *Biometrica* **42**, 425–440.

Simon, H. A. (1956). Dynamic programming under uncertainty with a quadratic criterion function. *Econometrica* **24**, 74–81.

Simon, H. A. (1957). "Models of Man, Social and Rational." Wiley, New York.

Sinha, S. M. (1966). A duality theorem for nonlinear programming. *Management Sci.* **12**, 385–390.

Slutsky, E. (1927). The summation of random causes as a source of cyclical processes. *Ecomometrica* **5**, 105–146.

Smirnov, N. (1948). Table for estimating the goodness of fit of empirical distributions. *Ann. Math. Statist.* **19**, 279–81.

Smith, H. W., and Man, F. T. (1969). Computation of suboptimal linear controls for nonlinear stochastic systems. *Int. J. Control* **10**, 645–655.

Smith, V. L. (1961). "Investment and production." Harvard Univ. Press, Cambridge, Massachusetts.

Solodovnikov, V. V. (1963). "Einfuehrung in die statistische Dynamik Linearer Regelungssysteme." Oldenbourg, Munich.

Solow, R. M. (1956). A contribution to the theory of economic growth. *Quart. J. Econ.* **70**, 65–94.

Solow, R. M. (1964). "Capital Theory and the Rate of Return." North Holland Publ., Amsterdam.

Solow, R. M., Tobin, J., von Weizsäcker, C. C., and Yaari, M. (1966). Neoclassical growth with fixed factor proportions. *Rev. Econ. Stud.* **33**, 79–116.

Soults, D. J. (1968). Asymptotic value distributions for matrix games. Unpublished Ph. D. Dissertation. Iowa State Uni, Amems.
Sprekle, C. (1961). Warrant prices as indicators of expectations and preferences. *Yale Economic Essays* **1**, 178-231.
Srinivasan, T. N. (1965). Optimal savings in a two-sector model of growth. *Econometrica* **32**, 358-73.
Steindl, J. (1965). "Random Processes and the Growth of Firms." Hafner, New York.
Stiglitz, J. E. (1970). A consumption-oriented theory of the demand for financial assets and the term structure of interest rates. *Rev. Econ. Studies* **37**, 321-351.
Stigum, B. P. (1969). Competitive equilibria under uncertainty. *Quart. J. Econ.* **83**, 533-561.
Stoleru, L. G. (1965). An optimal policy for economic growth. *Econometrica* **33**, 321-348.
Stone, B. K. (1970). "Risk return and equilibrium." MIT Press, Cambridge, Massachusetts.
Stone, R. (1962). Three models of economic growth. "Logic, Methodology and Philosophy of Science" (E., Nagel, P. Suppes, and A. Tarski, eds.). Stanford Univ. Press, Stanford, California.
Stone, R. (1965). A model of the educational system. *Minerva* **3**, No. 2, 172-186.
Stone, R., and Brown, A. (1962). "A Programme for Growth: A Computable Model of Economic Growth." Chapman and Hall, London.
Stone, R., Rowe, D.A., Corlett, W.J., Hurstfield, R., Potter, M. (1954). "The Measurement of Consumers Expenditure and Behavior in the United Kingdom." Cambridge Univ. Press, Cambridge, Massachusetts.
Struble, R. A. (1962). "Nonlinear Diflerential Equations." McGraw-Hill, New York.
Swan, T. (1956). Economic growth and capital accumulation. *Econ. Rebord* **32**, 334-61.
Sweezy. P. M. (1942). "The Theory of Capitalist Development." Oxford Univ. Press, New York.
Sworder, D. (1966). "Optimal Adaptive Control Systems." Academic Press, New York.
Takacs, L. (1960). "Stochastics Processes." Wiley, New York.
Takacs, L. (1962). "Introduction to the Theory of Queues." Oxford Univ. Press. London and New York.
Takashima, M. (1956). A note one evolutionary processes. *Bull. Math. Statist.* **7**, 18-24.
Talacko, J. (1959). On stochastic linear inequalities. Trabajos Estadist. **10**,
Tapley, B. D., and Lewallen, J. M. (1967). Comparison of several numerical optimization methods. *J. Optimization Theory* Appl. **1**, 1-32.
Theil, H. (1954). "Linear aggregation of economic relations." North-Holland Publ., Amsterdam.
Theil, H. (1957). A note on certainty equivalence in dynamic planning. *Econometrica* **25**, 346-349.
Theil, H. (1958; 2nd ed. 1961). "Economic Forecasts and Policy." North Holland Publ., Amsterdam.
Theil, H. (1961). Some reflections on static programming under uncertainty. *Weltwirtschaft.* **87**, 124-138.
Theil, H. (1964). "Optimal Decision Rules for Government and Industry." North Holland Publ., Amsterdam.
Theil, H. (1965). Linear desision rules for macrodynamic policy. Hickman, B. G. ed.
Theil, H. (1967). "Economics and Information Theory." North Holland Publ., Amsterdam.
Theil, H. and Nagar, A. L. (1961). Testing the independence of regression disturbances. *J. Amer. Statist. Assoc.* **56**, 793-806.
Thomas, D. R. (1967). Game value distributions: II. *Ann. Math. Statist.* **38**, 251-260.
Thomas, D. R., and David, H. (1967). Game value distributions: I. *Ann. Math. Statist.* **38**, 242-250.

Thompson. G. L., Cooper, W. W. and Charnes, A., eds. (1963). Characterizations by chance-constrained programming. Graves, R. L. and Wolfe, P. eds.
Tinbergen, J. (1939) "Statistical Testing of Business Cycle Theories." Business Cycles in the United States of America 1919-1929. Vol. 2. League of Nations, Geneva.
Tinbergen, J. (1952, 1955). "On the Theory of Economic Policy," Revised ed. North-Holland Publ., Amsterdam.
Tinbergen, J. (1954). "Centralization and Decentralization in Economic Policy." North-Holland Publ., Amsterdam.
Tinbergen, J. (1956). "Economic Policy, Principles and Design." North-Holland Publ., Amsterdam.
Tinbergen, J. (1960). Optimum savings and utility maximization over time. *Econometrica* **28**, 481-489.
Tinbergen, J., and Bos, H. C. (1962). "Mathematical Models of Economic Growth." McGraw-Hill, New York.
Tintner, G. (1940). "Variate Difference Method." Principia Press, Bloomington, Indiana.
Tintner, G. (1941). The pure theory of production under technological risk and uncertainty. *Econometrica* **9**, 305-312.
Tintner, G. (1942). The theory of production under nonstatic conditions. *J. Political Econ.* **50**, 645-667.
Tintner, G. (1942). A "simple" theory of business fluctuations. *Econometrica* **10**, 317-320.
Tintner, G. (1942). A contribution to the nonstatic theory of production. "Studies in Mathematical Economics and Econometrics" (O. Lange *et al.*, ed.). Univ. of Chicago Press, Chicago, Illinois.
Tintner, G. (1944). The "simple" theory of business fluctuations: A tentative verification. *Rev. Econ. Statisi.* **36**, 148-157.
Tintner, G. (1946). A note on welfare economics. *Econometrica* **14**, 69-78.
Tintner, G. (1948). Une theorie simple des fluctuations economiques. *Revue d'Econ. Polit.* **57**, 209 ff.
Tintner, G. (1949). Foundations of probability and statistical inference. *J. Roy. Statist. Soc.* **112**, 251 ff.
Tintner, G. (1952). "Econometrics." Wiley, New York.
Tintner, G. (1955). Stochastic linear programming with applications to agricultural economics. *Proc. 2nd Symp. Linear Programming* **1**, 197-228.
Tintner, G. (1957). Game theory, linear programming and input-output analysis. *Z. Nationaloekon.* **17**, 1-38.
Tintner, G. (1959). The application of decision theory of probability to a simple inventory problem. *Trabajos Estadist.* **10**, 239-247.
Tintner, G. (1960). "Handbuch der Oekonometrie." Springer, Berlin.
Tintner, G. (1960). A note on stochastic linear programming. *Econometrica* **28**, 490-495.
Tintner, G. (1960). Eine Anwendung der Wahrscheinlichkeitstheorie von Carnap auf ein Problem der Unternehmungsforschung. *Unternehmungsforschung* **4**, 164-170.
Tintner, G. (1961). A stochastic theory of economic development and fluctuations. "Money, Growth, and Methodology" (H. Hegeland, ed.), GWK Greerup, Lund. pp. 59-61.
Tintuner, G. (1961). The logistic law of economic development. *Arthaniti* **4**, 1-4.
Tintner, G. (1965). Stochastic linear programming with illustrations. Operations Research Verfaheren" (R. Henn, ed.), Vol. 2, pp. 108-121. Hain, Meisenheim am Glan.
Tintner, G. (1966). Modern decision theory. *J. Indian Sac. Agricultural Statist.* **18**.
Tintner, G. (1966). Some thoughts about the state of econnmetrics. "The Structure of Economic Science" (S. R. Krupp, ed.), pp. 114-128. Prentice Hall, Englewood Cliffs, New Jersey.

Tintner, G. (1967). Methodology of mathematical economics and econometrics. "Encyclopaeida of Unified Science." Univ. of chicago Press, Chicago, Illinois.

Tintner, G. (1968). "Methodology of Mathematical Economics and Econometrics." Univ. of Chicago Press, Chicago Illinois.

Tintner, G., and Davila, O. (1965). Aplicaciones de la econometria a la planificacion. *Fl Trimstre Econ.* **32,** 717-723.

Tintner, G., and Farghali, S. A. S. (1967). The application of stochastic programming to the UAR first five year plan. *Kyklos* **20,** 749-758.

Tintner, G., and Fels, E. (1967). Mathematical economics in the Soviet Union. *Communist Affairs* **5,** 3-8.

Tintner, G. and Millham C. (1970). "Mathamatics and Statistics for Economists." 2nd ed. Holt, New York.

Tintner, G., and Murteira, B. (1960). Um modelo "input-output" simplificado para a economic Portugesa. *Colectanea Estudos.* **8,** 1-14.

Tintner, G., and Narasimham, G. V. L. (1967). Lineare stochastische Differenzengleichungen fuer das deutsche Volkseinkommen 1851-1930. "Die Statistik in der Wirtschaftsforschung" (H. Strecker, and W. R. Bihn, eds.), pp. 451-452. Duncker & Humbolt, Festgabe fuer Rolf Wagenfuehr, Berlin.

Tintner, G., anp Narayanan, R. (1961). "An Econometric Model for India." Indian Statist. Inst. Calcutta (mimeographed).

Tintner, G., and Narayanan, R. (1966). A multi-dimensional stochastic process for the explanation of economic development. *Metrika* **11,** 85-90.

Tintner, G., and Patel, R. C. (1965). A log-normal diffusion process applied to the economic development of India. *Indian Econ. J.* **13,** 465-474.

Tintner, G., and Patel, R. C. (1966). Evaluation of Indian fertilizer projects: An application of consumer's and producer's surplus. *J. Farm Econ.* **48,** 704-710.

Tintner, G., and Patel, R. C. (1966). A log-normal diffusion process applied to the development of Indian agriculture with some considerations on economic policy. *J. Indian Soc. Agricultural Statist.* **13,** 465-467.

Tintner, G. and Rao, T. V. S. (1968). An optimal policy for economic growth in India. (mimeographed).

Tintner, G., and Sengupta, J. K. (1962). Some aspects of the design and use of a generalized growth model. *Indian Econ. Rev.* **6,** 1-21.

Tintner, G., and Sengupta, J. K. (1963). Ein verallgemeinerter Geburtenund Todesprozess zur Erklaerung der Entwicklung des deutschen Volk-seinkommens 1851-1939. *Metrika* **6,** 143-147.

Tintner, G., and Sengupta, J. K. (1964). Stochastic linear programming and its application to economic planning. "On Political Economy and Econometrics," Essays in honor of O. Lange, pp. 601-617. PWN Polish Sci. Publ., Warsaw.

Tintner, G., and Thomas, E. J. (1963). Un modele stochastique de développement economique avec application a l'industrie Anglaise. *Rev. Econ. Politique* **73,** 278-280.

Tintner, G., Narasimham, G. V. L., Patil, L., and Raghavan, N. S. (1961). A logistic trend for Indian agricultural income. *Indian J. Econ.* **42,** 79-84.

Tintner, G., Millham, C., and Sengupta, J. K. (1963). A weak duality theorem for stochastic linear programming. *Unternehmungsforschung* **7,** 1-8.

Tintner, G., Narasimham, G. V. L., Patil, L., and Raghavan, N. S. (1965). A simple stochastic process for the explanation of the trend of regional development. "Colloquium on the Application of Mathematics to Economics," pp. 355-358. Akademiai Kiado, Budapest.

Tintner, G., Patil, L., and Mukherjee, V. (1966). A multivariate exponential model of growth and transactions for international trade. *Opsearch* **3**, 63-70.

Tintner, G., Sengupta, J. K., and Thomas, E. J. (1966). Applications of the theory of stochastic processes to economic development. "The Theory and Design of Economic Development" (I. Adelman, and E. Thorbecke, eds.), pp. 99-110. Johns Hopkins Press, Baltimore, Maryland.

Tisdell, C. A. (1968). "The Theory of Price Uncertainty, Production and Profit." Princeton Univ. Press, Princeton, New Jersey.

Tobin, J. (1965). The theory of portfolio selection. "The Theory of Interest Rates" (F. H. Hahn, and F. P. R. Brechling, eds.). Macmillan, London.

Tobin, J. (1965). Money and economic growth. *Econometrica* **33**, 671-684.

Tobin, J. (1968). Liquidity preference as behavior towards risk. *Rev. Econ. Stud.* **25**, 65-86.

Tomovic, R. (1963). "Sensitivity Analysis of Dynamic Systems." McGraw Hill, New York.

Tukey, J. W. (1962). The future of data analysis. *Ann. Math. Statist.* **33**, 1-67.

Tustin, A. (1953). "The Mechanism of Economic Systems." Harvard Univ. Press. Cambridge, Massachusetts.

Uzawa, H. (1961). The stability of dynamic processes. *Econometrica* **29**, 617-631.

Uzawa, H. (1961). On a two-sector model of economic growth I. *Rev. Econ. Stud.* **29**, 40-47.

Uzawa, II. (1963). On a two-sector model of economic growth II. *Rev.Econ. Stud.* **30**, 105-118.

Uzawa, H. (1964). Optimal growth in a two sector model for capital accumulation. *Rev. Econ. Stud.* **31**, 1-24.

Uzawa, H. (1965). Optimum technical change in an aggregative model of economic growth. *Int. Econ. Rev.* **6**, 18-31.

Uzawa, H. (1966). An optimum fiscal policy in an aggregative model of economic growth. Adelman, I. and Thorbecke, E. eds.

Vajda, S. (1961). "Mathematical Programming." Addison-Wesley, Reading, Massachusetts.

Valentine, F. A. (1937). The problem of Lagrange with differential inequalities as added side conditions. "Contributions to the Calculus of Variations, 1933-1937." Univ. of Chicago Press, Chicago, Illinois.

Van de Panne, C., and Popp, V. (1963). Minimum-cost cattle feed under probabilistic protein constraints. *Management Sci.* **9**, 405-530.

Van Eijk, C., and Sandee, J. (1959). Quantitative determination of an optimum economic policy. *Econometrica* **27**, 1-13.

Van Praag, B. M. S. (1968). "Individual welfare functions and Consumer Behavior." North Holland Publ., Amsterdam.

Van Zwet, W. R. (1964). "Convex Transformations of Random Variables," Vol. 7. Mathematical Center Tracts, Amsterdam.

Veinstein, A. L., ed. (1963). "National Economic Models: Theoretical Problems of Consumption. (Russian) Moscow.

Verdoorn, P. J. (1956). Complementarity and long range projections. *Econometrica* **24**, 429-445.

von Neumann J. (1941). Distribution of the ratio of the mean square successive difference to the variance. *Annls of Mathatical Statistes*, **12**, 367-395.

von Neumann, J. (1945), A Model- of general economic economic equilibrium. *Revew of Econmic Studies* **13**, 1-9.

von Neumann, J., and Morgenstern, 0. (1944). "Theory of Games and Economic Behavior." Princeton Univ. Press, Princeton, New Jersey.

von Weizsäcker, C. C. (1965). Existence of optimal programs of accumulation for an infinite time horizon. *Rev. Econ. Stud.* **32**, 85-104.
von Weizsäcker, C. C. (1967). Lemmas for a theory of approximate optimal growth. *Rev. Econ. Stud.* **34**, 143-151.
Wagner, H. M. (1958), On the distribution of solutions in linear programming problems. *Amer. Statist. Assoc.* **53**, 161-163.
Wald, A. (1950). "Statistical Decision Functions." Wiley, New York.
Wald, A. (1951). On some systems of equations of mathematical economics. *Econometrica* **19**, 368-403,
Walras, L. (1954). "Elements of Pure Economics." Irwin, Homewood, Illinois.
Wang, P. K. C. (1964). Control of distributed parameter systems. "Advances in control Systems: Theory and Applications. "Academic Press, New York.
Wang, P. K. C. (1965). Invariance, uncontrollability and unobservability in dynamical systems. *IEEE Trance. Autmatic Control* **10**, 366-367.
Watts, D. G., ed. (1968). "The Future of Statistics." Academic Press, New York.
Weil, R. L. (1968). The decomposition of economic production systems. *Econometrica* **36**, 260-278.
Weiss, L. (1961). "Statistical Decision Theory." McGraw-Hill, New York.
Wessels, J. (1967). Stochastic programming. *Statist. Neerland.* **21**, 39-53.
Wetherill, G. B. (1966). "Sequential Methods in Statistics." Methuen, London.
Wets, R. J. B (1966). Programming under uncertainty: the equivalent convex program. *SIAM J. Appl. Math.* **14**, 89-105.
White, D. J. (1969). "Decision theory." Aldine Publ., Chicago.
Whittle, P. (1963). "Prediction and Regulation." English Univ. Press, London.
Wiener, N. (1949). "Extrapolation, Interpolation and Smoothing of Stationary Time Series." Wiley. New York.
Wiener, N. (1964). "God and Golem Inc." MIT Press, Cambridge, Massachusetts.
Wilde, D. J., and Beightler, C. L. (1967). "Foundations of Optimization." Prentice-Hall, Englewood Cliffs, New Jersey.
Williams, A. C. (1963). Marginal values in linear programming. *SIAM J.* **11**, No. 1.
Williams, A. C. (1965). On stochastic linear programming. *SIAM J.* **13**, 927-939.
Winckler, K. (1966). "*Multikollinearitaet.*" Schriften des Instituts fuer Statistik und Oekonometrie, Mannheim.
Within, T. M. (1957). "The Theory of Inventory Management." 2nd ed. Princeton Univ. Press, Princeton New Jersey.
Wold, H. (1954). Cansality and econometrics. *Econometrica* **22**, 62-174.
Wold, H., and Jureén, L. (1953). "Demand Analysis." Wiley, New York.
Wold, H. ed. (1965). "Bibliography on Time Series and Stochastic Processes." MIT Press, Cambridge, Massachusetts.
Wolfe, J. N. (1967). Planning by forecast. "Price Formation in Various Economies" (D. C. Hague, ed.). St. Martin's Press, New York.
Wolfe, P., and Cutler, L. (1963). Experiments in linear programming. Graves, R. L. and Wolfe, P. eds.
Wonham, W. M. (1968). Optimal stochastic control. "Stochastic Problems in Control: A Symposium of the American Automatic Control Council." Amer. Soc. of Mechanical Eng., New York.
Wonham, W. M., and Johnson, C. D. (1964). Optimal bang-bang control with quadratic performance index. *Trans. Amer. Soc. Meth. Eng. Ser. D* **1**, 107-115.
Yaari, M. (1964). On the existence of an optimal plan in a continuous time allocation process. *Econometrica* **32**, 576-590.

Yaari, M. (1965). Uncertain lifetime, life insurance and the theory of the consumer. *Rev. Econ. Studies.* **32**, 137-150.
Zangwill, W. I. (1965). Nonlinear programing via penalty functions. *Management Sci.* **13**, 344-358.
Zauberman, A. (1967). "Aspect of Planometrics." Yale Univ. Press, New Haven, Connecticut.
Zaycoff, R. (1937). Ueber die Ausschaltung der zufaelligen Komponente nach der Variate Diflerence Methode. Sofia Universitet. *Statisticheski Institut za Stopanski Prouchvanii Trudove* no. 1 pp. 1-46.
Zellner, A. (1969). Analysis of some control problems. Chapter of a forthcoming publication. (author, mimeographed).
Zellner, A., and Geisel, M. S. (1968). Sensitivity of control to uncertainty and form of the criterion function. "The Future of Statistics" (D. G. Watts, ed.).
Zeuthen, F. (1933). "Problem of Monopoly and Economic Welfare." Routledge and Kagan Paul, London.
Zoutendijk, G. (1960). "Methods of Feasible Directions." Elsevier, Amsterdam.

AUTHOR INDEX

Numbers in italics refer to the pages on which the complete references are listed.

A

Ackoff, R. L., 28, 34, *269, 273*
Adelman, I., *269*
Aitchison, J., 12, 65, 66, *269*
Aizerman, M. A., 98, 99, 103, 147, 160, 163, 164, 167, *269*
Akilov, G. P., *280*
Allen, R. G. D., 93, 94, 113, 142, 146, *269*
Alper, P., 95, 180, 182, *269*
Anderson, O. N., 17, *269*
Anderson, R. L., 15, 16, *269*
Anderson, T. W., 13, 15, 18, 19, *269*
Ando, A., *269*
Antosieswicz, H. A., *269*
Aoki, M., 27, 131, *269*
Aris, R., 108, 190, *269*
Armitage, P., 182, *270*
Arnoff, E. L., 34, *273*
Arrow, K. J., 21, 93, 112, 178, 188, 191, 202, 205, 210, 265, *270*
Aseltine, J. A., *270*
Athans, M., 155, 183, *270*
Aumann, R. J., 10, *270*

B

Babbar, M. M., 212, 243, 252, *270*
Bacharach, M. O. L., *282*
Bailey, N. T. J., 65, *270*
Balakrishnan, A. V., 100, 131, 148, 184, 185, 189, 229, 230, *270*
Balderston, F., 174, *278*
Balinski, M. L., 223, *270*
Balintfy, J. L., *284*
Bardhan, P. K., 95, 122, *270*
Barlow, R. E., 258, *270*

Barnett, S., *270*
Bartlett, M. S., 2, 15, 17, 20, 53, 65, *270, 271*
Basmann, R. L., 14, 23, *271*
Baumol, W., 11, 27, 28, *271*
Bausch, A., 188, 265, *281*
Beale, E. M. L., 208, *271*
Beckmann, M. J., 109, *271*
Beightler, C. L., 189, 195, 209, *295*
Bellman, R., 29, 100, 108, 109, 131, 184, 190, 229, 261, *271*
Beltrami, E. J., *271*
Bereanu, B., 210, 212, *271*
Bertram, J. E., 158, *271*
Bharucha-Reid, A. T., 21, 34, 45, 65, 78, 183, 262, *271*
Birnbaum, Z. W., 233, 234, 237, 267, *271*
Blaug, M., 9, 27, *271*
Blumen, L. M., 29, *271*
Boltyanskii, V. G., 100, 127, *286*
Boot, J. C. G., 229, *271*
Borch, K., 22, 28, 29, 186, 187, 207, 267, *271*
Bos, H. C., 190, *292*
Bose, S., 125, *271*
Box, G. E. P., 26, 131, 145, 204, 239, 258, *271, 272*
Boyce, W. E., 239, *272*
Bracken, J., 223, 226, *272*
Bradley, J. V., 239, 267, *272*
Brentani, P. B., 100, 103, 155, *278*
Brockett, R. W., 170, *272*
Brogden, H. E., 226, *272*
Brown, A., 169, 190, *291*
Brown, J. A. C., 12, 65, 66, *269*
Bruno, M., 94, 102, 103, 125, 127, 152, 166, 167, 169, 190, 194, *272, 273*
Buchanan, L., 185, 190, *272*

Bucy, R. S., 158, *272*
Burdick, D. S., *284*
Burmeister, E., 103, 152, *272*

C

Cairncross, A. K., 56, *272*
Candler, W., 108, *278*
Caratheodory, C., *272*
Carnap, R., *272*
Cass, D., 94, 191, 238, *272*
Chakravarty, S., 127, *272*
Chamberlin, E. H., 11, *272*
Champernowne, D. G., 35, *273*
Charnes, A., 28, 204, 205, 208, 239, 246, 260, *273*, *292*
Chenery, H. B., 95, 169, 171, 189, 190, 191, *273*
Chernoff, H., 149, *273*
Chipman, J. S., 267, *273*
Christ, C., 1, 12, *273*
Chu, K., *284*
Churchman, C. W., 34, *273*
Cochrane, D., 18, *273*
Connors, M., 95, *273*
Cooper, W. W., 28, 204, 205, 208, 246, *273*, *292*
Cootner, P. H., 29, 35, *273*
Corlett, W. J., 19, *291*
Courtillot, M., *273*
Cox, D. R., *273*
Cutler, L., 204, *295*

D

Dantzig, G. B., 12, 95, 108, 176, 178, 188, 204, 208, 225, 265, *273*, *274*
David, H., 213, *291*
Davila, O., *293*
Davis, H. T., 2, 15, 65, *274*
Day, R. H., 209, 210, *274*
D'Azzo, J. J., 131, *274*
Debreu, G., 9, 27, 188, 265, *274*
Derman, C., *274*
Dhrymes, P., 191, 199, 201, *274*
Dilberto, S. P., *274*
Dixit, A. K., 125, *274*
Dobell, A. R., *272*
Domar, E. D., 49, *274*
Dorfman, R., 28, *274*

Dostupov, B. G., 167, *274*
Dreyfus, S. E., 29, 101, 108, 132, 261, *271*, *274*
Duesenberry, J. S., 44, 50, 268, *274*, *275*
Duffin, R. J., 189, 195, 208, *275*
Durbin, J., 17, *275*
Dvoretsky, A., 21, *275*

E

Eckaus, R. S., 171, *275*
Eckstein, O., 268, *274*
Ellman, M., 179, 191, *275*
Elmaghraby, S. E., 204, 208, *275*
Enthoven, A. C., 202, *270*
Eppen, G. D., *277*
Evans, G. C., 93, *275*

F

Falb, P. L., 155, 183, *270*
Fama, E. F., 187, 264, *275*
Fan, L. T., 143, *275*
Farghali, S. A. S., 239, *293*
Farrar, D. E., 174, 205, *275*
Fei, J. C. H., *275*
Feldbaum, A. A., 147, 149, *275*
Feldstein, M. S., 174, *275*
Feller, W., 1, 12, 22, 52, 53, 54, 227, 233, *275*
Fellner, W., 186, *275*
Fels, E., 9, 23, 27, 108, *275*, *293*
Ferguson, R. O., *273*
Fiacco, A. V., 184, 189, 229, 230, *275*
Fishburn, P. C., 205, 267, *275*
Fisher, F., 13, 21, 99, *269*, *275*, *276*
Fisher, W., 99, *276*
Fisk, P. R., 12, *276*
Fisz, M., 2, 15, 19, 65, 78, *276*
Florentin, J. J., 138, *276*
Foley, D. F., 119, 122, *276*
Forstner, K., 34, 108, *281*
Fox, K. A., 24, 93, 94, 95, 96, 97, 98, 100, 104, 108, 122, 125, 127, 131, 137, 148, 154, 159, 168, 171, 175, 176, 178, 180, 182, 184, 186, 187, 189, 190, 191, 194, 198, 202, 203, 205, 208, 229, 230, 239, 262, 265, 266, *276*, *289*
Frank, M., *276*

Freund, R. F., 206, *276*
Frisch, R., 18, 51, 93, 97, 168, 181, 189, 190, 203, 246, *276*
Fromm, G., 44, 94, 268, *274, 275, 276*
Frost, P., 158, *279*
Funk, P., *276*

G

Galbraith, J. K., 11, *276*
Gale, D., 189, 194, *276*
Gamkrelidze, R. V., 100, 127, *286*
Geary, R. C., 239, *276*
Geisel, M. S., 173, 239, *296*
Georgescu-Roegen, N., 21, *276*
Geyer, H., *277*
Ghellinck, G. T., *277*
Gibrat, V., 12, 65, *277*
Gibson, J. E., 103, *279*
Gilbert, E. G., 99, *277*
Gnedenko, B. V., 12, *277*
Goldberger, A. S., 1, 12, 44, 190, *277, 280*
Goldman, S. M., 191, *277*
Goodwin, R. M., 94, 127, 152, 189, *277*
Granger, C. W. J., 2, 15, 29, 65, *277*
Graves, R. L., 28, *277*
Gray, P., *282*
Graybill, F. A., 1, *284*
Greenberg, B. G., 206, 213, *287*
Greenberger, M., 29, 35, 65, 268, *285*
Grenander, V., 2, 15, 22, *277*
Gruver, G., 217, 246, 255, *289*
Gumbel, E. J., *277*
Gupta, A. K., 209, *277*
Guttman, L., 237, *277*

H

Haavelmo, T., 1, 29, 34, 44, 161, *277*
Hader, J., 207, *277*
Hadley, G., *277*
Hahn, F. H., 10, 115, 181, 189, 191, 194, 265, *277*
Halkin, H., 104, *277, 279*
Hamada, K., 95, 122, *277*
Hannan, E. J., 2, 15, *277, 278*
Hanson, M. A., 211, *278*
Hanssmann, F., 24, 93, 95, *278*
Hardt, J. P., 107, *278*

Harris, T., 21, *270*
Harrod, R. F., *278*
Harsanyi, J. C., 11, *278*
Hart, B. I., 16, *278*
Hart, P. E., 190, *278*
Hartman, P., 118, *278*
Hass, J. E., 191, 199, *278*
Hatanaka, M., 2, 15, 29, *277*
Hayek, F. A., 11, *278*
Heady, E., 108, *278*
Heal, G. M., 191, *278*
Henn, R., 34, 108, *281*
Hestenes, M. R., *278*
Hickman, B. G., *278*
Hicks, J. R., 10, 189, 235, *278*
Hillier, F. S., 246, *278*
Ho, Y., 100, 103, 138, 150, 155, 186, *278*
Hoffmann, W. G., 44, *278*
Hoflenberg, M., 107, *278*
Hoggatt, A. C., 174, *278*
Hohenbalken, B., von, *278*
Holme, H., 93, *276*
Holt, C. C., 22, 28, 34, 96, 130, 256, 257, *278, 279*
Holtzman, J. M., 104, *279*
Hood, W. C., 13, *279*
Hotelling, H., 15, 16, *279*
Houpis, C. H., 131, *274*
Houthakker, H. S., 65, *286*
Howard, R. A., 29, 108, 131, 137, *279*
Hufschmidt, M. M., 204, 268, *279*
Hunter, J. S., 204, *271*
Hurstfield, R., 19, *291*
Hurwicz, L., 21, 178, 191, 210, *270*

I

Ichimura, S., 191, *279*
Inagaki, M., *279*
Intriligator, M. D., *279*
Isaacs, R., 150, 186, *279*
Isard, W., 190, *279*
Ito, K., 154, *279*

J

James, F. S., 17, *285*
Jenkins, G. M., 26, 131, 145, 239, 258, *271, 272, 279*
Johansen, L., 125, 168, 190, *279*

Johnson, C. D., 103, 148, *279, 295*
Johnston, J., 1, 12, 19, *279*
Jorgenson, D. W., 28, 127, *279*
Jureén, L., 1, 15, 18, *295*

K

Kailath, T., 158, *279*
Kalaba, R., 100, 108, 109, 131, 145, 184, *271, 279*
Kaldor, N., 189, 235, *279*
Kalecki, M., 93, *279*
Kalman, R. E., 97, 99, *279, 280*
Kantorovich, L. V., 12, 189, *280*
Kaplan, N., 107, *278*
Karlin, S., *270, 280*
Karreman, H. F., 131, *280*
Kataoka, S., 230, *280*
Kelley, H. J., 103, *280*
Kelley, J. E., *280*
Kemeny, J. G., 10, 207, *280*
Kendall, D., 161, 267, *280*
Kendall, M. G., 16, 23, 53, *280*
Keynes, J. M., 93, *280*
Kiefer, J., 21, *275*
Kingman, J. F. C., 206, *280*
Kirby, J. L., 239, 246, 260, *273*
Klein, L. R., 1, 12, 35, 44, 190, *275, 280*
Kogan, P. J., 29, *271*
Kolmogorov, A. N., 12, *277*
Koopmans, T., 13, 27, 94, 112, 188, 189, 194, 265, *279, 280, 281*
Kopp, R. E., 185, *281*
Korbel, J., 29, 35, 65, 268, *285*
Kornai, J., 95, 150, 176, 178, 186, 190, 191, 198, 199, 203, 213, 225, *281, 283*
Kozin, F., 158, *281*
Kreindler, E., 97, *281*
Kreko, B., 12, *281*
Kromphardt, W., 34, 108, *281*
Krylov, V. I., *280*
Kuenne, R. E., 10, *281*
Kuh, E., 44, *275*
Kuhn, H. W., 93, 108, 179, 185, 201, *281*
Kulikowski, R., 176, 198, *281*
Kumar, T., 243, 244, *289*
Kurz, M., 93, 103, 112, 113, 114, 115, 116, 181, 190, 191, 194, 265, *270, 281*

Kushner, H. J., 131, 138, 140, 143, 144, 158, 194, 229, *281*

L

Lange, O., 168, 190, 235, *281*
LaSalle, J. P., 170, *281*
Lasdon, L. S., 189, *281*
Lavi, A., *281*
Lee, E. B., 131, 177, 190, *281, 282*
Lefeber, L., 190, *282*
Lefkowitz, I., *282*
Lehmann, E. L., 209, *282*
Leitman, G., 131, 148, *282*
Leondes, C. T., 179, 183, *282*
Leontief, W., 190, *282*
Lerner, A. P., *282*
Leser, C. E. V., 12, *282*
Lesourne, J., 21, 34, *282*
Letov, A. M., 169, *282*
Levhari, D., 103, 190, 238, *282*
Levine, H. S., 107, *278*
Lévy, P., 22, *282*
Lewallen, J. M., 185, *291*
Lintner, J., 187, *282*
Liptak, Th., 95, 150, 176, 178, 186, 190, 191, 198, 199, 203, 213, 225, *281*
Lord, F. M., *282*
Luce, R. D., 29, 186, *282*

M

McCarthy, Av., 29, *271*
McCormick, G. P., 184, 189, 229, 230, *272, 275*
MacEwan, A., 95, 169, *273*
McFadden, D., 99, 176, 191, 194, *283*
McGill, R., 185, *281*
McGrath, R. J., 146, *283*
Machol, R. E., *282*
Macko, D., *283*
Madansky, A., 204, 205, *274, 282*
Mahalanobis, P. C., 171, 189, 190, 235, 237, *282*
Malinvaud, E., 1, 12, 191, 194, 265, *282*
Man, F. T., 155, 157, *290*
Mandelbrot, B., 12, 22, 29, 35, 65, 264, *282*
Mangasarian, O. L., 116, 202, *282, 283*
Mann, H. B., 21, *283*
Manne, A. S., 203, *283*

Marglin, S. A., 125, *283*
Markowitz, H., 174, 201, 205, 247, 260, *283*
Markus, L., 131, 177, 190, *281, 282*
Marschak, J., 9, 21, 265, 270, *283*
Marschak, T. A., 180, 191, *283*
Marshall, A., 10, *283*
Martos, B., 191, 198, *283*
Marx, K., *283*
Massé, P., 21, 34, *283*
Massera, J. L., 116, *283*
Matthews, R. C. O., 10, 189, *277*
Matusita, K., 227, *283*
Menges, G., 12, *283*
Merriam, C. W., 148, *283*
Merton, R. C., 161, 260, 261, *283*
Mesarovic, M. D., 176, *283*
Miller, H. D., *273*
Millham, C., 1, 22, 210, 220, 234, 239, 243, 246, *290, 293*
Mills, E. S., *283*
Mills, G., 190, *278*
Mills, H. D., *284*
Milnor, J., 204, *284*
Mirrlees, J. A., 189, 190, 239, 262, *279, 284*
Mishchenko, E. F., 100, 127, *286*
Mitter, S. K., 189, *281*
Modigliani, F., 22, 28, 96, 130, 256, 257, *278, 279*
Montias, J. M., 108, *284*
Mood, A. M., 1, *284*
Moran, P. A. P., 15, *284*
Morgenstern, O., 9, 10, 12, 29, 65, *277, 280, 284, 294*
Moriguti, S., *284*
Morishima, M., 10, 28, 103, 127, 159, 189, 190, 265, *284*
Morris, W. T., 177, *284*
Morrison, B., 22, 234, 239, 246, *290*
Morse, P. M., 21, *284*
Mossin, J., 187, *284*
Mukherjee, V., 39, *284, 294*
Murphy, R. E., 28, 29, 35, 174, 191, *284*
Murteira, B., *293*
Muth, J. F., 22, 28, 96, 130, 257, *279*

N

Nagar, A. L., 17, *291*
Narasimham, G. V. L., 16, 35, 53, *293*
Narayanan, R., 39, 59, *284, 293*
Naslund, B., 145, 190, 194, 203, 210, 239, 246, 250, 260, 263, 264, *284*
Naylor, T. H., 94, 143, 185, 186, 265, 268, *284*
Negishi, T., 191, 265, *277, 284*
Nehari, Z., 118, *284*
Nemhauser, G. L., 108, *284*
Nerlove, M., 15, *284*
Neustadt, L. W., 100, 131, 148, 184, 185, 189, 229, 230, *270, 285*
Newbold, P. M., 138, *278*
Newman, P., 112, *285*
Newton, R. H. C., 100, *285*
Nikaido, H., 191, *285*

O

O'Carroll, F. M., 35, *285*
Oldenburger, R., 96, 131, *285*
Oppelt, W., *277*
Orcutt, G., 17, 18, 29, 35, 65, 174, 185, 268, *273, 285*
Osborne, M., 263, *285*
Owen, D. B., 227, *285*
Ozga, S. A., 186, *285*

P

Paiewonsky, B., *285*
Pallu de la Barriere, R., 109, *285*
Pareto, V., 12, 22, 35, 65, *285*
Parikh, K. S., 171, *275*
Parzen, E., 47, *285*
Pasinetti, L. L., 95, *285*
Patel, R. C., 12, 65, 74, 78, 105, *293*
Patil, L., 16, 35, 53, *293, 294*
Pearson, J. D., 176, 199, *283, 285*
Peterson, E. L., 189, 195, 208, *275*
Pfanzagl, J., *285*
Pfouts, R. W., 201, *285*
Phelps, E. S., 94, 112, 113, 190, 238, 239, *285*
Phillips, A. W., 94, 96, 146, *285*
Pillai, K. C., *285*
Plant, J. B., 184, 185, *285*
Pontryagin, L. S., 98, 100, 127, 150, 190, *286*
Popp, V., 246, *294*
Portes, R. D., 191, *286*

Portillo-Campbell, J. H., 195, 205, 218, 239, 252, 258, 259, 260, 263, 264, 268, *289*
Potter, M., 19, *291*
Prais, S. J., 65, *286*
Pratt, J. W., 202, 204, 206, 226, *286*
Prekopa, A., 212, *286*
Proschan, F., 258, *270*
Pugachev, V. F., 108, *286*

Q

Quandt, R., 44, *286*
Quandt, R. E., 35, *286*
Quenouille, M., 2, 15, 17, 20, *286*

R

Radner, R., 95, 103, 171, 175, 180, 187, 189, 191, 192, 194, 265, 266, *286*
Radstrom, H., 213, *286*
Raghavan, N. S., 16, 35, 53, *293*
Raiffa, H., 29, 186, 204, *282*, *286*
Rajaraman, V., 146, *283*
Ramachandran, K. V., *285*
Ramsay, F. P., 94, 189, *286*
Ranis, G., *275*
Rao, J. N. K., 15, 18, *286*
Rao, T. V. S., 107, *293*
Rekasius, Z. V., 160, 163, 168, 229, *286*
Rhodes, D. D., 16, *286*
Rideout, V. C., 146, 163, *283*, *287*
Rivlin, A. M., 29, 35, 65, 268, *285*
Robbins, H., 265, *286*
Robinson, J., 11, *286*
Roos, C. F., 93, *286*
Rosen, J. B., 116, *287*
Rosenblatt, M., 2, 15, 22, *277*
Rosenfeld, J. L., 150, 212, *287*
Rosenstein-Rodan, P. N., 44, *287*
Rostow, W. W., 44, *287*
Rowe, D. A., 19, *291*
Roy, A. D., 175, 205, 232, 268, *287*
Rozonoer, L. I., 103, 106, 138, 143, *287*
Rubin, H., 13, *269*
Russell, W. R., 207, *277*
Ruubel, K. V., 167, *287*

S

Saaty, T. L., 34, 211, *287*
Sakawa, Y., 181, *287*
Samuels, J. C., 158, *287*
Samuelson, P., 28, 95, 112, 114, 115, 161, 181, 189, 260, *274*, *287*
Sandee, J., 182, 190, 191, *287*, *294*
Sanyal, B. C., 246, 267, *289*
Sarachik, P. E., 97, 158, *271*, *281*
Sarhan, A. E., 206, 213, *287*
Savage, L. J., 267, *287*
Scarf, H., *270*
Schlaifer, R., 204, *286*
Schultz, P. R., 149, *287*
Schultz, W. C., 163, *287*
Schumpeter, J., 9, *287*
Sen, A., *289*
Sengupta, J. K., 22, 24, 44, 46, 93, 94, 95, 96, 97, 98, 100, 104, 108, 122, 125, 127, 131, 132, 137, 142, 143, 145, 146, 148, 154, 158, 159, 165, 168, 171, 175, 176, 178, 179, 180, 182, 184, 186, 187, 189, 190, 191, 194, 195, 198, 202, 203, 204, 205, 206, 208, 209, 210, 211, 212, 215, 217, 218, 220, 226, 227, 229, 230, 232, 234, 235, 239, 243, 244, 246, 250, 252, 255, 258, 259, 260, 262, 263, 264, 265, 266, 267, 268, *276*, *287*, *288*, *289*, *290*, *293*, *294*
Sengupta, S. S., 28, *290*
Shackle, G. L. S., 267, *290*
Shapley, L. S., 204, *290*
Shell, K., 93, 95, 96, 108, 112, 115, 119, 122, 189, 238, *276*, *290*
Sheshinski, E., 103, 152, *272*
Shubik, M., 11, 186, 213, 265, *290*
Sidrauski, M., 119, 122, *276*, *290*
Simon, H., 22, 28, 29, 35, 96, 130, 131, 132, 142, 164, 205, 256, 257, *269*, *278*, *279*, *290*
Sinha, S. M., *290*
Slutsky, E., 16, *290*
Smirnov, N., 233, *290*
Smith, C., 95, 180, 182, *269*, *270*
Smith, H. W., 155, 157, *290*
Smith, V. L., 203, *290*
Soland, R. M., 223, 226, *272*
Solodovnikov, V. V., 131, *290*

Solow, R. M., 28, 95, 110, *274, 290*
Soults, D. J., 204, 213, *291*
Sprekle, C., 65, *291*
Srinivasan, T. N., 190, 238, *282, 291*
Steindl, J., 12, 22, 28, 29, 35, *291*
Stiglitz, J. E., 115, 187, *272, 290, 291*
Stigum, B. P., 187, 265, 266, *291*
Stoleru, L. G., 103, 107, *291*
Stone, B. K., 187, *291*
Stone, R., 19, 169, 182, 190, *291*
Struble, R. A., 99, *291*
Stuart, A., 16, 23, *280*
Stubberud, A. R., 185, *272*
Swan, T., *291*
Sweezy, P. M., *291*
Sworder, D., 172, 191, *291*
Szego, G. P., 93, 108, 185, *281*

T

Takacs, L., 2, 34, 50, 65, *291*
Takahara, Y., *283*
Takashima, M., 54, 161, *291*
Talacko, J., *291*
Tapley, B. D., 185, *291*
Taubman, P., 94, *276*
Teichron, D., 95, *273*
Theil, H., 1, 14, 17, 22, 51, 96, 97, 131, 132, 142, 164, 169, 182, 189, 191, 205, 208, 209, 258, *291*
Thomas, D. R., 213, *291*
Thomas, E. J., 44, 45, 49, *293, 294*
Thompson, G. L., 10, 204, 207, *280, 291*
Thorbecke, E., 93, 94, 96, 97, 98, 108, 125, 131, 159, 168, 171, 182, 184, 190, 191, 194, *269, 276*
Tinbergen, J., 44, 50, 96, 97, 188, 189, 190, *292*
Tintner, G., 1, 2, 9, 12, 13, 15, 16, 17, 18, 22, 23, 27, 29, 35, 39, 44, 45, 46, 49, 53, 59, 65, 74, 78, 105, 107, 108, 158, 191, 204, 210, 220, 230, 234, 235, 239, 243, 246, 252, 266, *275, 278, 284, 286, 289, 290, 292, 293, 294*
Tisdell, C. A., 265, *294*
Tobin, J., 119, 174, 205, 260, *290, 294*
Tomovic, R., 148, *294*
Tucker, A. W., 201, *281*

Tukey, J. W., 239, *294*
Tustin, A., 93, *294*

U

Uzawa, H., 21, 94, 95, 119, 121, 159, 189, 194, 195, 238, *270, 273, 294*

V

Vajda, S., *294*
Valentine, F. A., *294*
Van de Panne, C., 246, *294*
Van Eijk, C., 182, *294*
Van Praag, B. M. S., 266, *294*
Van Zwet, W. R., 226, *294*
Veinott, A. F., Jr., 203, *274, 283*
Veinstein, A. L., *294*
Verdoorn, P. J., 190, *294*
Vogl, T. P., *281*
von Neumann, J., 9, 10, 16, *294*
von Weizäcker, C. C., 189, 191, 194, *290, 295*

W

Wagner, H. M., 204, *295*
Wald, A., 21, 27, *283, 295*
Walker, D. A., *290*
Walras, L., 9, *295*
Wang, C. S., 143, *275*
Wang, P. K. C., 99, 181, *295*
Watson, G. S., 17, *275*
Watts, D. G., 26, *279, 295*
Waugh, F. V., 18, *276*
Webb, K. W., 211, *287*
Weil, R. L., 99, 209, *295*
Weiss, L., 172, 204, *295*
Wertz, K., 94, 143, 185, 186, 268, *284*
Wessels, J., 206, *295*
Wetherill, G. B., 172, 174, 179, 204, 229, *295*
Wets, R. J. B., 204, *295*
Whinston, A. W., 239, 246, 260, 263, *284*
Whitaker, J. K., 190, *278*
White, D. J., *295*
Whiten, T. M., 21, *295*
Whittle, P., 2, 27, 28, 34, 130, 131, *295*
Wiener, N., 34, 131, *295*

Wilde, D. J., 189, 195, 209, *295*
Williams, A. C., 208, 211, *295*
Winckler, K., 13, *295*
Winter, A., 118, *278*
Wold, H., 1, 15, 18, *295*
Wolfe, J. N., 179, *295*
Wolfe, P., 28, 176, 178, 204, 225, *274, 276, 277, 295*
Wolfowitz, J., 21, *275*
Wonham, W. M., 148, *295*
Wonnacott, T. H., 94, 143, 185, 186, 268, *284*

Y

Yaari, M., 190, 239, *290, 295, 296*

Z

Zangwill, W. I., *296*
Zauberman, A., 108, *296*
Zaycoff, R., 17, *296*
Zellner, A., 173, 239, *296*
Zener, C., 189, 195, 208, *275*
Zeuthen, F., 11, *296*
Zoutendijk, G., *296*

SUBJECT INDEX

A

Absolute (global) maximum, 176
Absolute stability, 170
Accuracy of estimate, 23, 153, 229
Activity analysis, 199, 251
  vector, 198
Adaptive control, 145
  methods and types, 146–150
  relation to game theory, 149
  applications, 162
Additive noise, 163
Adjoint (costate) variables (vectors), 101–104
Adjustment, 146–149
  of parameters, 146
  of errors, 167
Admissible control, 140, 155
  relation to singularity, 102
Aggregation related to
  decomposition, 176, 225
  production function, 114
  stability analysis, 178–180
  subunit models, 33, 179, 209
Aggregative models, 29, 46, 110, 119
Agricultural production model applications
  in stochastic control, 129
  in stochastic processes, 75
  in stochastic programming, 243
Algorithms
  control models and, 104, 183
  linear programming and, 189, 216
  nonlinear programming and, 189, 195, 206
  reliability programming and, 218, 249
  stochastic control and, 179
  stochastic processes and, 39, 59, 67, 229
  stochastic (probabilistic) programming and, 206, 216, 229

stopping rules and truncation and, 208, 213
Approach
  deterministic, 9, 24
  game-theoretic, 149, 212
  minimax, 150
  stochastic, 1, 9, 128
Approximations
  in decomposition methods, 176, 225
  in nonlinear programming, 157, 229
    by normal distribution, 155, 206
  in sensitivity analysis, 148, 210
  in stochastic control, 152
  in stochastic processes, 43, 54
  in stochastic (probabilistic) programming, 211
    by Taylor series expansion, 153, 206
A priori probability distribution, 149, 212
Arithmetic mean (expectation), 10, 50, 132
Artificial activities (variables), 249
Assets, 120
Asymptotic stability, 117
  in Lyapunov sense, 159
  in stochastic control, 163, 170
Autonomous system, 96, 184
Auxiliary (adjoint) variables, 101, 139
Average risk, 137–138

B

Balanced growth, 112, 121
  in aggregative models, 110, 161
  in disaggregative models, 125, 163, 167
Bang-bang control, 102, 127
  in relation to singular controls, 102
Best fit, 55, 73
Basic feasible solutions, 239
  first, second, and third best, 240
Basis, feasible, 222
Bayes' decision, 148, 172, 226

Bayes' estimates, 172
Bayes' formula, 173
  control rules, 149
  costs, 150
  criterion, 149
  in stochastic control, 172
  in stochastic (probabilistic) programming, 223
  theorem, 173
Bayesian system, 174, 226
Bienayme–Tchebycheff inequality, 206
Birth and death process, 35, 46
  applications
    to diffusion process, 65, 84
    to growth models, 48
  linear and nonlinear, 53–54
    homogeneous and nonhomogeneous, 51
Bliss level (point), 112
Boundedness assumptions, 112, 138
Branching (diffusion) process, 84
  applications, 65
Brownian motion process, 137

C

Calculus of variations, 100
Capital accumulation, optimal, 94, 112, 194, 260
Capital
  augmenting technological change, 113
  growth and, 121, 174, 235
  heterogeneity of, 114
  transferability of, 119, 168, 194
Capital budgeting (allocation), 168, 235, 263
Cauchy distribution, 23
Certainty equivalence, 131
  applications of, 130, 256
  theorem of, 164
Chance-constrained programming, 208
  applications
    to economic models, 213, 246
    to reliability, 248
    to other fields, 256, 263
  decision rules under, 235, 263
Chance constraints, 208
  in linear programming, 209
  in nonlinear programming, 214, 256, 263
  in reliability analysis, 216, 248

Chapman–Kolmogorov equation, 4, 66
Characteristic
  equation, 14
  function, 168
  roots, 14, 113, 167
χ^2 distribution
  central, 87, 214
  noncentral, 214
  in stochastic programming, 214
Classical optimization methods, 24, 94, 100
Closed loop control, 100
Closed sets, 104
Cobb–Douglas production function, 95, 175
Competitive equilibrium, 24
  market framework, 122, 265
  system and decomposition, 176, 225
Computing methods in
  stochastic control, 103, 143, 184
  stochastic processes, 60, 67
  stochastic programming, 204, 231
Complementarity, 122
Composite consumption good, 119
Concave (convex) functions, 101, 178
  convex set of feasible solutions and, 100
  nonlinear programming and, 121, 184, 196, 215
Conditional distribution, 149
  risk and, 172
Confidence (fiduciary) limits, 45, 77
Conflicting objectives, 179, 182
Conjugate gradient method, 184
Constant returns, 110
Constrained optimization, 100, 196
  stochastic (probabilistic) programming and, 205, 215, 258
Constraint qualifications, 104
  directional convexity and, 104
  Kuhn–Tucker theorem and, 100, 145
Consumer choice
  in normative theory, 111
  in stochastic allocation problems, 260
Consumption, 120, 260
  per capita, 111
  rate of growth, 235
  steady state, 112
Continuous maximum (Pontryagin's) principle, 101

Subject Index

Control theory, 94
  application of, 94, 150
  deterministic systems of, 94
  stochastic framework of, 128
  variational programming and, 100
Control systems
  with discrete time, 104, 143, 177
  with distributed parameters, 181
  with optimal and suboptimal controls, 154, 183
Control variables (vector), 95
Controllability, 98, 102
  in growth models, 115
  in stochastic models, 168, 179
Convergence
  in distribution, 213, 229
  in stability analysis, 117, 159
  in stochastic control, 145, 157, 179
  in stochastic processes, 30, 52, 87
  in stochastic (probabilistic) programming, 227
Convex functions, 101
  programming, 143, 178, 174, 249, 257
  sets, 100
Convexity (directional), 104
Convolution of random variables, 6, 263
Conner conditions, 112, 127
Correlation coefficient, 6
Co-state variables (vector), 101
Cost
  of control, 108,
  of control integral, 133
  of randomness (uncertainty), 148, 153, 208
Covariance (matrix)
  in stochastic control, 155
  in stochastic process, 61, 87
  in stochastic programming, 207
Criteria of best strategy, 149
Criterion, combined, 164
  generalized integral, 167
  optimality, 100, 110, 259
  primary, 243
  secondary, 243
  statistical, 173
  tertiary, 243
Criterion function, 99, 133
Cumulative distribution, 149
  in stochastic (probabilistic) programming, 215, 233

Curve, admissible, 102
  optimal, 125
  switching, 102, 127
Cybernetics, 130
Cycling behavior, 115, 167

D

Damping, 143
Data, qualitative, 10
  sampling, 23
  simulated, 22, 185
  statistical, 11
Decentralization related to
  computing algorithms, 98, 178, 224
  competitive equilibrium, 93, 265
  decomposition methods, 178
  dynamic control models, 96
  stochastic programming, 224
Decision functions, 148, 209
  minimax, 150, 173
  models, 95, 193, 219
Decision rules, 250
  applications
    to scheduling problems, 130, 258
    to stabilization policies 96, 129
    to stochastic programming, 250
Decomposition methods, 176
  applications of, 179, 224
  stochastic programming and, 224
Degeneracy (dingularity), 13, 102
Density, *a posteriori*, 172
  a priori, 149, 173
  conditional, 149, 172
  joint normal, 156, 172, 174, 206
  multi-dimensional, 84, 207
  one-dimensional, 158, 175, 207
  transition (stochastic process), 3, 35, 47, 153
Depreciation, 120
Derivative (left and/or right), 98, 211
Determinant, 14
  for testing stability, 116, 153
Deterministic transformations, 152, 162
  in stochastic control, 152, 167, 177
  in stochastic (probabilistic) programming, 204, 249
Deviation from desired levels, 169, 204
Differentiable functions
  in optimal control, 101

in stochastic control, 138, 183
 in stochastic programming, 238, 256
Difference equation (process), 65
Difference equations, 16, 29, 35
 in control theory, 104, 130, 143
Direct (sequential) search method, 179, 184
Directional, convexity, 104
Discontinuity, 211, 255
Discrete
 maximum principle, 104
 stochastic control, 135, 143
 stochastic process, 19, 35, 59
Dispersion (matrix), 85, 205
Distance function, 117
 in Lyapunov theory of stability, 117, 159
 in stochastic (probabilistic) programming, 227
Distribution (probability), 1, 11
 prior (*a priori*), 149, 172
Disturbance, random, 12, 29, 148, 183
Divergence
 between actual and nominal values, 163
 between private values and market behavior, 120
Domar (Harrod–Domar) type growth model, 48, 235
Dual constraints (variables), 126
 price system, 122, 127, 255
Duality theory, 101, 122, 197
 applications
  to control theory, 127
  to game theory, 150
  to stochastic programming, 201, 249
Dynamic models
 in economic growth, 94, 112
 in investment allocation and planning, 119, 125
 in reliability programming, 249
 in scheduling problems, 256
 in stochastic control, 146, 155, 162
 in stochastic programming, 235, 256, 264
Dynamic programming
 application, 134, 260
 formulation, 260
 Pontryagin's maximum principle and, 100

E

Economic growth, 29, 95
 applications of, 45, 59, 110, 235
 theory (models) of, 48, 119, 168, 171
Economic models
 applications
  to stochastic control, 150
  to stochastic processes, 35, 65
  to stochastic programming, 229
 firm behavior, 199
 growth and stabilization, 50, 165, 235
 operations research, 21, 130, 256
 planning (economic policy), 125, 235
Economic policy, 93
 analytical framework, 96
 economic growth and, 110, 121
 in stochastic control, 132
 in stochastic processes, 69
 in stochastic programming, 219
 operations research and, 130
 stability and optimality characteristics, 184
 theory of, 51, 94
Economietric methods, 13, 27
 models, 12
 systems, 21
Educational models, 182
Eigen (characteristic) root, 14, 113, 167
Efficiency, intertemporal, 115
 general, 177
Equations
 adjoint (canonical), 101, 104
 Bellman's, 100
 Euler–Lagrange, 100
 Euler's, 100
 finite difference, 16, 29, 104
 maximum likelihood, 19, 45, 67, 86
 of motion, 95
 normal (least squares), 37, 39
 Pontryagin's, 101, 104
  of trajectory, 112
 variational, 116
Equilibrium, long-run, general, 112, 158, 193, 265
 in economic models, 29, 112, 124
 game theory and, 213
 stability and, 116, 159
 in stochastic control, 163, 213
 in stochastic processes, 52

Equivalent systems, 54, 178
   nonlinear programs, 204, 207
Error analysis
   in stochastic control, 148, 153
   in stochastic processes, 19, 40, 72
   in stochastic (probabilistic) programming, 210, 249
Error
   mean square, 131, 147, 163, 167
   round-off, 131, 154
   steady state, 152, 167
   transient, 152
Estimation
   least squares, 14, 42, 61, 130
   maximum likelihood, 14, 19, 44, 67
   other types, 23, 43, 215
Estimation problems
   in stochastic control, 148, 154, 163
   in stochastic processes, 43, 67
   in stochastic programming, 215
Euclidean distance, 117
   space, 59, 220
Euler equations, 100
Euler–Lagrange conditions, 100
Excess demand, 255, 264, 267
Existence of solutions
   in control problems, 114
   in nonlinear programming, 184, 229
   in stochastic control and programming, 180, 229
   in variational programming, 99, 184, 229
Expanding economy (models of), 10, 113, 175
Expectation, mathematical, 37, 50, 130, 150
   conditional, 149, 249
Expected cost of uncertainty, 208
Expected value criterion, 143, 174, 206
Externalities, 177, 182
Extreme points, 212, 243
Extremum (maximum or minimum), 99, 138, 210

F

Factor market conditions, 119
Factor-augmenting technological change, 113
Factors of production, 110, 114, 119

Feasible control, 100, 148, 235
Feedback, 95
   positive (negative), 155, 165
   quadratic, 167
Feedback control, 95, 145, 167
   characteristics, 155, 263
   stability, 156
Filter, 7
   linear and nonlinear, 155, 180
   optimal, 157, 180
Firm behavior, theory of, 200
Fiscal policy, optimal, 119
Fluctuation, 118, 143
Forecast values, 7, 16, 77
Fractile programming, 205
   computing methods, 206
Frequency function (density), 2, 36, 226
Frontier, production (transformation), 114
Function
   auto-correlation, 6, 17
   criterion, 94, 99, 143, 208
   control, 96, 155
   correlation, 6, 17
   decision, 130, 136, 149
   distribution, 128
   likelihood, 19, 67, 215
   loss, 172
   optimal scalar, 154, 163, 235
   penalty, 103, 208
   spectral, 7, 26
   weighting, 147
Functional equation, 132
   application to dynamic programming, 134
Future divergence, 136, 167

G

Game theory, 10, 149
   decision criteria and, 149
   programming and, 212
Games, zero–sum two person, 212
General equilibrium, 187
   models, 122, 195
   limitations under uncertainty, 265
Geometric programming, use of, 197
Global maximum, 100
   stability, 115
Golden age growth, 112
   stability and optimality of, 112, 113

Golden rule of
  accumulation, 112
    duality relationships and, 126
  path, 112
Gradient methods, 179, 184
Growth, optimal economic, 110, 193
  in two-sector model, 119
  in multi-sector model, 114, 168
Growth models
  deterministic, 29, 48
  optimal, 110, 193
  probabilistic (stochastic), 29, 48, 235
Growth path
  balanced (equilibrium), 110
  neoclassical, 110, 193
  steady, 10, 52, 117

H

Hamiltonian, 101
  current value, 126
Harrod–Domar growth model, 48, 235
Harrod–neutral technical progress, 113
Hicks–neutral technical progress, 113
Horizon
  finite, 99, 104
  infinite, 111
Households, saving behavior of, 120
Hyperplane (hypersurface), 184

I

Identification problems, 13, 98, 122
Income
  disposable, 120
  national, 29
  real per capita, 46
Increasing returns, 113
Independent random variables, 217
Induced investment, 165
Inequality
  Jensen's, 229, 236
  probabilistic (stochastic), 220, 228
  saddle-point, 212
  Tchebycheff, 206, 227
Information, prior and posterior, 146, 149, 174
  cost (value), 213, 265
Initial conditions, 147
Input characteristics, 147

Instability
  in dynamic deterministic models, 116
  in stochastic models, 131, 147, 159
Instrument (control) variables, 97
Interest rate, 120
Intertemporal efficiency, 122, 152
Inventory models, 131, 256
Inverse, 13, 155
Investment allocation, theory of, 125, 171
Investment problems
  in allocation, 121, 171, 193, 263
  in optimal growth, 110, 235
  in planning, 126, 171, 235
Irreversibility, 200
Iterative computing methods, 184, 191

J

Jacobian matrix, 153
Jensen's inequality, 229
Jumps
  in optimal growth models, 127
  in singular control, 102
  in switching problems, 127, 184

K

Keynesian model, 93
Kolmogorov equations, 4, 66, 84
Kolmogorov–Smirnov statistic, 227, 233
Kuhn–Tucker theorem, 100, 201
  constraint qualifications of, 104

L

Labor-augmenting technical progress, 113
Labor supply, 110, 120
Lagrange multiplier, 122, 179
  interpretation of, 122, 126, 177
Lagrangean function, 202
Laplace transform, 212
Large numbers, law of, 10
Learning by doing, 96
Least squares estimates, 37, 131
  methods, 14, 39, 167
Leontief model, open dynamic, 168
Limitation on control variables
Linear
  decision rule, 131, 250, 263
  estimator, 7, 14
  filtering, 8, 180

Subject Index

time-varying (dynamic) system, 147, 184
Linear programming, 203
　applications of, 231, 235
　chance-constrained, 208, 216
　decomposition techniques of, 178, 223
　probabilistic (stochastic), 218
　sensitivity and stability of, 210
Lipschitz condition, 138
　in stochastic control, 140
Local maximum, 155, 264
　stability, 116, 211
Log–normal diffusion process, 77, 128
　distribution, 66
Loop
　closed, 100
　open, 100, 134
　transfer function, 131
Loss function, 149, 163, 172
　quadratic, 130, 258
Lyapunov function, 116
　in stability analysis, 158, 170

M

Marginal cost, 202
　equilibrium, 265
Marginal productivity, diminishing, 110, 123, 200
Marginal utility, diminishing, 112, 124
Markov chain (process), 2
　applications, 48
　types used, 35, 59
Mathematical properties
　of stochastic control, 137
　of stochastic processes, 2, 35
Matrix, 4
　characteristic roots of, 142, 168
　dispersion (covariance), 14
　information, 68
　Jacobian, 153
　nonsingular, 14
　transition probability, 65
Maximum
　likelihood estimate, 68, 86
　Pontryagin's principle, 100
Mean square error, 8
Measurable control, 144
Methods
　of classical variational calculus, 100
　of dynamic programming, 100
　of game theory, 149, 212
　of statistical estimation, 8, 12
　of stochastic control, 128
　of stochastic processes, 35
　of stochastic (probabilistic) programming, 203
Minimax strategy (solution), 150
　theorem, 212
Minimum risk, 174
Minimum-time control model, 107
Mixed economy, 119
Models, *see also* specific types
　of dynamic investment allocation, 119, 235, 263
　of economic growth and stabilization, 93, 110, 168
Moments, 42
　methods of, 244
Money, role in growth models, 119, 123, 265
Monetary policy, 119
Monte Carlo techniques, 185, 223
Multiplier–accelerator models, 96, 165
Multi-product firms, models of, 201
Multisector models of economic growth, 114, 126, 168
Multivariable system, 98, 168
Multivariate normal, 85, 155
Myopic foresight (discounting), 112

N

Necessary conditions
　of existence of stochastic control, 139, 144
　of identification, 13
　of the extremum (optimality), 100
Neighborhood of solutions, 153
　first, second, third best, 240
Neoclassical production function, 110, 114
　one-sector model, 110
　two-sector model, 119
Neumann (von) model, 10, 103
Neutral technological change, 113
　types, 113
Noise
　additive, 12
　autocorrelated, 16

-to-signal ratio, 183
Nominal path (trajectory), 167
Nonlinear
  programming, 184
  stochastic control, 153, 179
  stochastic processes, 53, 65
  stochastic programming, 205
Nonmarket framework, 182
Nonnegativity conditions, 121, 203
Nonparametric methods, 185, 227
  applications of, 233
Nonsequential decision approaches, 5, 145
Nonstationary stochastic process, 34, 184
Normal distribution, 13, 66, 154
  convergence to, 206

O

Objective function
  linear, 96, 203, 231, 249
  nonlinear, 111, 249, 256
  stochastic, 205
Objectives and subgoals, 176
Observability condition (theorem), 99
One-sector growth model, 111
  neoclassical, 113
  optimal path in, 112
One-stage optimal allocation, 126, 219
Open-loop, 100
  control, 100
  filters, 180
  –optimal feedback control, 157
Operational methods (models), 93, 98
Operations research models, 21
  applications of, 130, 258
Opportunity cost, 202
Optimal control
  applications, 110
  feedback and stochastic control and, 155
  necessary conditions, 100
  nonlinear programming and, 100, 104
  sensitivity and stability, 148, 155
Optimal decision rule, 130, 250
Optimality principle (dynamic programming), 100
Optimality
  of control rule, 100
  of decisions, 149
  of feedback control, 155
  of path (intertemporal), 114, 148
  of performance, 147
  of policy, 142, 226, 235
  of principal, 100
  of trajectories, 114
  of trade-off with stability, 155, 245
Optimum (optimal) economic growth, 111
  bliss point (level), 112
  canonical form, 126
  criterion functional, 113, 169, 174
  heterogenous capital goods, 114
  maximum principle, 111
  multiple subsystems, 176
  in one-sector model, 111
  overtaking principle, 194
  robust controls, 173, 184
  solution by variational calculus, 100
  sufficient conditions for an optimal path, 100
  with surplus labor, 125
  in two-sector model, 121
  with technical change, 113
Optimum (optimal) subsystem, 176
Oscillations
  in control processes, 155, 166
  in dynamic models, 169, 184
  in optimal growth models, 116

P

Parametric programming, 209
Pareto optimality, 10, 265
Pay-off function, 149, 212
Penalty function, 103, 208
  in stochastic programming, 208
Perfect foresight, 112
Performance measure, 103, 143, 177
Period of adjustment, 147, 179, 267
Periodicity, 6, 184
Perturbation
  analysis in control, 148, 153, 163
  in stochastic programming, 211, 250
Phase plane, 124
  stability analysis in, 125, 127
Poisson distribution, 36
  process, 37, 41
Policy, 27
  centralized, 49, 76, 94
  decentralized, 97, 176

Subject Index

dynamic investment, 125, 238
economic, theory of, 94
optimal, 95, 101, 104
Policy models, 110
  applications, 119
  types of, 110, 128, 148, 171, 235
Pontryagin's maximum principle, 100
Population growth, 110
Portfolio models
  in dynamic allocation problems, 174, 264
  in static stochastic programming, 207, 238
Positive (negative) definite form, 117, 155
Posterior distribution, 149, 172
Prediction methods, 7, 14, 18, 20, 36
  problems, 13
  use in planning, 51, 75, 130
Preference function; *see* Objective function
Price adjustment, 124
Pricing
  implicit, 111, 239
  shadow, 124, 151, 254
Principle components, 56
Principle, Pontryagin's maximum, 100
Prior distribution, 172
Private sector, role of, 120
Probability distribution, 2, 36, 91
  for state variables, 148, 168
  in stochastic control, 144, 149, 157
  in stochastic processes, 6, 36, 52, 84
  in stochastic programming, 206, 215
  of errors, 13, 167
Production function, 167
  aggregate, 110
  neoclassical, 120
Production scheduling, models of, 256
Profit function
  expected value, 143, 264
  maximization criterion, 201, 243
  variance of, 201
Programming
  dynamic, 100, 184
  linear, 230, 235, 249
  nonlinear, 230, 235, 249
  stochastic (probabilistic), 205
  under uncertainty, 208
  variational, 100
Proportional control, 96, 155, 184

derivative control and, 96
integral control and, 96
Public good, 120
Pure strategy, 213

Q

Quadratic criterion, 134, 258
  performance measure, 130, 258
  programming, 201, 206
Qualification, constraint, 104
Quantitative economic policy model, 94, 119
Quasi-concave (convex) function, 184, 201
Queuing theory, 21
  applications of, 21

R

Ramsay problem, 112
Random variables, 2, 12, 46
Rate
  of discount, 112
  of growth, 110
  of return, 263
  of saving, 123
  of subscription, 202
Real disposable income, 120
Recursive equations, 98
  programming, 191
Reliability analysis, models of, 249
Renewal problems, 21
Rental on capital, 121
Resource allocation, 120, 171, 193
Response function, 167
Riccati equation, 142, 155
Risk-analysis
  in portfolio allocation, 174, 260, 264
  stochastic (probabilistic) programming, 205, 267
Risk aversion, 206
  applications, 215, 218
  different measures, 216
Robust control, 173
  estimate, 227

S

Saddle point, 150
Saddle value problem, 212

Sample estimate, 19, 37
  function, 7, 23
Saving policy, 110
  optimal, 113, 126
Savings ratio, variable, 123
Second-best policy, 157, 243
Sensitivity analysis, 151
  in stochastic control, 148, 154, 168
  in stochastic programming, 210
Sequential decisions, 172
  revision of estimates, 177
  policy, 173
Shadow prices
  of consumption and investment, 112, 126
  of private and public capital, 122
  of system reliability (components), 253
Signal-to-noise ratio, 184
Signal frequency, 8
Simplex and related methods, 223, 249
Simulation methods
  in stochastic control models, 184
  in stochastic processes, 28, 92
Stability, 95
  asymptotic, 116
  dual, 126
  global, 116
  in growth models, 113, 174
  in stochastic vs. deterministic systems, 159, 179
  local, 116
  stochastic, 157, 174
Stability theorems, deterministic (growth)
  models, 115
  stochastic systems, 157, 220
Stabilization models, 142
  policy, 143, 155
State
  density function, 47
  space (system), 96
  vector (variables), 99
Static optimization criteria, 174, 201, 205, 249
Stationary equilibrium, 112, 163
  process, 7, 22, 131
  state, 163
Statistical movements, 66
Steady growth path, 112, 163
Steady-state characteristics
  of consumption pattern, 112
  of equilibrium prices, 127
  of saddle-point solution, 113, 127
  of uniqueness, 112
Stochastic control
  applications, 128, 165
  characteristics, 151
  models of, 137
  theory of, 97, 137, 168
  types used, 129, 146
Stochastic process
  applications, 56, 61
  characteristics, 3, 25, 66
  models of, 2, 35
  theory of, 26, 65, 84
  types used, 25, 65
Stochastic programming
  applications, 235, 246
  models of, 203
  characteristics, 203
  theory of, 203
  types used, 203
Strategy, 150
Suboptimal policy (solution), 154, 243
Sufficiency conditions
  in stochastic control, 100, 145
  in stochastic process, 7
Sufficient conditions for optimality, 100
Superposition, 98, 147
Switching phenomena, 102, 127
Synthesis of controls, 102, 124
Systems approach, 95

T

Tangent hyperplane (gradient), 139, 153
Target variables (vector), 148, 163, 193
  desired value, 103
  role in theory of economic policy, 94
Taxes in economic models, 120
Taylor series, 113, 116, 153, 206
Tchebycheff inequality, 175, 206
Technological change (progress), 113
  applications, 45, 113
  various types, 113
Terminal conditions (control), 101, 112
Theory, *see* Specific theory
Time
  continuous, 5, 35, 98
  discrete, 3, 29, 35, 104, 175
Time discount, 111
  preference, 114

Subject Index

Time-varying matrix (system), 148
Tinbergen-type policy model, 97
Totally unstable (stable) system, 170
Transfer function, 131
Transient solution, 115, 158
Transition matrix, 4, 84, 154
 probability, 2, 35, 135
Transversality condition, 111, 127
Truncated distribution, 178, 208
 maximand, 243
Two-sector growth model, 119, 235
Two-stage least squares, 14
 programming, 208

U

Uniform distribution, use of, 213
Unstable equilibrium, 117
Uniqueness of solution, 13, 112, 116
Upper bound, 62, 150, 171, 179, 210, 229
Utility function (functional), 111, 121, 206
 in reliability programming, 248
Utility maximizing criterion, 121, 206

V

Value
 of a game, 150, 213
 of extra information, 213
Variable savings ratio, 123
Variance minimization, 132, 174, 207
Vector maximum problem, 265
von Neumann model, 10, 103

W

Wald's (minimax) criterion, 150
Weights in objective function, 163, 168, 248
Wealth, 260
White (Gaussian) noise, 154
Wiener process, 155
Wronskian matrix, 98

Z

Zero-sum game, 212
 Bayes' principle, 173
 randomized strategy, 213, 263